Photo Acoustic and Optical Coherence Tomography Imaging, Volume 1

Diabetic retinopathy

Online at: https://doi.org/10.1088/978-0-7503-2052-8

Photo Acoustic and Optical Coherence Tomography Imaging, Volume 1

Diabetic retinopathy

Edited by
Ayman El-Baz
University of Louisville, Louisville, KY, USA

Jasjit S Suri
AtheroPoint LLC, Roseville, CA, USA

IOP Publishing, Bristol, UK

ISBN 978-0-7503-2052-8 (ebook)
ISBN 978-0-7503-2050-4 (print)
ISBN 978-0-7503-2053-5 (myPrint)
ISBN 978-0-7503-2051-1 (mobi)

DOI 10.1088/978-0-7503-2052-8

Version: 20231201

IOP ebooks

British Library Cataloguing-in-Publication Data: A catalogue record for this book is available from the British Library.

Published by IOP Publishing, wholly owned by The Institute of Physics, London

IOP Publishing, No.2 The Distillery, Glassfields, Avon Street, Bristol, BS2 0GR, UK

US Office: IOP Publishing, Inc., 190 North Independence Mall West, Suite 601, Philadelphia, PA 19106, USA

With love and affection to my mother and father, whose loving spirit sustains me still.

—Ayman El-Baz

To my late loving parents, immediate family, and children.

—Jasjit S Suri

Contents

Preface

This book covers the state-of-the-art techniques of optical coherence tomography (OCT) imaging for the diagnosis of retinal diseases. Clinical disorders of the retina have been attracting the attention of researchers, aiming at reducing the blindness rate. This includes uveitis, diabetic retinopathy, macular edema, endophthalmitis, proliferative retinopathy, age-related macular degeneration and glaucoma. Currently, most ophthalmologists perform diagnosis by visual observation and interpretation. Treatment is significantly dependent on having an early and accurate diagnosis, which can be significantly improved by employing disease-specific computer-aided diagnostic (CAD) systems based on different image modalities such as: OCT, fundus imaging, and optical coherence tomography angiography (OCTA). This book will focus on OCT imaging for the diagnosis of retinal diseases. Among the topics discussed in the book are computerized tools for the automatic segmentation of diffuse retinal thickening edemas using OCT scans; recent developments in OCTA imaging for the diagnosis and assessment of diabetic retinopathy; multimodal photoacoustic microscopy; identification and measurement of abnormal retinal fluid; comparison of ocular ultrasound with OCT in the evaluation of diabetic retinopathy; OCT biomarkers in diabetic macular edema; deep learning-based multi-class retinal fluid segmentation and detection in OCT images; OCT and OCTA for the diagnosis and treatment of diabetic macular edema; eye sicknesses diagnosis using OCT and fundus imaging techniques; and early identification of diabetic retinopathy through a higher-order spatial 3D-OCT appearance model CAD system.

In summary, the main aim of this book is to help advance scientific research within the broad field of OCT imaging for the diagnosis of retinal diseases. The book focuses on major trends and challenges in this area, and it presents work aimed to identify new techniques and their use in biomedical analysis.

Ayman El-Baz
Jasjit S Suri

Acknowledgements

The completion of this book could not have been possible without the participation and assistance of so many people whose names may not all be enumerated. Their contributions are sincerely appreciated and gratefully acknowledged. However, the editors would like to express their deep appreciation and indebtedness particularly to Dr Ali H Mahmoud and Dr Yaser Elnakieb for their endless support.

Ayman El-Baz
Jasjit S Suri

Editor biographies

Ayman El-Baz

Ayman El-Baz is a Distinguished Professor at University of Louisville, Kentucky, United States and University of Louisville at Alamein International University (UofL-AIU), New Alamein City, Egypt. Dr El-Baz earned his BSc and MSc degrees in electrical engineering in 1997 and 2001, respectively. He earned his PhD in electrical engineering from the University of Louisville in 2006. Dr El-Baz was named as a Fellow for IEEE, Coulter, AIMBE and NAI for his contributions to the field of biomedical translational research. Dr El-Baz has almost two decades of hands-on experience in the fields of bio-imaging modeling and non-invasive computer-assisted diagnosis systems. He has authored or coauthored more than 700 technical articles.

Jasjit S Suri

Jasjit S Suri is an innovator, scientist, visionary, industrialist and an internationally known world leader in biomedical engineering. Dr Suri has spent over 25 years in the field of biomedical engineering/devices and its management. He received his PhD from the University of Washington, Seattle and his Business Management Sciences degree from Weatherhead, Case Western Reserve University, Cleveland, Ohio. Dr Suri was crowned with the President's Gold medal in 1980 and made Fellow of the American Institute of Medical and Biological Engineering for his outstanding contributions. In 2018, he was awarded the Marquis Life Time Achievement Award for his outstanding contributions and dedication to medical imaging and its management

List of contributors

Waleed H Abdulla
Department of Electrical, Computer, and Software Engineering, The University of Auckland, New Zealand

Josh Agranat
Boston Medical Center/Boston University School of Medicine, Boston, MA, USA

Marah Alhalabi
Electrical and Computer Engineering Department, Abu Dhabi University, UAE

Nurettin Bayram
Etlik City Training
and
Research Hospital, Department of Ophthalmology, University of Health Sciences, Ankara, Turkey

Pablo Carnota-Méndez
Centro de Ojos de La Coruña, La Coruña, Spain

Renoh Johnson Chalakkal
oDocs Eye Care Ltd, New Zealand
and
Department of Electrical, Computer, and Software Engineering, The University of Auckland, New Zealand

José Cunha-Vaz
AIBILI, Coimbra, Portugal and University of Coimbra, Coimbra, Portugal

Vani Damodaran
Department of Biomedical Engineering, SRM Institute of Science and Technology, Kattankulathur, Tamil Nadu, India

Joaquim de Moura
Department of Computer Science and Information Technology, CITIC, University of A Coruña, Spain

Ahmed El-Baz
duPont Manual, Louisville, KY, USA
and
Bioimaging Laboratory, Department of Bioengineering, University of Louisville, Louisville, KY, USA

Ayman El-Baz
Bioimaging Laboratory, Department of Bioengineering, University of Louisville, Louisville, KY, USA

Mohamed Elsharkawy
Bioimaging Laboratory, Department of Bioengineering, University of Louisville, Louisville, KY, USA

Mohammed Ghazal
Electrical and Computer Engineering Department, Abu Dhabi University, UAE

María Gil-Martínez
Complejo Hospitalario Universitario de Santiago de Compostela (Spain). Instituto Oftalmológico Gómez-Ulla, Santiago de Compostela, Spain

Guruprasad Giridharan
Bioimaging Laboratory, Department of Bioengineering, University of Louisville, Louisville, KY, USA

Eugene Hsu
Boston University School of Medicine, Boston, MA, USA

Pauline John
Laboratory for Advanced Bio-Photonics and Imaging, Division of Engineering, New York University Abu Dhabi, Abu Dhabi, UAE

Fahmi Khalifa
Department of Electrical and Computer Engineering, Morgan State University, Baltimore, MD, USA

Prakash Kumar Karn
Department of Electrical, Computer, and Software Engineering, The University of Auckland, New Zealand

Conceição Lobo
AIBILI, Coimbra, Portugal
and
University of Coimbra, Coimbra, Portugal

Ali Mahmoud
Bioimaging Laboratory, Department of Bioengineering, University of Louisville, Louisville, KY, USA

João Heitor Marques
Institute for the Biomedical Sciences Abel Salazar - University of Porto, Centro Hospitalar Universitário de Santo António, Porto, Portugal

Carlos Méndez-Vázquez
Centro de Ojos de La Coruña, La Coruña, Spain

Sandra R Montezuma
University of Minnesota, Munneapolis, MN, USA

Hossein Nazari
University of Minnesota, Munneapolis, MN, USA

Van Phuc Nguyen
Department of Ophthalmology and Visual Sciences, University of Michigan, Ann Arbor, MI 48105, USA

Jorge Novo
Department of Computer Science and Information Technology, CITIC, University of A Coruña, Spain

Nuria Olivier-Pascual
Complejo Hospitalario Universitario de Ferro, Spain

Marcos Ortega
Department of Computer Science and Information Technology, CITIC, University of A Coruña, Spain

Pawarissara Osathanugrah
Boston Medical Center/Boston University School of Medicine, Boston, MA, USA

Yannis Paulus
Department of Ophthalmology and Visual Sciences, University of Michigan, Ann Arbor, MI 48105, USA

Bernardete Pessoa
Unit for Multidisciplinary Research in Biomedicine - Institute for the Biomedical Sciences Abel Salazar - University of Porto, Centro Hospitalar Universitário de Santo António, Porto, Portugal

Amber Piazza
University of Minnesota Medical School, Minneapolis, MN, USA

Sara Rubio-Cid
Complejo Hospitalario Universitario de Ferrol, Spain

Harpal Sandhu
Bioimaging Laboratory, Department of Bioengineering, University of Louisville, Louisville, KY, USA

Nayan Sanjiv
Boston Medical Center/Boston University School of Medicine, Boston, MA, USA

Torcato Santos
AIBILI, Coimbra, Portugal

Ahmed Sharafeldeen
Bioimaging Laboratory, Department of Bioengineering, University of Louisville, Louisville, KY, USA

Ahmed Soliman
Bioimaging Laboratory, Department of Bioengineering, University of Louisville, Louisville, KY, USA

Manju Subramanian
Boston Medical Center/Boston University School of Medicine, Boston, MA, USA

N Sujatha
Biophotonics Laboratory, Department of Applied Mechanics, Indian Institute of Technology, Madras, India

Meysam Tavakoli
Department of Radiation Oncology and Winship Cancer Institute, Emory University, Atlanta, Georgia, USA

Aristomenis Thanos
Legacy Devers Eye Institute, Portland, OR, USA

Carlos Torres-Borrego
Centro de Ojos de La Coruña, La Coruña, Spain

Nilesh J Vasa
Opto-mechatronics Laboratory, Department of Engineering Design, Indian Institute of Technology, Madras, India

Daniel Velázquez-Villoria
Hospital POVISA, Vigo (Spain). Clínica Villoria, Vigo, Spain

Plácido L Vidal
Department of Computer Science and Information Technology, CITIC, University of A Coruña, Spain

Xueding Wang
Department of Biomedical Engineering, University of Michigan, Ann Arbor, MI 48105, USA

Wei Zhang
Department of Biomedical Engineering, University of Michigan, Ann Arbor, MI 48105, USA

IOP Publishing

Photo Acoustic and Optical Coherence Tomography Imaging, Volume 1

Diabetic retinopathy

Ayman El-Baz and Jasjit S Suri

Chapter 1

Computerized tool for the automatic segmentation of DRT edemas using OCT scans

Joaquim de Moura, Plácido L Vidal, Jorge Novo and Marcos Ortega

Diabetic macular edema (DME) is a complication of diabetes mellitus that results from the formation of intraretinal leakage in the macular region. This relevant eye disorder is recognised as a leading cause of visual loss among the industrialized world, as reported in the statistics of the World Health Organization guidelines. This chapter presents a software tool for the automated segmentation of diffuse retinal thickening regions from optical coherence tomography (OCT) images. For this purpose, two retinal regions were defined and extracted: the inner retina and the outer retina. Then, a learning process was used to analyze a comprehensive and heterogeneous subset of relevant patterns in the OCT scans. Finally, two complementary post-processing stages were applied to improve the obtained performance and the overall efficiency of the presented tool. The presented tool achieved satisfactory performance, achieving a Jaccard of 0.6625 and a Dice of 0.7899, which demonstrates the suitability of the adopted solution.

1.1 Introduction

Image processing, analysis and computer vision represent very interesting, interdisciplinary and dynamic scientific fields of computer science [1]. In particular, these relevant areas provide different computational tools that are commonly employed in many technological domains to solve different real-world problems [2]. In this context, after an explosion of interest during the 1980s and 1990s, the last three decades have been characterised by the maturity of these areas and a notable growth in different active applications from different domains of knowledge, such as industry [3], medicine [4], finance [5], engineering [6], agriculture [7] and education [8], among others [9, 10]. Therefore, as a result of this considerable technological advance, we can observe a significant increase in emerging computational solutions that include hardware,

doi:10.1088/978-0-7503-2052-8ch1

software, services and many automatic technologies that have the main objective of improving and facilitating the daily work of specialists and professionals [11, 12].

In particular, in the field of medicine, clinical experts often use different computer-aided diagnosis (CAD) systems for automatic or semi-automatic processing, analysis and recognition of medical images of different types, such as conventional x-ray [13, 14], magnetic resonance [15], computerised tomography [16] or ultrasound scans [17], among others. Therefore, the use of CAD solutions has grown in importance in recent years, facilitating the work of clinicians in diagnostic procedures, avoiding tedious and time-consuming manual procedures.

Specifically in the field of ophthalmology, CAD tools spread rapidly over the years, progressively being integrated into the clinical workflow to assist the clinical specialists in diagnostic, prognostic and therapeutic tasks in daily practice. In this context, these computational tools use the clinical information obtained through different imaging modalities, such as classical retinography [18], fluorescein angiography [19], optical coherence tomography (OCT) [20–22] and optical coherence tomography angiography (OCTA) [23], among others.

OCT is a non-invasive imaging examination widely used in ophthalmology for retinal imaging as well as for morphological analysis of different healthy or pathological structures [24, 25]. This well-established imaging technique uses low-coherence (high-bandwidth) interferometric technology to provide, in real-time, a set of two-dimensional scans of the histological structures of the main ocular tissues via sequential gathering of longitudinal and lateral reflections. In figure 1.1, we can see a representative illustration of a spectral domain OCT system.

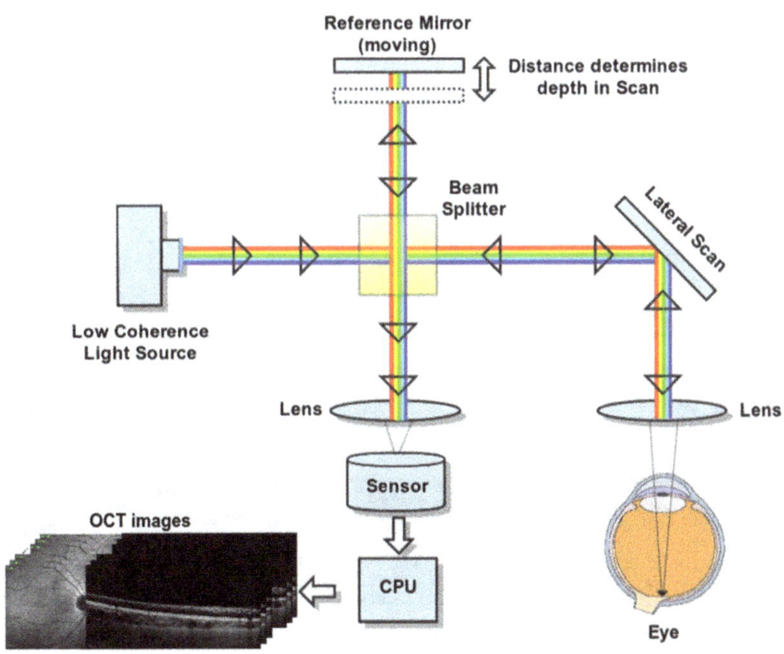

Figure 1.1. Representative illustration of a spectral domain OCT system.

The OCT scans allow a direct high-resolution visualization of the morphology and architecture of the retina and their corresponding histopathological properties. Consequently, these images provide a valuable resource for the detection, diagnosis, and treatments of several eye disorders [26, 27] such as, for example, glaucoma, central serous chorioretinopathy, pigment epithelium detachment, age-related macular degeneration, epiretinal membrane or diabetic macular edema (DME).

With regards to DME, this serious ocular disease is considered a worldwide health concern, in accordance with the World Health Organisation (WHO) guideline statistics [28]. In particular, DME is one of the most important consequences associated with diabetes mellitus, being considered a major cause of vision loss and affects mainly the developed countries. Specifically, figure 1.2 illustrates 6 OCT images showing the presence and absence of DME disease.

Using the OCT imaging as a reference, Otani *et al* [29] proposed a categorisation of DME disease according to three classes: diffuse retinal thickening (DRT), serous retinal detachment (SRD) and cystoid macular edema (CME). To do so, the authors analysed several imaging characteristics of the OCT scans. Subsequently, Panozzo *et al* [30] expanded the existing clinical categorisation by defining new characteristics that can be seen on OCT images and that better characterise this relevant eye disorder. To do this, the authors included information on the volume, diffusion, morphology and presence of the epiretinal membrane. Figure 1.3 illustrates an OCT scan with the three clinical categories of DME analysed.

Regarding the DRT, this type of DME is typically defined by a sponge appearance as a consequence of fluid leakage with restricted reflectivity in the retinal tissues. In addition, since this type of DME usually appears before the SRD and CME regions, it is frequently considered by the clinical experts as a valuable

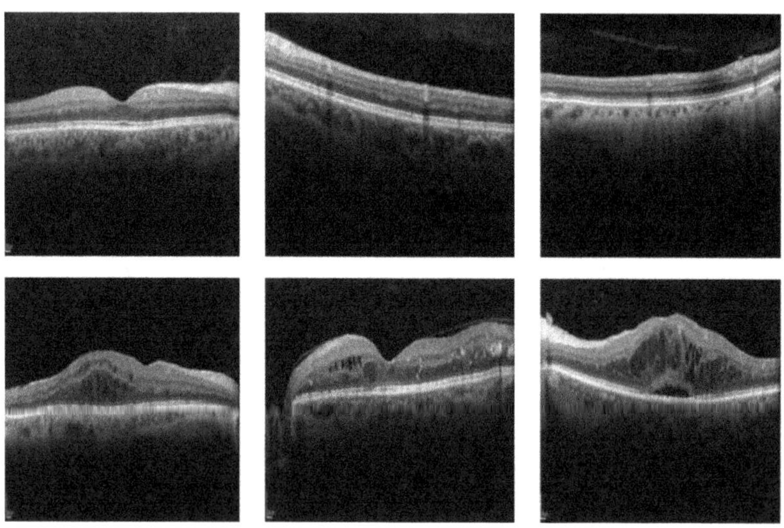

Figure 1.2. Examples of OCT scans. First row, OCT scans of patients without DME disease. Second row, OCT scans of patients with DME disease.

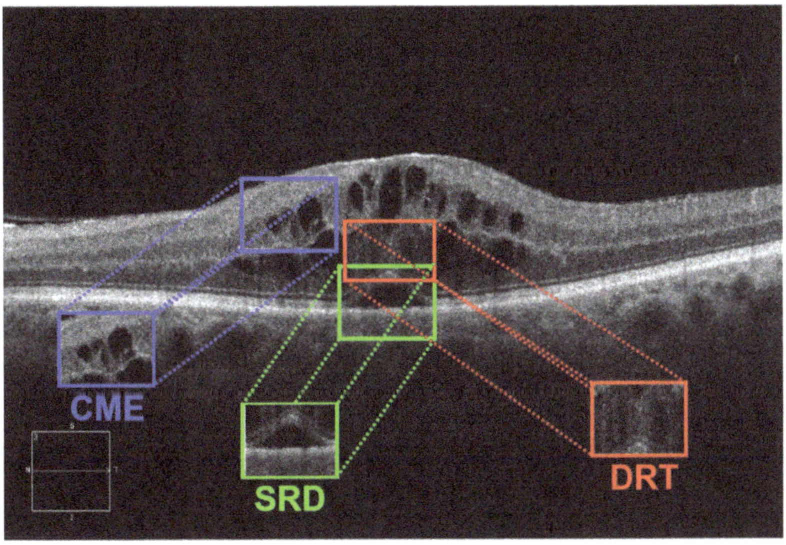

Figure 1.3. OCT image showing the presence of all classes of DME: DRT, CME and SRD.

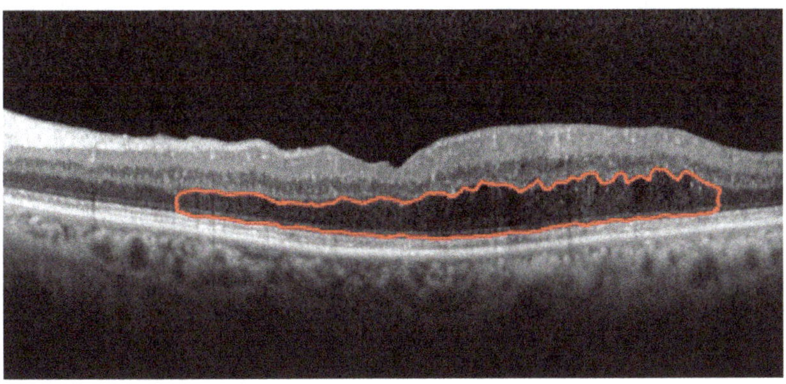

Figure 1.4. OCT scan with the manual delineation of the pathological DRT region.

marker for the diagnosis of this relevant eye disorder [31]. In figure 1.4, we can see an OCT scan with the manual delineation of the pathological DRT region.

Some proposals using OCT scans for the identification, segmentation or characterisation of intraretinal fluid regions associated with DME disorder have been published in recent years. As reference, Gopinath *et al* [32] proposed a strategy for the segmentation of macular edemas in OCT scans. To achieve this, the authors use a convolutional architecture to train a mapping function that captures the output of multiple motions to generate a probability map of the locations of pathological fluids in a given OCT scan. Following a similar strategy, Schlegl *et al* [33] proposed an automatic tool for the quantification of fluid regions in OCT scans by means of different machine learning models. In the work of de Moura *et al* [34], the authors proposed a comprehensive analysis of representative descriptors for the intraretinal fluid characterization in OCT images. In another proposal [35], the authors presented a novel paradigm to identify fluid accumulations in the retina using intuitive heat maps. Roy *et al* [36] presented a CNN architecture for the segmentation of pathological fluid regions in OCT scans. Samagaio *et al* [37] proposed a novel approach to classify the presence of macular edemas using OCT scans. In another proposal [38], the authors presented an automatic system for the segmentation and characterization of the DME regions in OCT scans. In the work of de Moura *et al* [39], the authors proposed a deep features analysis in a transfer learning-based process for DME screening using OCT scans. Similarly, Chan *et al* [40] proposed a framework based on a transfer learning approach for DME recognition on OCT scans. As we can see from the existing studies, the proposed systems only aimed at identifying areas of intraretinal fluid and, therefore, did not address the identification or segmentation of DRT regions. In this sense, at present, only the works [41, 42] addressed the precise segmentation of DRT regions by OCT scans.

In this chapter, we describe a fully automatic system that identifies and segments DRT edemas from OCT images, following the reference clinical classification in the field of ophthalmology. Firstly, two regions of the retina were automatically delineated: one corresponding to the ILM/OPL region (inner retina) and other to the OPL/RPE region (outer retina). Then, a learning strategy was adopted, analyzing a set of samples of a specific size to extract different feature descriptors. And finally, a post-processing step was applied to improve the overall performance of the presented system.

The chapter is structured as follows: Section 1.2 contains a detailed explanation of the methodology presented. Section 1.3 presents and discusses the obtained results with a brief explanation on their significance. Finally, section 1.4 includes a series of final notes drawn for this research and a commentary on future lines of work.

1.2 Computational identification and segmentation of DRT edemas

The presented pipeline, illustrated in figure 1.5, consists of three main stages: a first stage, in which the main layers of the retina are segmented and two retinal regions are delimited: inner retina and outer retina; a second stage, a set of features within the outer retina is extracted and a machine learning strategy is adapted for the DRT segmentation; a third step, in which a post-processing stage was applied to refine the DRT segmented regions. Each of these stages will be discussed below.

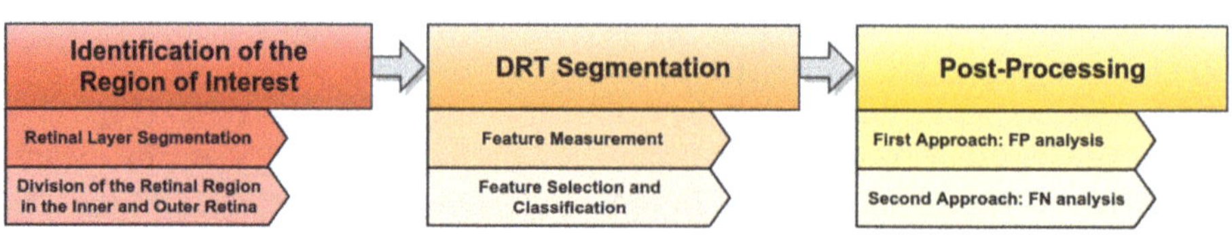

Figure 1.5. Main stages of the automatic segmentation of DRT regions.

1.2.1 Identification of the region of interest

The different types of pathological fluid accumulations are normally found in typical relative positions within the layers of the retina. Specifically, DRT edemas usually occur in the OPL/RPE region. In this way, two regions of the retina were identified, facilitating the subsequent segmentation of this relevant DME type. The following subsections describe this entire process in more detail.

1.2.1.1 Retinal layer segmentation

Regarding the automatic segmentation of the main retinal layers, we followed the previous study of González-López *et al* [43]. In particular, we segment four retinal layers: the inner limiting membrane (ILM), the inner/outer segments (ISOS), the outer plexiform layer (OPL) and the retinal pigment epithelium (RPE). For this purpose, firstly, we used a denoising algorithm based on the Butterworth Fourier filter to mitigate the speckle noise [44], preserving the information contained in the OCT scans. Next, an active contour-based strategy was used to delineate the retinal boundaries. As said, these retinal regions correspond to the region of the retina where the DRT edema usually appears (figure 1.6).

1.2.1.2 Segmentation of the inner/outer retinal regions

Using the segmented retinal layers as reference, two regions are segmented: the ILM/OPL region (inner retina) and the OPL/RPE region (outer retina), as represented in figure 1.7. Based on clinical knowledge, these retinal regions were identified in order to simplify the subsequent DRT segmentation stage.

1.2.2 DRT segmentation

In order to accurately segment the DRT regions, a machine learning algorithm was employed to characterize the pathological regions only in the restricted search space (outer retina). To achieve this, a set of windows of a given size was analyzed, thereby extracting a comprehensive subset of features. Finally, a post-processing step was carried out to improve the results obtained in the segmentation stage. The following subsections describe this entire process in more detail.

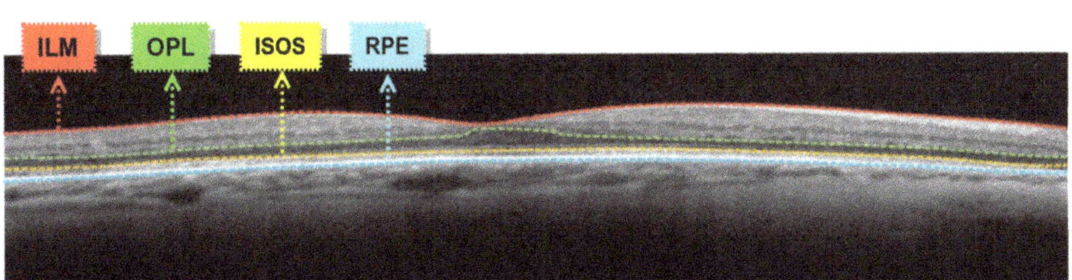

Figure 1.6. Representative example of the retinal layer segmentation stage.

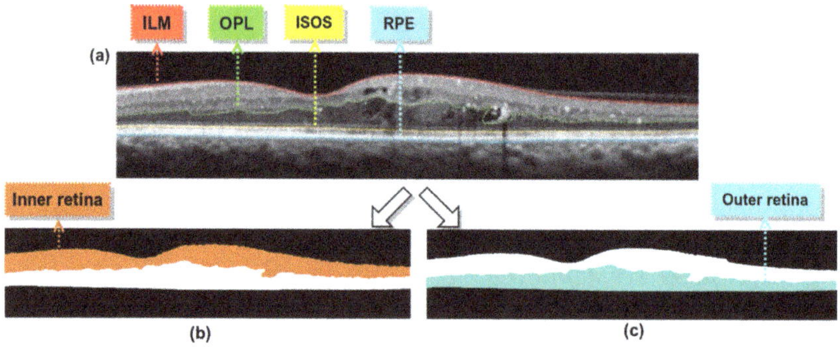

Figure 1.7. Representative example of the segmentation of the inner/outer retinal regions. (a) OCT scan with the main retinal layers segmented. (b) The inner retinal region. (c) The outer retinal region.

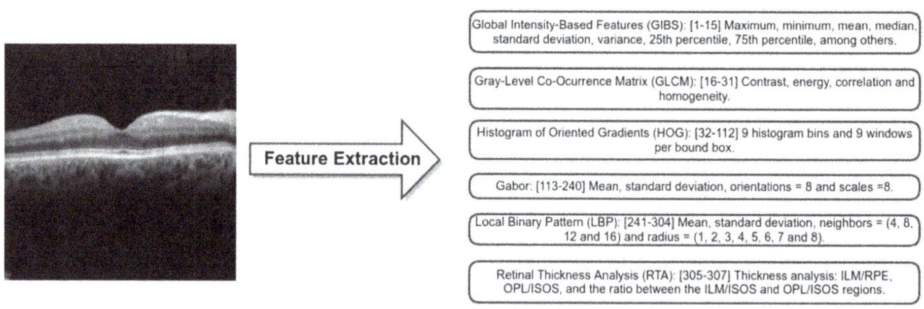

Figure 1.8. Schematic representation of the feature extraction.

1.2.2.1 Feature extraction

To characterize the pathologic patterns of DRT-type edema, a comprehensive set of 307 features was extracted from the outer retinal region, as represented in figure 1.8. In particular, this set of features includes characteristics of intensity, texture and knowledge of the domain.

1.2.2.2 Feature selection and classification

The 307 extracted features were posteriorly analyzed to obtain the subset that maximizes the separability between the DRT and non-DRT regions and, therefore, facilitating the classification process. To do this, we use the well-known Sequence Forward Selection (SFS) [45] algorithm. In particular, this feature selector employs a strategy in which features are sequentially added to a subset of empty candidates until the addition of more features does not decrease the given selection criteria. A machine learning technique is then used to evaluate different prediction models using the previously chosen subset of features. To this end, four classifiers were used to measure the performance of the presented methodology: the Naive Bayes, the Parzen, the Quadratic Bayes Normal Classifier (QDC) and the k-Nearest Neighbors (kNN) for three different k values (k = 3, 5 and 7).

As training details, the initial image dataset was partitioned into two smaller datasets with 50% for training and 50% for testing. In addition, a 10-fold cross-validation with 50 repetitions was performed. As a final result of this classification stage, all columns in the outer retinal region were categorized into DRT or non-DRT categories.

1.2.3 Post-processing

In this stage, two independent post-processing approaches were designed to improve the results obtained by the presented system. In the following subsections, this whole process is described in more detail.

1.2.3.1 First approach: FP analysis

The first post-processing approach focuses on the analysis and subsequent reduction of the false positive rates. In this sense, these false detections of DRT columns usually occur due to the presence of other pathological structures of similar appearance that can be observed in the outer retinal region. To do this, we implemented a strategy that calculated the minimum width of each segmented region with respect to the nearest corresponding region, thus eliminating small isolated regions, as represented in figure 1.9.

1.2.3.2 Second approach: FN analysis

The second post-processing approach focuses on the analysis and subsequent mitigation of the false negative rates. In particular, these misclassified regions are mainly derived from the presence of speckle noise and/or vascular shadows in the outer region of the retina. To achieve this, we implemented a strategy based on an aggregation factor (*d*). Specifically, this strategy joins two contiguous regions if the distance between them is less than a predefined aggregation factor, as represented in figure 1.10.

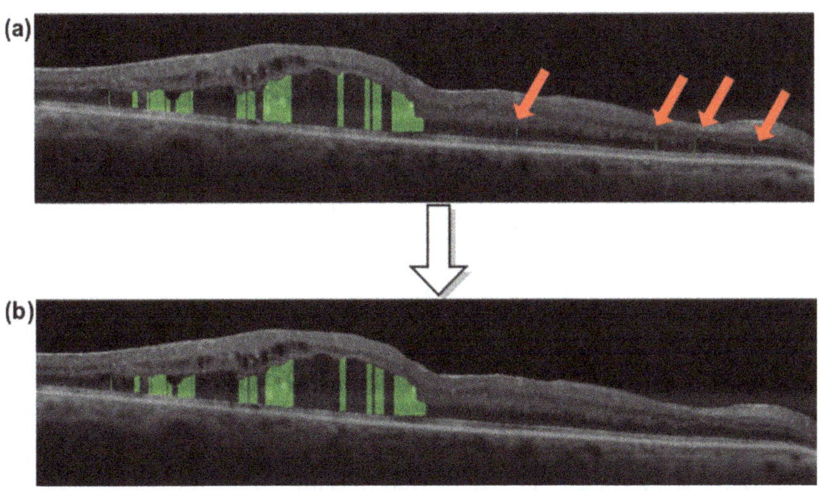

Figure 1.9. Representative example of the first post-processing step. (a) DRT regions without the post-processing step. (b) DRT regions with the post-processing step.

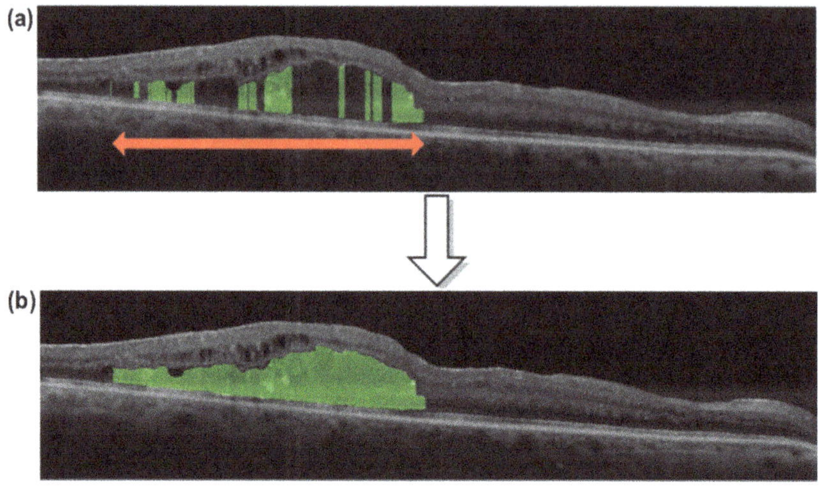

Figure 1.10. Representative example of the second post-processing step. (a) DRT regions without the post-processing step. (b) DRT regions with the post-processing step.

1.3 Results and discussion

The presented method was validated using an image dataset consisting of 70 scans. These scans were obtained using an OCT confocal scanning laser ophthalmoscope (cSLO) imaging device from Heidelberg Spectralis. All the OCT scans were obtained focusing on the macular region of patients diagnosed with DME. In addition, this dataset has a variable resolution that ranges from 401×1015 to 481×1521 pixels. To ensure the complete anonymity and confidentiality of participants in this study, we used anonymised data images available for research purposes.

To validate the presented methodology, an expert clinician labeled 560 samples to represent the presence of DRT edemas, including 280 for each category, DRT and non-DTR columns. As said, the used dataset was partitioned into 2 subsets, 50% for training and 50% for testing. In addition, we performed a 10-fold cross-validation with 50 repetitions without any pre-processing stage on the input OCT images. In particular, the presented system was evaluated by means of the Accuracy, Jaccard and Dice coefficients, described in equations (1.1), (1.2) and (1.3), respectively.

$$\text{Accuracy} = \frac{\text{TP} + \text{TN}}{\text{TP} + \text{TN} + \text{FP} + \text{FN}} \tag{1.1}$$

$$\text{Jaccard} = \frac{\text{TP}}{\text{TP} + \text{FP} + \text{FN}} \tag{1.2}$$

$$\text{Dice} = \frac{2 \times \text{TP}}{2 \times \text{TP} + \text{FP} + \text{FN}} \tag{1.3}$$

Firstly, we analyze the subset of features that maximizes the separability between the DRT and non-DRT regions. To do this, we use an SFS algorithm to analyze the initial set of 307 features. As a result of this feature analysis, we can conclude that most of the features were selected from the HOG, Gabor and LBP. Figure 1.11 shows the results obtained with different classifier configurations using a subset of features that was obtained by the SFS algorithm.

Once the best subset of features has been determined, we analyze the different classifiers considered in this work to determine which best discriminates between DRT and non-DRT regions. Figure 1.12 presents the results obtained by each classifier using the most relevant subset of features. As we can see, the best results

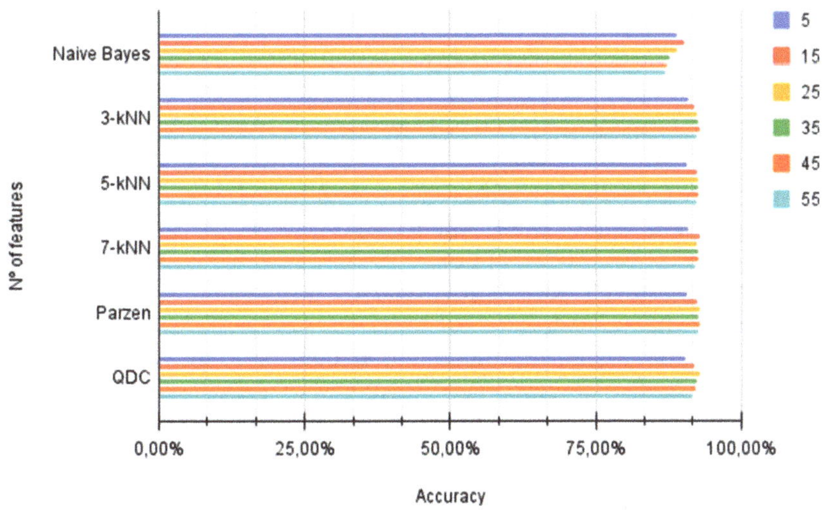

Figure 1.11. Analysis of different classifiers using larger progressive feature sets.

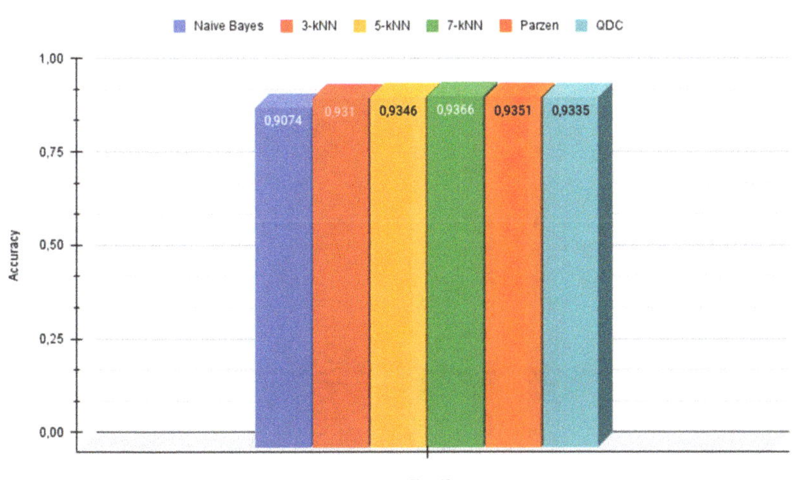

Figure 1.12. Summary of the accuracy results obtained from each classifier using the most relevant subset of features.

were obtained with the kNN algorithm with $k = 7$, reaching an accuracy value of 93.66% with only 21 features.

Using the best classifier configuration as a reference, we analyze different window sizes to determine the best way to distinguish the texture patterns that are present in the DRT edema regions of the surrounding healthy tissues. Each window has a variable height value (h) centered on the analyzed column. These h values are calculated by the distance between the ISOS and the OPL retinal layers. Figure 1.13 presents the performance of the 7-kNN learning strategy for different window sizes. Once again, satisfactory results were obtained, reaching accuracy values over 88.66%. In particular, the best values were achieved with a window size of ($h \times 23$) pixels, which resulted in an accuracy of 93.66%.

To evaluate the segmentation process, DRT identifications (SFS feature selector + 7-kNN algorithm) and their respective height values of the outer retina (distance between the ISOS and the OPL retinal layers) were used. The presented method obtained satisfactory results, achieving a 0.8381, 0.6106 and 0.7480 in Accuracy, Jaccard and Dice coefficients, respectively, without any post-processing step.

Using the segmentation of DRT regions as a reference, we tested the capabilities of both designed post-processing approaches. To do so, firstly, we analysed the first post-processing step for the reduction of the FP rates, eliminating the isolated DRT regions. As mentioned above, these false detections are generally produced by the existence of other pathological structures that may be found within the analyzed region of the retina. In particular, we analyzed the best combination between the width of the DRT regions (w_{min}) and the distance to the closest DRT columns (d_{min}). The results obtained provided a reasonable balance between Jaccard and Dice (0.6162 and 0.7516, respectively) using values of w_{min} and d_{min} (16 and 10, respectively).

The second post-processing step aims at the analysis of DRT columns and subsequent reduction of the FN rates, connecting nearby pathological regions using

Figure 1.13. Accuracy results obtained from the 7-kNN algorithm for different window sizes.

the aggregation factor (*d*). As indicated, these misclassified DRT regions are mainly derived from the presence of speckle noise and/or vessel shadows. Consequently, the presented method obtained satisfactory results using an aggregation factor of 34, reaching values of 0.6625 and 0.7899 for the Jaccard and Dice coefficients, respectively.

1.4 Conclusions

This chapter presents a fully automatic system for segmentation of DRT edema in OCT images, following the reference clinical classification in ophthalmology. For this purpose, two retinal regions were defined and extracted for subsequent analysis: the ILM/OPL region (inner retina) and the OPL/RPE region (outer retina). A learning process was then applied using different classifiers to validate the appropriateness of the selected features in the segmentation of these ocular lesions. In addition, two complementary post-processing stages were designed to improve the results obtained by the presented system. This system was validated using 70 OCT scans, being 560 samples labeled to represent the presence of DRT edemas, including 280 samples for each category, DRT and non-DTR. The best result was obtained by the 7-kNN, using 21 features and a window size of ($h \times 23$) pixels, according to Jaccard and Dice (0.6625 and 0.7899, respectively) and with a combination of post-processing stages. Therefore, the presented system has demonstrated its suitability in the automatic segmentation of DRT regions in OCT scans.

Acknowledgments

This research was funded by Instituto de Salud Carlos III, Government of Spain, DTS18/00136 research project; Ministerio de Ciencia e Innovación y Universidades, Government of Spain, RTI2018–095894-B-I00 research project, Ayudas para la formación de profesorado universitario (FPU), grant reference FPU18/02271; Ministerio de Ciencia e Innovación, Government of Spain through the research project with reference PID2019-108435RB-I00; Consellería de Cultura, Educación e Universidade, Xunta de Galicia, Grupos de Referencia Competitiva, grant reference ED431C 2020/24 and through the postdoctoral grant contract reference ED481B 2021/059; Axencia Galega de Innovación (GAIN), Xunta de Galicia, grant reference IN845D 2020/38; CITIC, Centro de Investigación de Galicia reference ED431G 2019/01, receives financial support from Consellería de Educación, Universidade e Formación Profesional, Xunta de Galicia, through the ERDF (80%) and Secretaría Xeral de Universidades (20%).

References

[1] Sonka M, Hlavac V and Boyle R 2007 *Image Processing, Analysis and Computer Vision* (Berlin: Springer) 3rd edn

[2] Umbaugh S E 2010 *Digital Image Processing and Analysis: Human and Computer Vision Applications with CVIPtools* (Boca Raton, FL: CRC Press)

[3] Gunasekaran S 1996 Computer vision technology for food quality assurance *Trends Food Sci. Technol.* **7** 245–56

[4] de Moura J, Novo J, Charlón P, Barreira N and Ortega M 2017 Enhanced visualization of the retinal vasculature using depth information in OCT *Med. Biol. Eng. Comput.* **55** 2209–25

[5] Hassanein K S, Wesolkowski S, Higgins R, Crabtree R and Peng A 1997 Integrated system for automated financial document processing *25th AIPR Workshop: Emerging Applications of Computer Vision* vol 2962 (Bellingham, WA: International Society for Optics and Photonics) 202–12

[6] Salgado L, Menendez J M, Rendon E and Garcia N 1999 Automatic car plate detection and recognition through intelligent vision engineering *Proceedings IEEE 33rd Annual 1999 Int. Carnahan Conf. on Security Technology (Cat. No. 99CH36303)* (Piscataway, NJ: IEEE) 71–6

[7] Brosnan T and Sun D W 2002 Inspection and grading of agricultural and food products by computer vision systems—a review *Comput. Electron. Agric.* **36** 193–213

[8] Bebis G, Egbert D and Shah M 2003 Review of computer vision education *IEEE Trans. Educ.* **46** 2–21

[9] Schadt E E, Linderman M D, Sorenson J, Lee L and Nolan G P 2010 Computational solutions to large-scale data management and analysis *Nat. Rev. Genet.* **11** 647–57

[10] Fernández A, Ortega M, de Moura J, Novo J and Penedo M G 2019 Automatic evaluation of eye gestural reactions to sound in video sequences *Eng. Appl. Artif. Intell.* **85** 164–74

[11] Novo J, Barreira N, Penedo M G and Santos J 2012 Topological active volume 3D segmentation model optimized with genetic approaches *Nat. Comput.* **11** 161–74

[12] Novo J, Penedo M G and Santos J 2010 Evolutionary multiobjective optimization of topological active nets *Pattern Recognit. Lett.* **31** 1781–94

[13] de Moura J, Ramos L, Vidal P L, Cruz M, Abelairas L, Castro E, Novo J and Ortega M 2020 Deep convolutional approaches for the analysis of Covid-19 using chest x-ray images from portable devices *IEEE Access* **8** 195594–607

[14] de Moura J, Novo J and Ortega M 2020 Fully automatic deep convolutional approaches for the analysis of Covid-19 using chest x-ray images *Appl. Soft Comput.* **115** 108190

[15] El-Dahshan E S A, Mohsen H M, Revett K and Salem A B M 2014 Computer-aided diagnosis of human brain tumor through MRI: a survey and a new algorithm *Expert Syst. Appl.* **41** 5526–45

[16] Javaid M, Javid M, Rehman M Z U and Shah S I A 2016 A novel approach to CAD system for the detection of lung nodules in CT images *Comput. Methods Programs Biomed.* **135** 125–39

[17] Thomaes T, Thomis M, Onkelinx S, Coudyzer W, Cornelissen V and Vanhees L 2012 Reliability and validity of the ultrasound technique to measure the rectus femoris muscle diameter in older CAD-patients *BMC Med. Imaging* **12** 7

[18] Ortega M, Barreira N, Novo J, Penedo M G, Pose-Reino A and Gómez-Ulla F 2010 Sirius: a web-based system for retinal image analysis *Int. J. Med. Inform.* **79** 722–32

[19] Tavakoli M, Toosi M B, Pourreza R, Banaee T and Pourreza H R 2011 Automated optic nerve head detection in fluorescein angiography fundus images *2011 IEEE Nuclear Science Symp. Conf. Record* (Piscataway, NJ: IEEE) 3057–60

[20] de Moura J, Novo J, Rouco J, Charlón P and Ortega M 2019 Artery/vein vessel tree identification in near-infrared reflectance retinographies *J. Digit. Imaging* **32** 947–62

[21] Cabaleiro P, de Moura J, Novo J, Charlón P and Ortega M 2019 Automatic identification and representation of the cornea–contact lens relationship using AS-OCT Images *Sensors* **19** 5087

[22] de Moura J, Novo J, Ortega M and Charlón P 2016 3D retinal vessel tree segmentation and reconstruction with OCT images *Int. Conf. on Image Analysis and Recognition* (Cham: Springer) 716–26

[23] Baamonde S, de Moura J, Novo J, Charlón P and Ortega M 2019 Automatic identification and characterization of the epiretinal membrane in OCT images *Biomed. Opt. Express* **10** 4018–33

[24] Baamonde S, de Moura J, Novo J, Charlón P and Ortega M 2019 Automatic identification and intuitive map representation of the epiretinal membrane presence in 3D OCT volumes *Sensors* **19** 5269

[25] Díaz M, de Moura J, Novo J and Ortega M 2019 Automatic wide field registration and mosaicking of OCTA images using vascularity information *Procedia Comput. Sci.* **159** 505–13

[26] Huang D, Swanson E A, Lin C P, Schuman J S, Stinson W G, Chang W and Puliafito C A 1991 Optical coherence tomography *Science* **254** 1178–81

[27] Puliafito C A, Hee M R, Lin C P, Reichel E, Schuman J S, Duker J S and Fujimoto J G 1995 Imaging of macular diseases with optical coherence tomography *Ophthalmology* **102** 217–29

[28] Romero-Aroca P 2011 Managing diabetic macular edema: the leading cause of diabetes blindness *World J. Diabetes* **2** 98

[29] Otani T, Kishi S and Maruyama Y 1999 Patterns of diabetic macular edema with optical coherence tomography *Am. J. Ophthalmol.* **127** 688–93

[30] Panozzo G, Parolini B, Gusson E, Mercanti A, Pinackatt S, Bertoldo G and Pignatto S 2004 Diabetic macular edema: an OCT-based classification *Seminars in Ophthalmology* vol 19 (Milton Park: Taylor and Francis) pp 13–20

[31] de Moura J, Samagaio G, Novo J, Charlón P, Fernández M I, Gómez-Ulla F and Ortega M 2019 Automatic identification of diabetic macular edema biomarkers using optical coherence tomography scans *Int. Conf. on Computer Aided Systems Theory* (Cham: Springer) pp 247–55

[32] Gopinath K and Sivaswamy J 2018 Segmentation of retinal cysts from optical coherence tomography volumes via selective enhancement *IEEE J. Biomed. Health Inform.* **23** 273–82

[33] Schlegl T, Waldstein S M, Bogunovic H, Endstraßer F, Sadeghipour A, Philip A M and Schmidt-Erfurth U 2018 Fully automated detection and quantification of macular fluid in OCT using deep learning *Ophthalmology* **125** 549–58

[34] de Moura J, Vidal L, Novo P, Rouco J, Penedo J, Ortega M G and M 2020 Intraretinal fluid pattern characterization in optical coherence tomography images *Sensors* **20** 2004

[35] Vidal P L, De Moura J, Novo J, Penedo M G and Ortega M 2018 Intraretinal fluid identification via enhanced maps using optical coherence tomography images *Biomed. Opt. Express* **9** 4730–54

[36] Roy A G, Conjeti S, Karri S P K, Sheet D, Katouzian A, Wachinger C and Navab N 2017 ReLayNet: retinal layer and fluid segmentation of macular optical coherence tomography using fully convolutional networks *Biomed. Opt. Express* **8** 3627–42

[37] Samagaio G, Estévez A, de Moura J, Novo J, Fernández M I and Ortega M 2018 Automatic macular edema identification and characterization using OCT images *Comput. Methods Programs Biomed.* **163** 47–63

[38] de Moura J, Samagaio G, Novo J, Almuina P, Fernández M I and Ortega M 2020 Joint diabetic macular edema segmentation and characterization in OCT images *J. Digit. Imag.* **33** 1–17

[39] de Moura J, Novo J and Ortega M 2019 Deep feature analysis in a transfer learning-based approach for the automatic identification of diabetic macular edema *2019 Int. Joint Conf. on Neural Networks (IJCNN)* (Piscataway, NJ: IEEE) 1–8

[40] Chan G C, Muhammad A, Shah S A, Tang T B, Lu C K and Meriaudeau F 2017 Transfer learning for diabetic macular edema (DME) detection on optical coherence tomography

(OCT) images *2017 IEEE Int. Conf. on Signal and Image Processing Applications (ICSIPA)* (Piscataway, NJ: IEEE) 493–6

[41] de Moura J, Novo J, Charlón P, Fernández M I and Ortega M 2019 Retinal vascular analysis in a fully automated method for the segmentation of DRT edemas using OCT images *Procedia Comput. Sci.* **159** 600–9

[42] Samagaio G, de Moura J, Novo J and Ortega M 2018 Automatic segmentation of diffuse retinal thickening edemas using optical coherence tomography images *Procedia Comput. Sci.* **126** 472–81

[43] González-López A, de Moura J, Novo J, Ortega M and Penedo M G 2019 Robust segmentation of retinal layers in optical coherence tomography images based on a multistage active contour model *Heliyon* **5** e01271

[44] Samagaio G, de Moura J, Novo J and Ortega M 2017 Optical coherence tomography denoising by means of a fourier butterworth filter-based approach *Int. Conf. on Image Analysis and Processing* (Cham: Springer) 422–32

[45] Berk K N 1980 Forward and backward stepping in variable selection *J. Stat. Comput. Simul.* **10** 177–85

IOP Publishing

Photo Acoustic and Optical Coherence Tomography Imaging, Volume 1
Diabetic retinopathy
Ayman El-Baz and Jasjit S Suri

Chapter 2

Recent developments in optical coherence tomography angiography imaging for the diagnosis and assessment of diabetic retinopathy

Pauline John, Vani Damodaran, N Sujatha and Nilesh J Vasa

This chapter provides a comprehensive view of the recent advancements in optical coherence tomography (OCT) imaging techniques for the detection and assessment of diabetic retinopathy (DR). DR is a serious complication that arises due to the poor control of blood glucose levels, which causes extensive damage to retinal blood vessels and eventually blindness. Various pathological conditions due to DR are discussed in brief and the different imaging techniques currently used in diagnosis are described. OCT is one such technique that has been a gold standard for imaging in ophthalmology mainly due to its high-resolution imaging capability and non-invasiveness. The various types of OCT configuration, commercial systems and its ability to detect pathological conditions are described in detail. A functional extension known as OCT angiography, capable of imaging retinal microvasculature, and its implementation are described. The role of deep learning in enhancing imaging and leading towards real-time imaging is also highlighted.

2.1 Introduction

2.1.1 Diabetic retinopathy

A steady rise in the diabetes rate worldwide has led to an increase in several systemic complications. Diabetic retinopathy (DR), a microvascular disease that affects the retinal circulation, is one of the complications caused as a result of chronic poor glycemic control [1, 2]. Subjects with a chronic diabetic condition for 15 years or longer have a 90% chance of developing retinopathy [2]. Hence, the duration of diabetes is also a significant factor that influences the progress of retinopathy [3]. DR progresses from non-proliferative abnormalities characterized by (i) vascular

doi:10.1088/978-0-7503-2052-8ch2

permeability related micro-infarction (fluffy white swellings or cotton wool spots appearing in the retina), (ii) hard exudates (yellow eosinophillic masses which contain foamy macrophages with lipids), (iii) microaneurysms (bulged walls of retinal blood vessels), (iv) haemorrhage (leakage of red blood cells into the retina as a result of vessel wall breakdown), to (v) proliferative neovascularization (newly formed vessels) caused as a reaction to ischemia [1, 2, 4]. Contraction of the accompanying fibrous tissue by proliferative DR leads to tractional retinal detachment or vitreous haemorrhage, eventually leading to irreversible vision loss [2].

One of the severe complications of DR is neovascular glaucoma, which can lead to painful blindness and eye shrinkage [2]. Other non-proliferative complications of DR include diabetic macular edema (DME), which causes visual impairment in diabetic subjects and is characterised by the leakage of fluids, proteins and lipids into the sensory retina as a result of defective inner blood retinal barriers, leading to retinal thickening and formation of hard exudates, leading to fluid accumulation in the macula and diabetic macular ischemia (DMI), which causes occlusion of the macular capillary network [2, 5, 6]. Development of an efficient diagnostic system to detect DR at an early stage is crucial to avoid serious risks such as permanent vision loss. The following section describes various conventional and advanced screening techniques utilized in the diagnosis of several complications of DR.

2.1.2 Diagnostic techniques for screening DR

The accepted methods for screening diabetic retinopathy are as follows [4, 7]:

- Direct and indirect ophthalmoscopy;
- Color fundus film photography;
- Digital mydriatic and non-mydriatic photography;
- Fluorescein angiography (FA); and
- Optical coherence tomography (OCT).

Traditional screening methods performed by ophthalmologists include direct examination of the retina with a limited field of view, as well as an indirect examination providing a wide field of view of the entire retina by dilating the pupil using an ophthalmoscope [1, 4, 8]. It can be challenging for the human experts to detect diabetic retinopathy in its early stage. Alternatively, the seven-standard field stereoscopic 30° color film fundus photography, which aids in viewing the major structures of the fundus such as the retina, optic disc, macula, fovea and blood vessels, is considered to be superior over direct and indirect ophthalmoscopy in terms of its sensitivity and specificity [4, 9, 10]. A typical fundus camera has an approximately 30° field of view for viewing the entire optic disc and temporal side of the macula simultaneously [10]. Current colour fundus photography methods use either a stereotypic or nonstereotypic camera, which can take seven 30°, three wide angle 60° or nine overlapping 45° fields [10]. Curvature of the fundus and aberrations caused by the patient's eye results in poor image quality for cameras with wider fields as high as 150°. Other limitations of dilated (mydriatic) fundus or retinal photography are the requirement of a professional photographer as well as an interpreter, high cost of the instrument, time consumption and uncomfortable pupillary dilation [4, 7, 10]. Comparatively, digital non-mydriatic fundus or retinal photography is advantageous as it neither requires dilation of the pupil nor requires a skilled operator [4, 11]. In addition, digital photography assists in documenting the progression of DR, aids in remote analysis and allows implementation of image processing methods [7]. However, complete evaluation of this method is still required as its senstivity has not been sufficient enough to detect DME [4, 12].

Another well-established method is the retinal fluorescein angiogram (FA), used for detecting the presence of neovascularisation in the retina, which is challenging to identify using clinical examinations [10]. FA is capable of providing ultra-wide field imaging of 200°, wherein the entire posterior eye segment can be visualized extending beyond the equator of the eye. However, it does not provide cross-sectional information of retinal layers as it is not a depth-resolved approach [13]. Although automated computerised methods using FA have been developed for the detection and quantification of microaneurysms, it is not considered as an ideal method of detection due to the following limitations, such as the inability to repeat the examination immediately or after a short term, requirement of invasive venipuncture and intravenous injection of sodium fluorescein or indocyanine green dye into the body [14]. In addition, though extremely rare, the injected dye might cause other life threatening reactions such as bronchospasm, anaphylaxis and cardiac arrest [10, 14].

Compared to FA, which necessitates injection of fluorescein, a non-invasive method of detection based on optical coherence tomography (OCT) techniques is capable of providing high-resolution (order of micrometers) and depth-resolved imaging, is considered to be more suitable for routine screening and early detection of retinal diseases with high specificity. OCT technique has also been reported to have a higher sensitivity of >90% when compared to retinal thickness analyser in the clinical diagnosis of macular edema [15]. In addition to microaneurysms developing in the early stages of diabetic retinopathy, neovascularizations, intraretinal microvascular abnormalities and non-perfusion areas are the other morphological changes of diabetic retinopathy, detectable by optical coherence tomography angiography (OCT-A), which are subtle to be detected by other methods such as clinical dilated examinations using FA and color fundus photography [13, 16, 17], as shown in figure 2.1 (reprinted from [6]). Figure 2.1(a) shows the fundus photography, figure 2.1(b) shows the FA image, figures 2.1(c) and (d) show the 3 × 3 mm image obtained using FA and the OCT-A image, respectively, with white arrows denoting superficial collaterals and orange arrows denoting non-perfusion areas. Figure 2.1(e) shows the corresponding B-scan obtained using spectral domain OCT. Automated early detection and grading of DR have also been possible using OCT-A imaging modality incorporated with machine learning techniques [18, 19].

A detailed explanation of the OCT technique can be seen in the forthcoming sections. Section 2.2 comprises of a brief discussion on the optical properties of eye, followed by section 2.3, which describes the basic principle of the OCT technique, the wavelengths used and a detailed discussion on different OCT imaging modalities

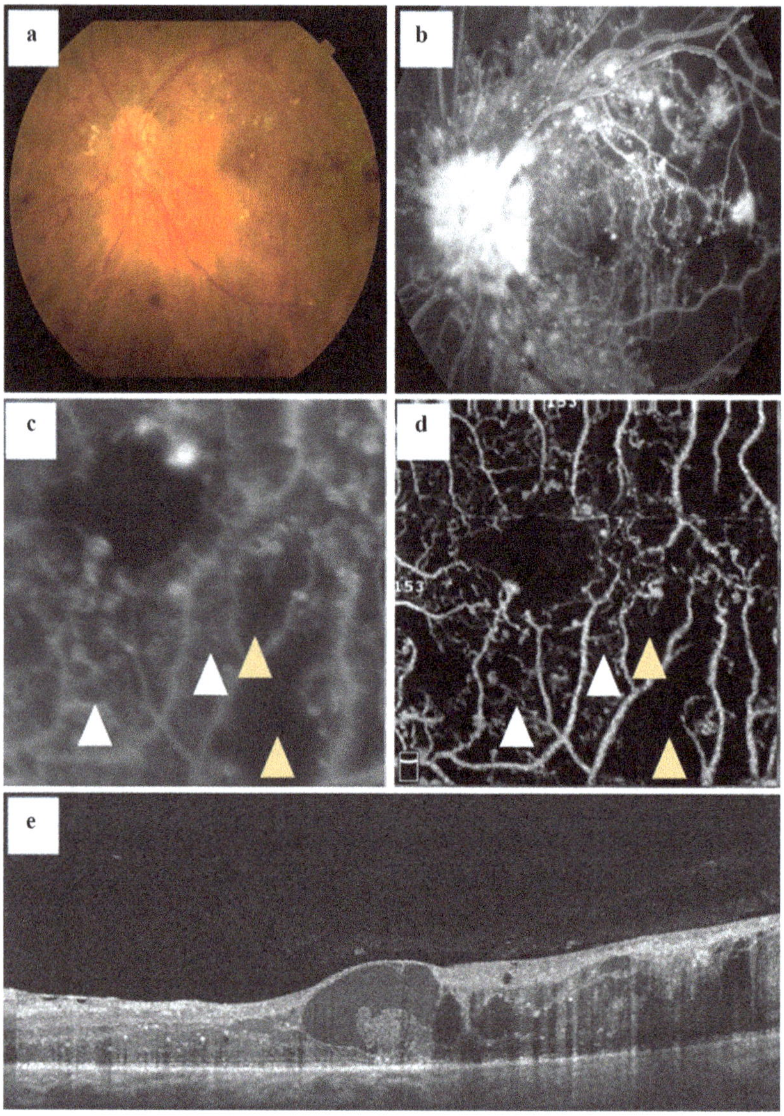

Figure 2.1. Results of intraretinal microcirculation and capillary non-perfusion in proliferative diabetic retinopathy condition (a) Fundus photography, (b) FA image, (c) 3 × 3 mm image of FA, (d) 3 × 3 mm OCT-A image, showing superficial collaterals and non-perfusion areas with white and orange marks and (e) Corresponding cross-sectional B-scan of the vasculature obtained using SD-OCT. [Reprinted with permission from Springer Nature from [6] under the Creative Commons Attribution 4.0 International License (http://creativecommons.org/licenses/by/4.0/).]

such as spectral domain OCT (SD-OCT) and swept source OCT (SS-OCT) systems used in the detection of DR. Section 2.4 includes the milestones of the OCT technique in the field of ophthalmology. Section 2.5 comprises a detailed discussion on the assessment of pathological conditions related to DR using OCT-A, which includes the principle, commercial systems available, limitations, algorithms implemented to obtain artifact-free OCT-A images and also highlights the applications of deep learning in OCT-A imaging.

2.2 Optical properties of human eye tissue

Figure 2.2 shows the major structures of the eye, which includes the cornea, lens, retina, anterior and posterior chambers of the eye filled with aqueous humor and vitreous humor, respectively, with their refractive indices and the transmission of different wavelengths of light passing through the eye structures. The refractive index of the aqueous and vitreous humor of the eye is 1.336, while the refractive index of the cornea is 1.376 and the lens is 1.386–1.406 [9]. The refractive index of fibers and the surrounding ground substance of the cornea are 1.47 and 1.354, respectively [9].

Absorption properties of the ocular media is an important factor to be considered, as an increased absorption would lead to increased attenuation of the light resulting in a decrease in the backreflected signal (due to the roudtrip of light from the source reaching the retina and then reflecting back to the detector) [20]. Water being the major component of the eye strongly influences the spectral absorption of the ocular media. Apart from water, proteins and other cellular components also influence the absorption process. Strong absorption of light by the cornea and lens occurs below 300 nm and between 300 and 400 nm, respectively [9]. The layers of the retina in front of the cones and rods are considered to be highly transparent and absorption is attributed to the visual pigments such as retinal macular pigment, visual pigments in the photoreceptors, haemoglobin in the choroid and melanin in the pigment epithelium [9]. The amount of pigments, particularly

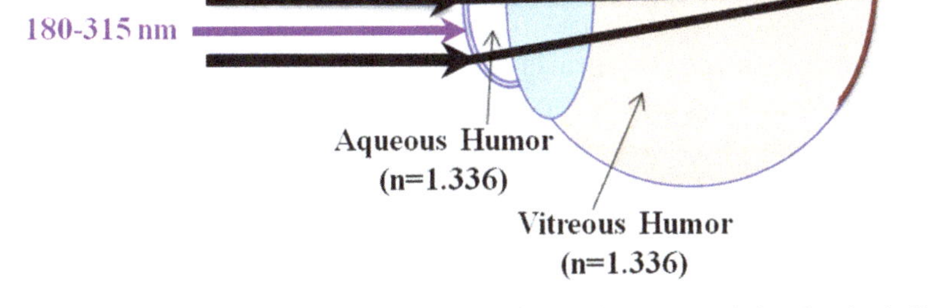

Figure 2.2. Schematic representation of the major structures of a human eye, their refractive indices and the transmission wavelengths of light passing through the eye.

melanin, vary with individuals. The absorption of these pigments are prominent between 400–600 nm and is less significant beyond 600 nm, hence illumination band is selected beyond 500 nm in visible source based OCT system.

Choroid is a thin, highly vascularized layer present between the retina and sclera [9]. Choroid contains the absorbers melanin and haemoglobin, which strongly absorb short wavelength light and backscatter longer wavelengths. A small amount of light penetrates further into schlera, a dense, whitish layer, from which light gets backscattered strongly and reaches the retina. Fundus reflectance is the light that specularly gets reflected and scattered at the fundus and travels back out of the eye. It is low at shorter wavelengths (below 600 nm) and high at longer wavelengths (beyond 600 nm), attributing to the blood in the choroid [9]. Wavelengths around 850, 1050 and 1300 nm are commonly used in ophthalmic OCT techniques, omitting the prominent water absorption bands around 1400 and 1900 nm [21], to achieve an adequate penetration of light through the retinal layers to image choroid [20].

It is essential to understand the optical properties of various structures in the eye and the maximum permissible exposure limit, in order to choose the light source with a suitable wavelength and power level to avoid thermal damages to the structures of the eye. The visible OCT (Vis-OCT) retinal imaging system uses a maximum illumination power of 226 μW and an illumination band beyond 500 nm, while the near infrared OCT (NIR-OCT) retinal imaging system uses a maximum illumination power of 1 mW [22, 23]. The following section includes the basic principle of the OCT technique and a brief discussion on the advantages and challenges of Vis-OCT and NIR-OCT systems.

2.3 OCT technique

2.3.1 Basic principle

OCT is a non-invasive interferometric technique and is considered to be an important diagnostic tool in ophthalmology [1]. It is similar to the pulse-echo ultrasonography technique, which is based on 'time of flight' [24]. OCT is advantageous in providing cross-sectional images of better resolution, around 10 μm, compared to other imaging techniques such as magnetic resonance imaging and ultrasound imaging [24]. In addition, unlike ultrasound techniques which require probe tissue contact, in the OCT technique the light beam can pass through the air-tissue interface without the need for any immersion fluid or contact of the probe with the tissue to be investigated. The use of near infrared light, allows probing of depths from 1–2 mm beneath the tissue surface, better than confocal microscopy [1].

The OCT system is generally operated in a Michelson interferometer configuration as shown in figure 2.3. In the interferometer setup, a part of the light is directed to the sample and the rest to a reference arm of known length. The backscattering (amplitude and phase information) from within the tissue is made to interfere with the backreflected reference beam and the resultant interferogram is analysed to form high resolution cross-sectional images of the tissue. The sample is illuminated at one spot and the depth profile at that spot is obtained by analysing the

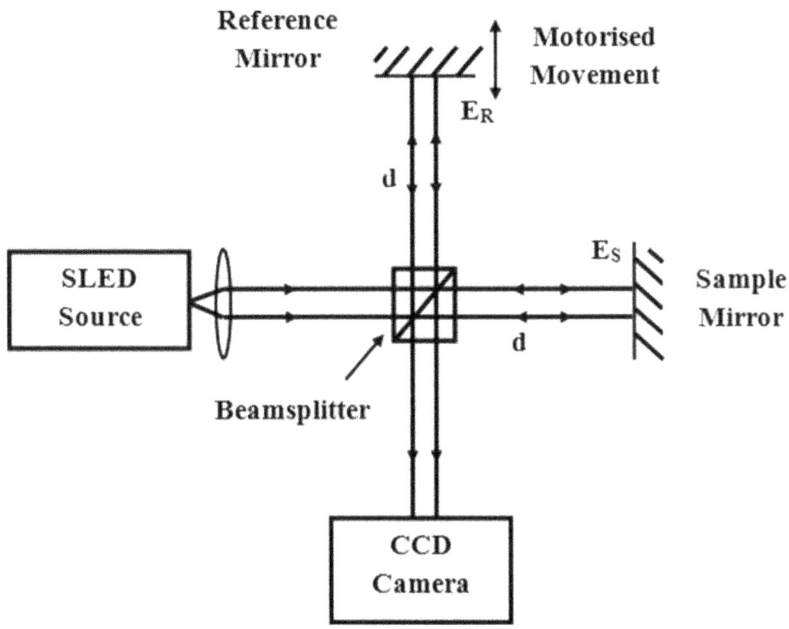

Figure 2.3. Optical coherence tomography configuration.

backscattered light. By scanning the spot on the tissue surface, 2D and 3D images can be obtained.

The amplitude of the electric field of the light source is expressed as E_{SR}. The source beam is split into a reference beam and a sample beam with amplitude of E_R and E_S, respectively, as shown in figure 2.3. The reflectivity values of the mirrors are assumed to be 100%. The backreflected light from the two arms combine at the beamsplitter and the electric field detected by the detector E_D, is shown in equation (2.1) [25],

$$E_D = \frac{1}{\sqrt{2}}E_R + \frac{1}{\sqrt{2}}E_S \tag{2.1}$$

$$E_R(x) = \frac{1}{\sqrt{2}}E_{\mathrm{SR}}e^{i(\omega x_r)/c} \tag{2.2}$$

$$E_s = \frac{1}{\sqrt{2}}E_{\mathrm{SR}}e^{i(\omega x_s)/c} \tag{2.3}$$

where ω is the angular frequency, x_r is the displacement of the reference mirror, x_s is the displacement of the sample mirror, c is the speed of light. Considering a phase difference being introduced in the sample arm, a difference in phase is introduced between the two beams and E_D is as shown in equation (2.4).

$$E_D(x) = \frac{1}{\sqrt{2}}E_{\mathrm{SR}}e^{i(\omega x_r)/c} + \frac{1}{\sqrt{2}}E_{\mathrm{SR}}e^{i(\omega x_s + a)/c} \tag{2.4}$$

The detector generally measures the irradiance and not the electric field, which is the time average of the square of the amplitude of the electric field as shown in equation (2.5).

$$I_D(x) = \langle E_D E_D^* \rangle = \frac{1}{2}\langle E_R E_R^* \rangle + \frac{1}{2}\langle E_S E_S^* \rangle + \frac{1}{2}\langle E_S E_R^* \rangle + \frac{1}{2}\langle E_R E_S^* \rangle \tag{2.5}$$

where E_R^* and E_S^* are the complex conjugate of the electric field in the reference and sample arm, respectively. Equation (2.5) is expressed as follows:

$$I_D(x) = \frac{1}{2}I_R(x) + \frac{1}{2}I_S(x) + \frac{1}{4}\langle E_{\mathrm{SR}}^* e^{-i(\omega x_r)/c} E_{\mathrm{SR}} e^{i(\omega x_s)/c} + E_{\mathrm{SR}} e^{i(\omega x_r)/c} E_{\mathrm{SR}}^* e^{-i(\omega x_s)/c}\rangle \tag{2.6}$$

The first two terms in equation (2.6) represent constant DC terms, which do not carry any information relevant for imaging. The cross-correlation terms carry the interference information. By using the identity, $2\cos\theta = e^{i\theta} + e^{-i\theta}$, equation (2.6) becomes

$$I_D(x) = \frac{1}{2}I_R(x) + \frac{1}{2}I_S(x) + \frac{1}{2}\left\langle \mathrm{Re}(\frac{1}{2}E_{\mathrm{SR}}E_{\mathrm{SR}}^* e^{i(\omega(x_r - x_s)/c)})\right\rangle \tag{2.7}$$

$$I_D(x) = \frac{1}{2}I_R(x) + \frac{1}{2}I_S(x) + \frac{1}{4}\langle E_{\mathrm{SR}}E_{\mathrm{SR}}^*\rangle \cos\theta \tag{2.8}$$

When θ is 0 or $\pm 2\pi$, the cosine value is at a maximum of 1, denoting constructive interference, and when θ is $\pm\pi$, the cosine value is at minimum of −1, denoting destructive interference. Coherence length (l_c) of a source helps in defining the axial resolution (δz) of imaging in OCT systems. Considering a source with a Gaussian shaped spectrum with center wavelength λ_0, spectral bandwidth $\Delta\lambda$, and a medium with refractive index n, the axial resolution δz can be estimated as follows [26]:

$$l_c = \delta z = \frac{2\ln 2}{n\pi}\left(\frac{\lambda_0^2}{\Delta\lambda}\right) \tag{2.9}$$

The focusing optics in the sample arm play an important role in determining the lateral resolution of imaging. For a Gaussian beam with diameter D, focused by a lens with focal length f, the diameter of the focal spot δx is estimated as follows [26]:

$$\delta x = \frac{4\lambda_0 f}{\pi D} \tag{2.10}$$

2.3.2 Visible and NIR light sources used in OCT systems

Light sources mostly used in commercial OCT systems, which are based on low coherence interferometry are (i) inexpensive superluminscent light emitting diodes (SLEDs) with a typical bandwidth of less than 100 nm and centre wavelength ranging from 670 to 1600 nm. Other sources include (ii) supercontinuum sources (SCs) in which nonlinear effects of photonic crystal fibers result in spectral broadening of the pump laser, which leads to the emergence of ultra broadband (900–2800 nm) and

Table 2.1. Comparison of visible and NIR OCT systems [22].

Vis-OCT	NIR-OCT
Strong attenuation coefficient of biological tissue - limits illumination power into deeper tissues	Deeper penetration depth achievable upto schlera compared to Vis-OCT
Increased image contrast due to higher scattering coefficient of biological tissue in the visible region	Limited axial resolution of 5 to 7 μm compared to 2 μm possible using Vis-OCT in imaging retinal pigmented epithelium
Large variation in refractive indices of glass and chromatic aberration of optical lenses - makes ultrabroadband Vis-OCT challenging	
Lower reflectivity of gold and aluminium coated mirrors in the visible region compared to NIR region	
Lower A-line rate of Vis-OCT makes it more susceptible to motion artifacts compared to NIR-OCT	

multiband OCT systems capable of providing sub-micrometer resolutions, (iii) ultrafast Ti:sapphire laser sources operating at 800 nm and (iv) swept sources which sweep through continuous wavelengths [21, 22, 27].

Swept sources used in OCT systems replace spectrometers with single element photodetectors and are capable of providing increased A-line rates and effective penetration depth [22]. Earlier, free-space OCT systems were adopted to avoid excessive loss and imbalanced dispersion caused by fiber coupler [28]. However, recently, compact fiber-based OCT system incorporated with better performance fiber couplers are established, providing ease in alignment and maintenance [29].

Table 2.1 lists the comparison of Vis-OCT system and NIR-OCT systems [22]. Vis-OCT system is advantageous in retinal imaging and studies based on oxygen-related hemodynamics, as oxygenated hemoglobin and reduced hemoglobin are distinctive in the visible wavelength region compared to NIR spectral range. Whereas, NIR-OCT systems are generally used in imaging human retina, as they provide improved imaging depth as well as due to the availability of low cost light sources in the NIR spectral range. The rapid development of light sources and detection systems in OCT have not only enabled morphological imaging, but also have advanced towards functional imaging, using optical microangiography [22].

2.3.3 Different types of OCT systems

OCT systems are broadly classified into time domain (TD-OCT), spectral domain OCT (SD-OCT) and swept OCT (SS-OCT) systems based on the source, detector and the scanning mechanism incorporated in the system.

2.3.3.1 Time domain OCT system

In the TD-OCT type, the scanning of the biological sample is done by varying the reference arm length as shown in figure 2.4. As the path difference Δz is changed by moving the reference mirror, the backscattering from the corresponding depth is picked up [30]. This allows scanning of the sample in the axial or depth direction. Such a scan is often referred to as an A-scan. The mode of scanning can be in the form of mechanical stepper motor systems, piezoelectric, galvano scanners, etc.

The interference pattern is recorded using a photodetector or a camera at various points during scanning (steps of coherence length). A simple DC subtraction procedure is performed in order to see the final image [31]. The main advantages of the TD-OCT system are a lower post-processing procedure and the depth of imaging is dependent on the attenuation properties of the sample [32]. As this system uses a photodetector and requires a scanning mechanism in order to acquire the cross-sectional information, the acquisition time is longer, which makes it difficult in ophthalmology applications. The imaging resolution is dependent on the scanning mechanism.

2.3.3.2 Spectral domain OCT system

SD-OCT provides depth-resolved tissue structure information by spectral analysis of the interference fringe pattern. Unlike TD-OCT, the reference arm is stationary in SD-OCT systems which allows for high-speed imaging. SD-OCT can be further divided into two, based on the light source being used. The first one uses a broadband light source and a spectrometer and the interference pattern is recorded as a function of wavelength.

The captured spectral information on post-processing using a Fourier transform gives the cross-sectional information and the resolution of the system is dependent on the spectrometer resolution. The speed of imaging is dependent on the readout

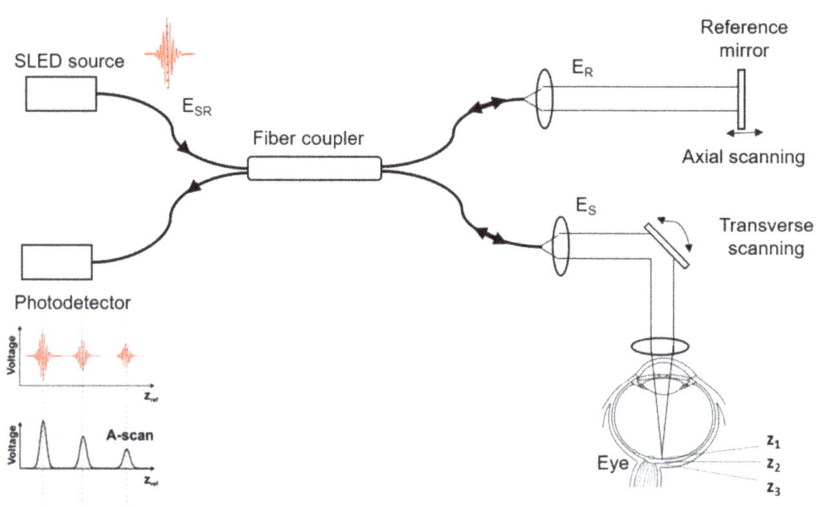

Figure 2.4. Schematic of a time domain optical coherence tomography system.

time of the sensor in the spectrometer [33]. In this system, all reflections from within the sample (due to the presence of multiple scatters) is available at once in the backreflected spectrum as shown in figure 2.4. Each of these reflectors has to be separated out to form the reflectivity versus depth profile or the A-scan. The zero in the A-scan represents the reference arm position and the sample reflectors (with respect to the reference arm) appear at double the distance owing to the round trip distance of each reflector as shown in figure 2.5. Hence, a new single pass depth variable is accommodated in the post-processing algorithm to correct this term. The post-processing performed on the spectrometer output is as shown in figure 2.6.

The time spent to collect photons for OCT signal reconstruction is dependent on the total depth scanned x_{depth}, total scan time T and coherence length l_c as shown in the expression below [34]. SD-OCT completely makes use of the detected photons and hence more sensitive than TD-OCT.

$$\frac{l_c}{x_{\text{depth}}}T \tag{2.11}$$

In k-space, the measured signal is proportional to $x_r - x_s$. Smaller $x_r - x_s$ value produces a slower sinusoidal oscillation than larger $x_r - x_s$ where x_r is the reference arm length and x_s is sample arm length. The scanning depth achievable in a SD-OCT system is based on the spectral resolution of the spectrometer used [34]. For a spectrometer with N pixels, the highest detectable spectra is N/2 beyond which aliasing takes place.

$$x_{\text{depth}} = \left(\frac{N}{2}\right)l_c \tag{2.12}$$

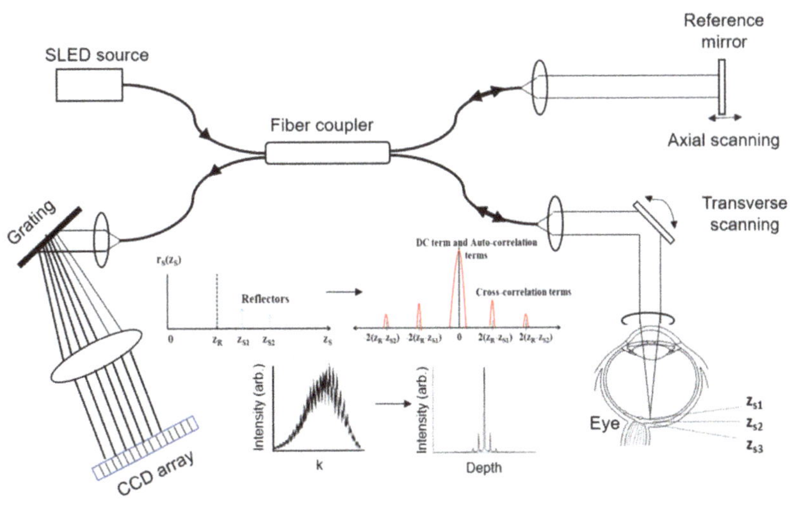

Figure 2.5. Schematic of a spectral domain optical coherence tomography system.

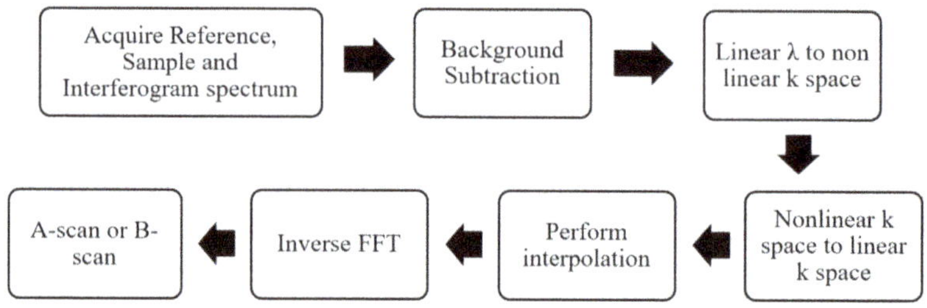

Figure 2.6. FD-OCT post-processing algorithm flowchart.

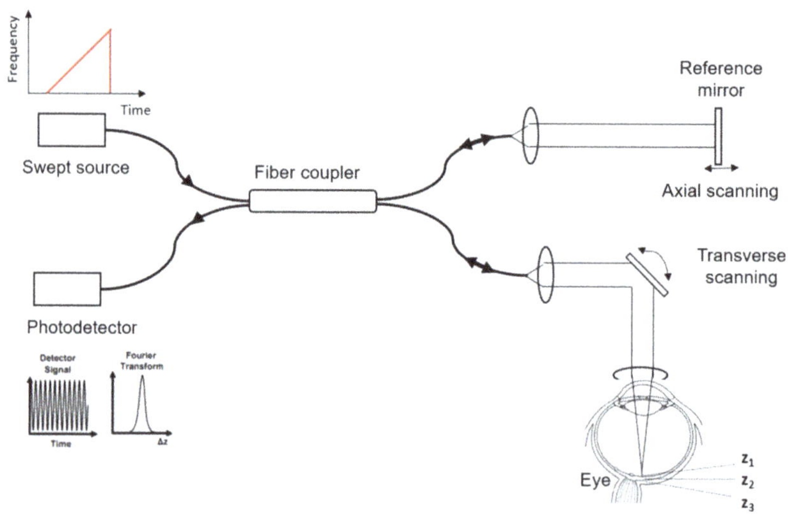

Figure 2.7. Schematic of a swept source optical coherence tomography system.

2.3.3.3 Swept source OCT system

SS-OCT uses an optical source which rapidly sweeps a narrow line-width over a broad range of wavelengths and a photodetector similar to that of TD-OCT, as shown in figure 2.7 [35]. The sweeping rates are in the order of a hundred kilohertz with ultrafast detection and analog-to-digital conversion in the order of gigahertz. The spectral interferogram for each wavelength is recorded at each sample point and Fourier transformation is used to extract their individual contribution as a function of depth position.

The speed of imaging is dependent on the sweep rate of the swept source. This high speed has resulted in volumetric real-time imaging in the field of ophthalmology leading to better diagnosis. In the rapidly tunable narrowband laser source, the source $k(t)$ is swept linearly,

$$k(t) = k_0 + \delta k \times t \tag{2.13}$$

$$\delta k = \Delta k / \Delta t \tag{2.14}$$

where Δk is the total bandwidth swept and Δt is the total sweep time. The signal is evenly spaced at M such that $M \times \delta k = \Delta k$ [34]. The depth scan range can be given by the following equation.

$$x_{\text{depth}} = \frac{M}{2} l_c \tag{2.15}$$

The signal being evenly spaced, a discrete Fourier transform (DFT) on the interference signal will help derive the depth scan of the tissue.

2.4 Milestones in the advancement of OCT in ophthalmology

The first OCT studies on *ex-vivo* bovine eyes was conducted in 1991, using a low-coherence diode source at wavelength ~800 nm to obtain the A-scans of the anterior eye chamber of the bovine eye, with 10 μm axial resolution and a detection sensitivity of −100 dB [36]. In 1993, the first human retinal images were obtained *in vivo* at 840 nm wavelength with an axial resolution of 15 μm [36]. In 1996, the first commercial ophthalmic OCT device was introduced. The second and third generation OCT instruments in ophthalmology were developed in 2000 and 2002, respectively. Commercially available time domain OCT (TD-OCT) 3000 (Stratus OCT) has been a standard diagnostic tool for detecting posterior segment retinal tomography. Compared to TD-OCT, which involves translation of the reference mirror pathlength in time, frequency domain OCT (FD-OCT) in the spectral or Fourier domain enables extraction of spectral information from the sample in one exposure with the help of a spectrometer with a high signal-to-noise ratio (SNR) and also provides 40 000 axial scans per second [5, 9]. The potential of TD-OCT and FD-OCT in the evaluation and treatment of DME and diabetic retinopathy has been discussed in several reviews [5, 36]. Figure 2.8 shows the milestones of OCT development in the field of ophthalmology since 1991.

In 2006, commercial spectral domain OCT (SD-OCT) was introduced into the ophthalmic market [36]. SD-OCT aids in quantification of fluid accumulation, identification of different edematous patterns, and identification of microstructural changes in the retina and fovea [5, 37]. The ability to detect alterations in the microvasculature using fluorescein angiography is limited because of the superposition of the capillary network and leakage. Alternatively, the three dimensional imaging capability of SD-OCT provides vascular mapping of the macular perfusion [16]. Accurate *in vivo* measurement of retinal thickness for segmentation of the retinal layers has also been possible using SD-OCT technique.

Faster imaging speeds and longer imaging range of the swept laser based swept source OCT (SS-OCT) system, without the need of spectrometers and line scan cameras compared to SD-OCT, have made SS-OCT favourable for retinal imaging with an axial resolution of ~5 to ~6 μm at a wavelength of 1050 nm, resulting in 70 000 and 312 500 axial scans per second, thereby reducing the exposure time and eliminating motion artifacts [38]. Compared to SD-OCT systems, a longer wavelength of 1050 nm used in the SS-OCT system provides improved imaging depth because of less attenuation from ocular structures [39]. The SS-OCT system with a

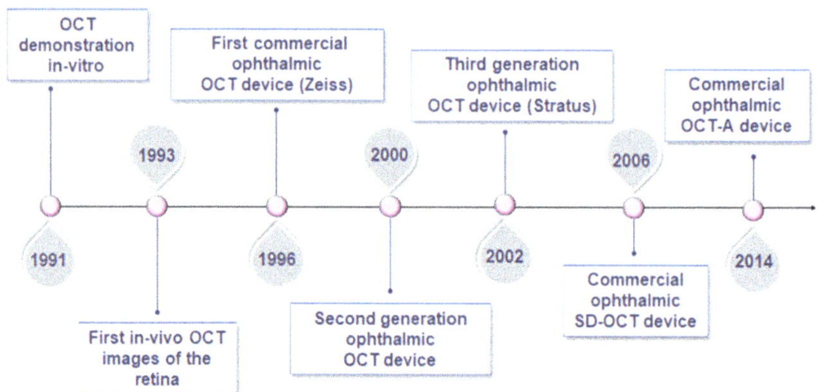

Figure 2.8. Schematic representation of different stages of OCT development in the field of ophthalmology.

vertical cavity surface emitting laser (VCSEL) source providing wide field images of retina and choroid at 580 000 axial scans per second has been demonstrated [35, 39]. The high speed capability of SS-OCT coupled with an eye tracking algorithm allows for a wider field of imaging in addition to deeper imaging, thereby producing denser scans [40]. However, SD-OCT has better resolution compared to SS-OCT and the usage of shorter wavelength gives much more detail on the layers of the retina when compared to SS-OCT [41]. SD-OCT is still the most widely used OCT system clinically due to its low cost.

Optical coherence tomography angiography (OCT-A) was introduced in 2014; is a label-free and a functional modality of the OCT technique. It is a high-resolution imaging technique based on the fast SD-OCT or SS-OCT systems, which provides two- and three-dimensional volumetric angiograms of choroidal blood circulation within the eye and intraretinal microvascular abnormalities, allowing repeatable precise visualization, without any risk to the patients, overcoming all the drawbacks of fluorescein angiography [16]. OCT-A also helps in differentiating microaneurysms in both superficial and deep vascular plexus, which is challenging to visualize by fluorescein angiography [42]. Most of the commercial OCT-A systems are SD-OCT based. SS-OCT based angiography can reliably detect microvascular anomalies and neovascularisation.

2.5 OCT-A technique

2.5.1 Principle of OCT-A technique

The underlying principle of OCT-A is to develop contrast between static blood vessels and flowing blood cells (intrinsic contrast) by means of Doppler shift or speckle variance [36]. Unlike Doppler, OCT-A helps in separating out the scatterers from the static blood vessels in order to create angiograms. The principle of OCT-A depends on repeated detection of the OCT signal at the same location that varies with time, with respect to the movement caused by the constantly moving red blood cells which act as a contrast mechanism for imaging the flow in the blood vessels compared to the static signal reflected from the static tissue, thereby aiding in the

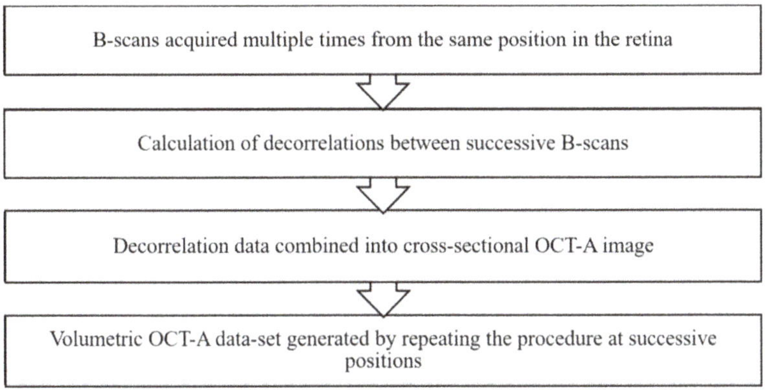

Figure 2.9. Flowchart representing the steps performed in obtaining volumetric OCT-A image.

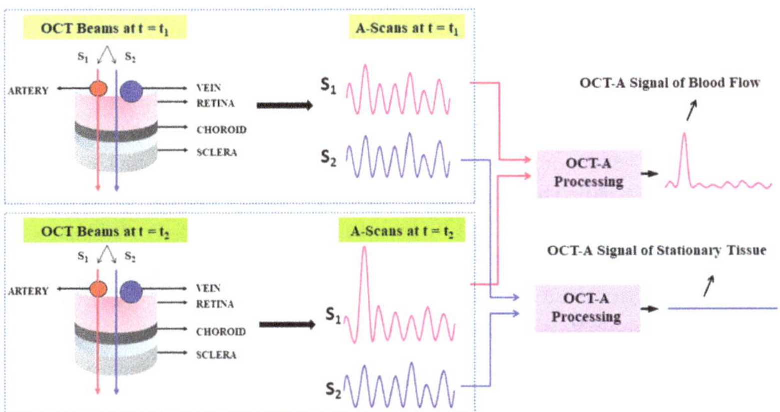

Figure 2.10. A simplified schematic representation of the OCT-A principle: light beams striking the retinal tissue interacting with the blood vessel (S_1—pink arrow) and the stationary tissue (S_2—blue arrow) at time $t = t_1$ and $t = t_2$, and the OCT A-scans generated from the light signals backscattered from the blood vessel and the stationary tissue represented as pink (S_1) and blue (S_2) traces at time t_1 and t_2. The averaged signals after processing form composite signals representing increased variability of the OCT signal from beam S_1 as a result of movement of red blood cell (pink) compared to the signal obtained from stationary tissue (blue).

visualization of the microvasculature [36, 43]. A simplified flowchart representing the steps performed in obtaining a volumetric OCT-A image is shown in figure 2.9 and a simplified illustration of the OCT-A principle is shown in figure 2.10, representing the averaged OCT-A signals obtained after processing the OCT signals S_1 and S_2 backscattered from the blood vessel and the stationary tissue of the retina, respectively, at time t_1 and t_2.

2.5.2 Assessment of various pathological conditions related to DR using OCT-A

OCT-A has also been utilized as an alternative non-invasive approach to FA in detecting abnormalities in the intraretinal microvasculature, diabetic macular ischemia and diabetic macular edema [44]. Traditional FA shows only the superficial plexus whereas OCT-A helps image retinal and choroidal microvasculature, thereby finding many retinal diseases such as age-related macular degeneration, arterial and venous occlusion of the retina and uveitis [6] accurately with precise anatomical depth information [37]. The changes induced in the intensity and phase of the backscattered light signal due to the blood flow in the blood vessels is detected without the need for any fluorescein dye. Grading the severity of DR could be assessed by detecting precisely the extent of non-perfusion areas of capillary in the macular region using OCT-A compared to FA [16]. This technique also helps in post-treatment monitoring of proliferative diabetic retinopathy condition [13].

2.5.2.1 Quantitative measurement of features for early diagnosis of DR using OCT-A

Quantitative measurements of the following features aid in understanding the pathophysiology and thereby determining the progression of DR: (i) area density of the vessel, (ii) spacing of the vessel, (iii) length density of the vessel, (iv) diameter index of the vessel, (v) total length of the vessel, (vi) tortuosity or fractal dimension of the vessel and (vii) area of the avascular zone in the fovea [13, 45–51]. Change in the vessel density in the deep capillary plexus is considered to predict the severity of DR. Decrease in the vessel density in both the superficial as well as the deep capillary plexus enables monitoring follow-up of treatment in patients [13]. On the other hand, early detection of DR using OCT-A may be possible by detecting the inter-capillary spacing for non-perfusion as it tends to occur even before the manifestation of changes in the vessel area density., whereas increase in the vessel tortuosity has been considered as an early indicator of damage to the retinal microvasculature. The measurement of vessel tortuosity features may be used to differentiate non-proliferative diabetic retinopathy (NPDR) from proliferative diabetic retinopathy (PDR) in the superficial capillary plexus, as the vessel tortuosity increases in the case of NPDR and decreases in the case of PDR [52]. Another early indicating feature of DR is the fractal dimension, which measures the complexity of the branching pattern of blood vessels, which was found to reduce in the superficial as well as the deep capillary plexus as a result of diabetes [53]. The foveal avascular zone (FAZ) is another feature which tends to increase during NPDR conditions [54]. Compared to the commonly used descriptive scaling system based on error-prone qualitative grading analysis to predict the severity of DR, OCT-A can provide improved complementary quantitative analysis by providing measurements of the values of features, including the FAZ [17]. These features are extracted from the acquired OCT-A images and further compiled for enabling early diagnosis and classification of diabetic retinopathy using computer-aided diagnosis (CAD) methods [19, 51, 55].

2.5.3 Advancements in the OCT-A imaging modalities

2.5.3.1 SD-OCT based OCT-A

RTVue XR Avanti (Optovue) with the AngioVue software is the first commercially available OCT-A system with a proprietary split-spectrum amplitude decorrelation

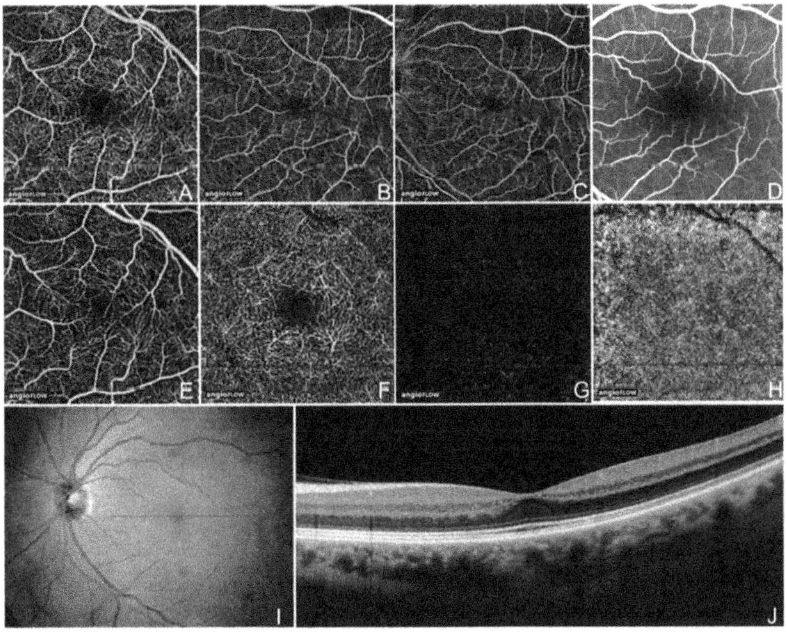

Figure 2.11. Images of a normal eye obtained using RTVue XR Avanti (Optovue) and processed using AngioVue software (A) 3 × 3 mm OCT-A image, (B) 6 × 6 mm OCT-A image, (C) 8 × 8 mm OCT-A image, (D) Fluorescein angiography image cropped to 8 × 8 mm, (E) 3 × 3 mm OCT-A image of the 'Superficial' inner retina, (F) 3 × 3 mm OCT-A image of the 'Deep' inner retina, (G) 3 × 3 mm OCT-A image of the outer retina (no vasculature), (H) 3 × 3 mm OCT-A image of the choriocapillaris, (I) En-face intensity OCT image. (J) Highly-sampled OCT B-scan image. [Reprinted with permission from Springer Nature from [56] under the Creative Commons Attribution 4.0 International License (http://creativecommons.org/licenses/by/4.0/).]

angiography (SSADA) software to produce detailed images by improving signal-to-noise ratio and helps minimize scan time [56]. It uses a light source centered at 840 nm with a spectral width of 45 nm. The imaging resolution axially is 5 μm and laterally is 15 μm and can image a wide field of view of 6 × 6 mm. The imaging speed is 70 000 A-scans per second and each B-scan contains 216 A-scans. The system takes approximately 3 s to produce a 3D scan. The system is inbuilt with a motion correction technology to reduce motion artifacts. The images taken with this system is shown in figure 2.11 [56].

Topcon's 3D OCT-1 (Type: Maestro2) OCT system enables OCT-A imaging with a scanning speed of 50 000 A-scans per second with resolutions of 6 μm axially and 20 μm laterally. Zeiss's Cirrus 6000 system employs a 840 nm SLED source and has a high scan speed of 100 000 A-scans per second with resolutions of 5 μm axially and 15 μm laterally with a field of view of upto 12× 12 mm for OCT-A imaging [6].

2.5.3.2 SS-OCT based OCT-A

SPECTRALIS from Heidelberg engineering is another commercially available OCT system with OCT-A module. The imaging resolution of this system is 3.9 μm axially and 5.7 μm laterally with a field of view of 2.9 × 2.9 mm. The proprietary TruTrack active eye tracking system enables detailed visualization of capillary network. The equipment uses a probabilistic OCT-A algorithm to separate out the stationary tissue and moving blood accurately [57].

Topcon DRI Triton is an SS-OCT based angiography system with a wide field of view upto 12 × 9 mm and scans at the rate of 100 000 A-scans per second and uses the OCT-A ratio analysis (OCT-ARA) which keeps the whole spectrum intact in order to preserve the axial resolution. The axial resolution obtained is around 2.6 μm. The volumetric scan is acquired over a 3 × 3 mm field of view in 4 s [58, 59].

AngioPlex OCT angiography equipment from Carl Zeiss has a scanning rate of 68 000 A-scans per second and covers a field of view of about 3 × 3 mm and 6 × 6 mm. This system has a tracking software known as FastTrac retinal tracking. This system uses a complex algorithm known as OCT microangiography-complex (OMAG) which analyses the intensity and phase differences between consecutive B-scans at the same anatomical location to generate the image [6].

Zeiss PLEX Elite 9000 is an SS-OCT angiography equipment which provides a scan speed of 100 000 kHz with a center wavelength of 1060 nm, with resolution of 6 μm axially and 14 μm laterally enabling visualization of deeper (3 mm) and wider range from vitreous to sclera in the posterior segment [60]. The image of an eye with retinal vein occlusion is shown in figure 2.12 along with the corresponding FA image. The SS-OCT image covered more area compared to FA image with greater detail.

Diabetic retinopathy is characterised by the microaneurysms, ischemia and capillary non-perfusion in the retina. A study conducted by Federico Corvi and

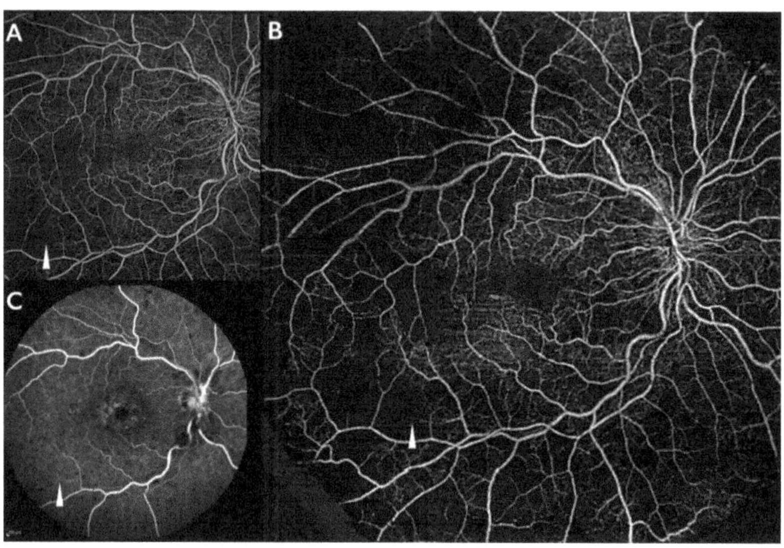

Figure 2.12. Images of an eye with retinal vein occlusion: (A) SS-OCT-A image, (B) SS-OCT-A extended wide field image and (C) FA image. The white arrow shows interrupted vessels in OCT-A image and filled vessel in FA. The slow flow in vessel influence OCT-A image signal leading to over estimation of non-perfusion. [Reprinted with permission from PLOS ONE from [59] under the Creative Commons Attribution 4.0 International License (http://creativecommons.org/licenses/by/4.0/).]

group compared five different OCT-A systems to evaluate the ability to identify retinal microaneurysms (MA) in the superficial capillary plexus (SCP) and deep capillary plexus (DCP) [59]. It was found that Plex Elite and AngioPlex showed higher number of MA's in SCP. DRI OCT Triton, Spectralis and Avanti showed better in DCP [61]. Figure 2.13 as reprinted from [6] shows that OCT angiography imaging can show the origin of the microaneurysms as demarcated saccular or fusiform shapes in the dilated capillaries in the retinal layers and the lesions can be precisely localised in terms of depth.

Canon has recently introduced the Xephilio OCT-S1, a large wide field SS-OCT system capable of imaging high resolution images up to 23 × 20 mm in a single scan. The system also has an artificial intelligence (AI) based intelligent denoise technology which helps to obtain high quality OCT-A image with reduced noise, increased detail and visibility. This system enables deeper and high resolution imaging of the eye. Thus, OCT-A is a promising imaging technique based on SD or SS-OCT for volumetric visualization of the retinal and choroidal microcirculation without the need of a contrast agent. The volumetric detail available helps in the assessment and treatment of diabetic retinopathy. Future scope would be in the direction of real

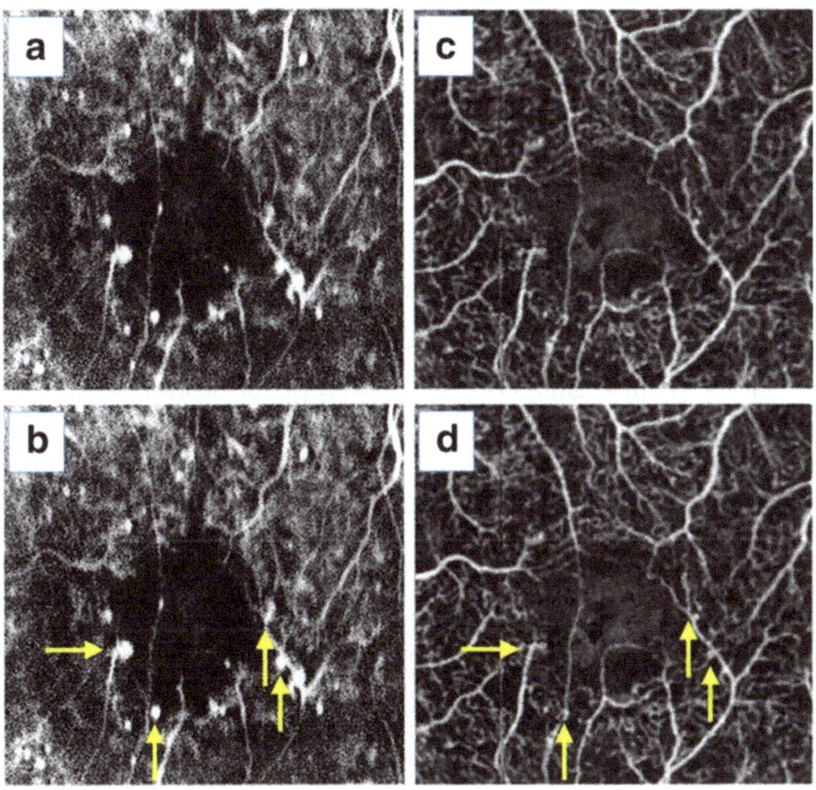

Figure 2.13. Image from a diabetic patient (a, b) FA image and (c, d) OCT-A image from SCP. Yellow arrows indicate MAs. [Reprinted with permission from Springer Nature from [6] under the Creative Commons Attribution 4.0 International License (http://creativecommons.org/licenses/by/4.0/).]

Table 2.2. OCT-A vs traditional angiography.

OCT-A	Angiography
Non-invasive	Invasive—dye injection
Imaging time—4–5 s	Imaging time—minimum of 10 min
Field of view—12 × 12 mm (max)	Field of view—200 degrees
Lesion depth can be identified and boundaries can be marked	Lesion depth cannot be identified
Multiple images at different depths	Depth information not available
Repeatable	Not repeatable
Quantitative measurement	Qualitative measurement

time monitoring of blood flow in the microvasculature and imaging larger area with shorter acquisition time.

A comparison of OCT-A based diagnosis of DR with that of conventional approaches have been reviewed by several research groups [6, 16, 17, 62]. The benefits of OCT-A over Angiography is tabulated in table 2.2.

2.5.4 Challenges of OCT-A imaging technique

Although OCT-A is advantageous over FA in many ways, one of the limitations of OCT-A is the narrow field of view compared to FA which limits generation of peripheral retinal images of good quality [56]. Another limitation of OCT-A for clinical application includes motion artifacts caused due to the movement and blinking of the eye, which may lead to appearance of white and black lines across the scan [62]. In the case of en face OCT-A, formation of shadow artifacts by moving blood cells in the superficial layers onto deeper layers lead to misinterpretation of results which is one of the major drawbacks in the detection of choroidal neovascularization. Another limitation is the flow projection artifacts in cross-sectional OCT-A, caused by superficial vessels of the retina, which limits the visualization of deeper vessel layer [63]. Development and incorporation of eye-tracking device and optimal algorithms to resolve and correct issues related to these artifacts is crucial [17, 64]. Despite these artifacts, as manual identification of neovascular tissue is time consuming, development of algorithms to perform automated repeatable segmentation to identify particular retinal vascular layer is another challenge related to OCT-A [62]. The following section describes various algorithms implemented based on the intensity, phase and a combination of both intensity and phase information of the OCT signal to overcome these challenges.

2.5.5 Algorithms used in OCT-A imaging for artifact removal

2.5.5.1 Split-spectrum amplitude-decorrelation angiography (SSADA) algorithm

Extraction of flow signal by distinguishing vessels from stationary tissues with an improved SNR and elimination of artifacts caused by the bulk motion of tissue has been provided by the split-spectrum amplitude-decorrelation angiography (SSADA) algorithm, wherein decorrelation in the intensity of the signal between the two consecutive B-scans that has been split into sub-bands from the full OCT spectrum, is computed and thereafter averaged to obtain the final flow signal as shown in equation (2.16) [43].

$$\mathrm{Flow}_{\mathrm{SSADA}}(x, z) = 1 - \frac{1}{N-1}\frac{1}{M}\sum_{i=1}^{N-1}\times\sum_{m=1}^{M}\frac{I_{im}(x, z)I_{(i+1)m}(x, z)}{\left[\frac{1}{2}I_{im}(x, z)^2 + \frac{1}{2}I_{(i+1)m}(x, z)^2\right]} \tag{2.16}$$

where M denotes the number of split-spectrums, N denotes the number of repetition of B-scans at the same location, $I_{im}(x, z)$denotes the value of intensity measured in the ith B-scans of mth split-spectrum at lateral location denoted as x and depth position denoted as z. Splitting of the spectrum into sub-bands is considered to increase the SNR without increasing the acquisition time of scan. However, the axial resolution for imaging flow gets reduced to ~15 μm, limiting the detection of smaller vessels [43].

2.5.5.2 Optical microangiography (OMAG) algorithm

OMAG algorithm includes both intensity and phase information of the OCT signal in calculating the flow signal with an increased sensitivity by subtracting the consecutive complex signals after performing phase compensation. The following equation shows the calculation based on OMAG algorithm [43],

$$\mathrm{Flow}_{\mathrm{OMAG}}(x, z) = \frac{1}{N-1}\sum_{i=0}^{N-1}\left| C_{i+1}(x, z) - C_i(x, z) \right| \tag{2.17}$$

where N denotes the number of repetition of B-scans performed at the same location, $C_i(x, z)$ denotes the complex signal consisting of both intensity and phase values in the ith B-scans at lateral location x and depth position z. Averaging the absolute values of the differences of complex signal in each pair of B-scan results in the final flow intensity. OMAG algorithm is also capable of providing information regarding the direction of the moving blood cell relative to the direction of the incident OCT beam, using Hilbert transformation. This method is considered to be efficient in detecting retinal and choroidal capillaries with high sensitivity and also provide information on the progression of eye disease. Advanced OMAG method on multiple signal classification based on the principle of orthogonality has also been proposed to provide estimation of super-resolution spectral information by eigen-decomposition, thereby suppressing static tissue signal without the need of phase compensation [43].

2.5.5.3 Slab subtraction (SS) algorithm

Slab subtraction method in OCT-A has been introduced to effectively remove projection artifacts caused by superficial layers, by subtracting the superficial slab with maximum projected flow from the current slab. The decorrelation value of the slab subtraction of the OCT-A image, denoted as C_{slab2}, is obtained using the following equation [65],

$$C_{\text{slab2}} = \begin{cases} D_{\text{slab2}} - D_{\text{slab1}}, \text{ if } D_{\text{slab2}} > D_{\text{slab1}} \\ 0, \text{ otherwise} \end{cases} \tag{2.18}$$

where, D_{slab2} denotes the decorrelation value within the slab of interest from maximum flow projection, while D_{slab1} denotes the maximum flow projection of all layers above D_{slab2}. The drawback of this method is the appearance of shadowing artifacts and tail artifacts on cross-sectional OCT-A image as it fails to suppress flow projections within the slab [65].

2.5.5.4 Projection resolved OCT-A (PR-OCT-A) algorithm

PR algorithm has been implemented to remove both projection and tail artifact from the vessels of the retina in OCT-A image. The theory of PR algorithm is explained as follows [53], where, the normalised flow signal F, is represented as [65],

$$F = \frac{D}{S} \tag{2.19}$$

where D is the decorrelation and S is the log amplitude OCT signal. The PR-corrected values of decorrelation C_n, is expressed as,

$$C_n = \begin{cases} D_n, \text{if } F_n > (1 + \alpha) \max (F_i), \ 1 \leqslant i \leqslant n - 1 \\ 0, \text{ otherwise} \end{cases} \tag{2.20}$$

where, i and n represent the index of a voxel in an A-line from the shallow inner and proximal end, respectively, and α denotes noise. The process of PR-OCT-A has been illustrated in figure 2.14 as reprinted from [65].

The comparison of choroidal neovascularization (CNV) acquired as en face and cross-sectional OCT-A images before and after processing with PR and SS algorithms has been studied [66]. Compared to the result obtained using SS algorithm, PR-OCT-A has been observed to provide clearest angiograms in the outer retina, greater vessel area of CNV and vascular connectivity. Whereas, the result obtained using SS algorithm retains fewer regions and the true vessels are interrupted or erased, disrupting the vascular morphology (figure 2.15). Common algorithms used in OCT-A and their outcomes have been listed in table 2.3.

2.5.6 Application of deep learning in OCT-A imaging

OCT-A image helps visualize retinal and choroidal microvasculature with high resolution along with depth information. An early detection and assessment of diabetic retinopathy from OCT-A image is essential for successful clinical outcomes. By automating the detection and assessment of the eye disease, precise treatment can be planned. Deep learning (DL) based techniques are the key to achieve it. Deep learning has been applied to fundus images and OCT images for classification and detection of various diseases such as DR, retinopathy of prematurity, glaucoma, age related macular degeneration etc. Deep learning technique can play a very important role in screening of ophthalmic disease and ease the burden of shortage

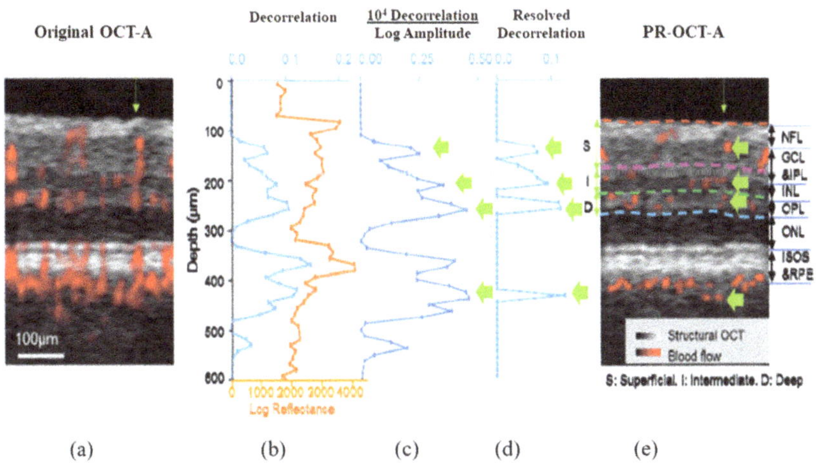

Figure 2.14. Illustration of Steps performed in obtaining PR-OCT-A. (a) original OCT-A image before projection suppression taken from parafoveal region, (b) original decorrelation and log amplitude values of the A-line pointed by the green arrow in (a), (c) decorrelation normalized by log amplitude according to equation (2.19), the *in situ* flow in real vessels is represented by the four successive higher peaks marked by green arrows, (d) decorrelation plot obtained by applying PR algorithm by setting to zero the decorrelation values representing projection artifacts outside the successive peaks by equation (2.20) and (e) composite cross-sectional OCT-A image obtained after clean-up of projection artifacts using PR algorithm showing 4 vessels marked using green arrows. S, I and D represent the superficial, intermediate and deep vascular plexus in the inner retina. NFL - nerve fiber layer, GCL—ganglion cell layer, IPL—inner plexiform layer, INL—inner nuclear layer, OPL—outer plexiform layer, ONL—outer nuclear layer ISOS—inner and outer segments of photoreceptors and RPE—retinal pigment epithelium. (Reprinted with permission from [65] © The Optical Society.)

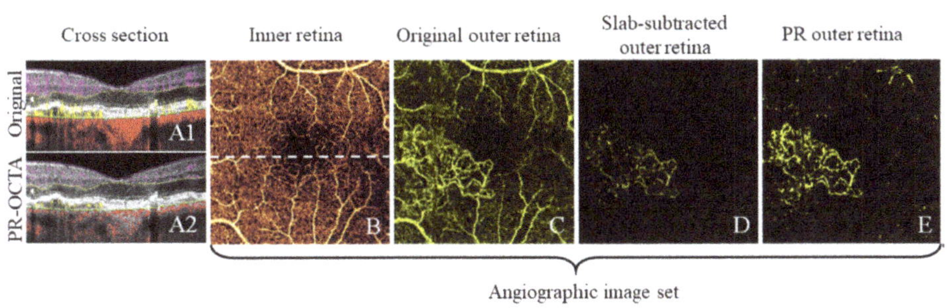

Figure 2.15. (A1) Original angiographic input image set, (A2) projection-resolved (PR) OCTA with inner retinal with inner retinal (violet), choroidal (red), and outer retinal (yellow) flow overlaid on structural OCT, (B) inner retinal angiogram, with white dotted line indicating the position of the B-scans in (A), (C) outer retinal angiogram generated from the original OCTA demonstrated in (A1), (D) outer retinal angiogram processed by slab-subtraction approach, (E) PR outer retinal angiogram, with the entire CNV preserved but with some residual projection artifacts. (Reprinted with permission from [66] © The Optical Society.)

of healthcare professionals, ophthalmologists, optometrists, etc. Researchers around the world have applied deep learning based techniques to publicly available OCT datasets and reported promising results.

Table 2.3. List of various algorithms implemented in OCT-A and their outcomes.

S. No.	Algorithm/reference	Outcome
1	Montaging algorithm	Generates wider field of view [56]
2	Eye tracking algorithm	Overcomes motion artifacts [64]
3	Split-spectrum amplitude decorrelation angiography (SSADA)	Assists in visualizing retinal vasculature with improved SNR [43]
4	Optical microangiography (OMAG) algorithm	Provides suppression of background static tissue from contributing to the angiographic signals [43]
5	Slab subtraction (SS) algorithm	Effective removal of flow projection artifacts from deeper vessels but leaves shadowing and tail artifacts on cross-sectional OCT-A image [65]
6	Projection resolved OCT-A (PR-OCT-A) algorithm	Resolves projection and tail artifacts caused due to the flickering shadow formed by the blood flow in the superficial capillary plexus onto the deeper layers and enhances depth resolution of retinal vasculature [65]
7	Inter-capillary space (ICS) based algorithm	Early detection of capillary dropout or non-proliferative areas [67]

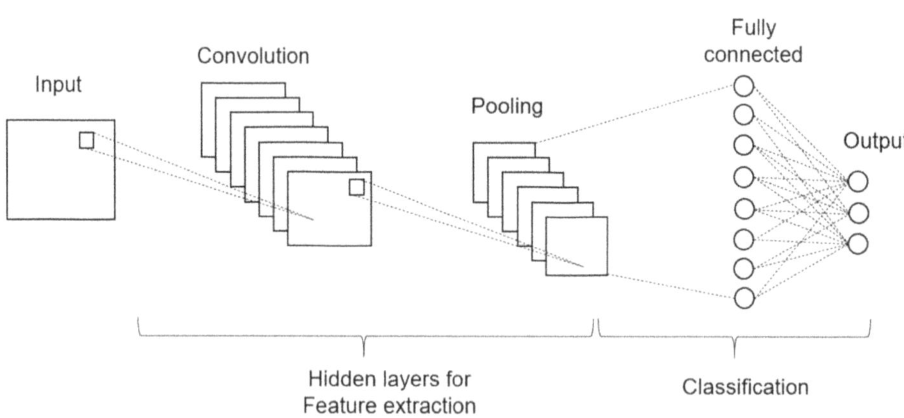

Figure 2.16. CNN architecture.

An important deep learning technique used for in OCT imaging is convolution neural network (CNN) and widely used for retinal layer segmentation. CNN is comprised of a huge number of artificial neurons tasked to process and extract image features both simple and complex in order to classify the images as shown in figure 2.16 CNN network has two parts—convolution mask base for feature extraction and a classifier. The input images are passed through a number of convolution masks where the initial masks extract simple features and later masks extract complex features. The extracted feature maps are then subjected to pooling operation where the most prominent features (max pooling) or average of all features (average pooling) are generated. This step also reduces the dimensionality of the feature maps and thereby reduces amount of computation. The generated features are then flattened and fed into the fully connected layer which consists of multiple hidden layers. Using activation function such as sigmoid or softmax the outputs are classified into various classes.

CNN requires good amount of training data in the form of labelled data/ground truth images. An important issue with this data is disagreement or variability among clinicians around the world on the labelling of various diseases and its stages. This makes it difficult for OCT-A images due to the unavailability of such a database. Research groups around the world have used transfer learning based technique to overcome this issue. Transfer learning technique uses pretrained CNN networks (model trained on large datasets to solve similar problems) and modifies the weights for the task in hand. A number of pretrained CNN networks are available, a few of them are VGG, GoogLeNet, ImageNet, Inception, MobileNet, etc. The first step in transfer learning is to modify the classifier layer based on the output expected from OCT-A images (based on presence or absence of disease and stage of the disease). The OCT-A images to be fed into the network should be modified based on the pretrained networks requirement. The whole idea is to use an appropriate pretrained network as it is for OCT-A image classification. If required, the learning rate and weights in the network can be adjusted.

In case of DR assessment, the international clinical diabetic retinopathy disease severity scale has classified it into non-proliferative DR (NPDR) with subclassifications of mild, moderate and severe based on the number of microaneurysms present and changes observed in vision and proliferative diabetic retinopathy (PDR) (presence of neovascularisation) [58]. Another technique used for OCT-A image classification supervised machine learning approach for screening of retinopaties—normal and diseased condition, different ocular diseases and severity of the diseases. Although the available OCT-A images are small, the features are quite distinct and hence helpful in classification. For DR assessment, various features such as blood vessel density, blood vessel caliber, blood vessel tortuosity, vessel perimeter index, foveal avascular zone (area or contour) etc are used for classification and identify the stage of DR.

Xincheng and group [68] used a backward elimination technique where the classifier which performs least is eliminated and the network is retrained on training data. After the training is complete, the test data is fed for classification. This work also used optimal feature selection for improving classification accuracy. Each of the features plays a significant role in the disease condition or staging and the selection of the relevant features for each of this is important. The study has shown promising results. Blood vessel density, foveal avascular zone area and contour are useful for disease classification and blood vessel density and foveal avascular zone area was found to play a significant role in DR severity staging. OCT-A imaging takes around

4–5 s for acquisition and trained classifier would take 1 or 2 s to analyse and classify the input image thus making the screening process real-time.

Deep learning can be employed to denoise the OCT-A images and enhance microvasculature. Akihito and group [69] employed a U-net based DL technique to denoise the OCT-A images. U-net is a symmetric architecture which consists of an encoder (down sampling) and decoder (up sampling). The encoder takes one image as input and computes feature maps on different scales as shown in figure 2.17. These feature maps are under further convolution iteratively until deep features are extracted. The decoder deconvolves by rearranging features and not performing convolution operation while upscaling along with concatenation with the image of the same resolution in the encoding stages and generates the final denoised image. The study showed that the averaging technique usually employed in OCT-A image generation takes more time than the deep learning based denoising technique. The hyperparameter tuning in deep learning plays an important role in avoiding artifacts (capillary over drop out or pseudo capillary generation) generated in the up sampling step in U-net. A larger database is necessary to test the efficiency of any developed DL model.

As OCT-A images are found to be less in number, Aaron Y Lee and group [70] have used a DL based technique to generate OCT-A images from standard OCT images. This is done by identifying structure and vasculature in B-scan OCT image to generate OCT-A image. A U-net based DL architecture was used with multiple blocks of convolution (deepest block with 18 convolution filters) and concatenation. The DL model identified the large and medium size blood vessels and micro-vasculature with better detail as shown in figure 2.18. The speckle pattern varies for static structure and moving scatters (finer speckle for faster movement). This property is used in the DL network to infer the dynamic information from B-scan images. This study helps in generating OCT-A like images from the available OCT databases to study the history of vasculature in patients and integration with existing OCT machines to produce OCT-A images of patients being scanned.

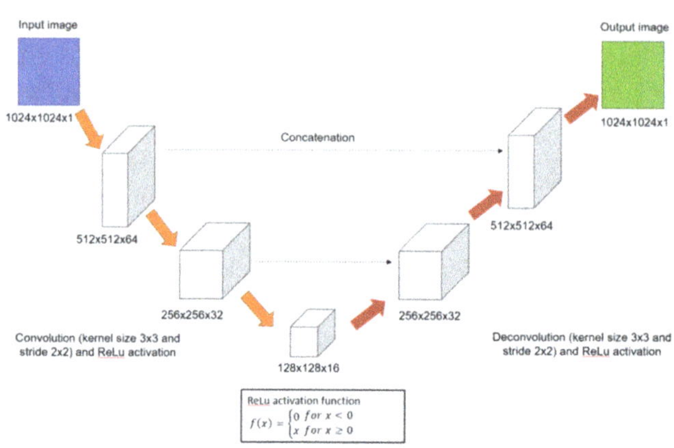

Figure 2.17. U-net architecture.

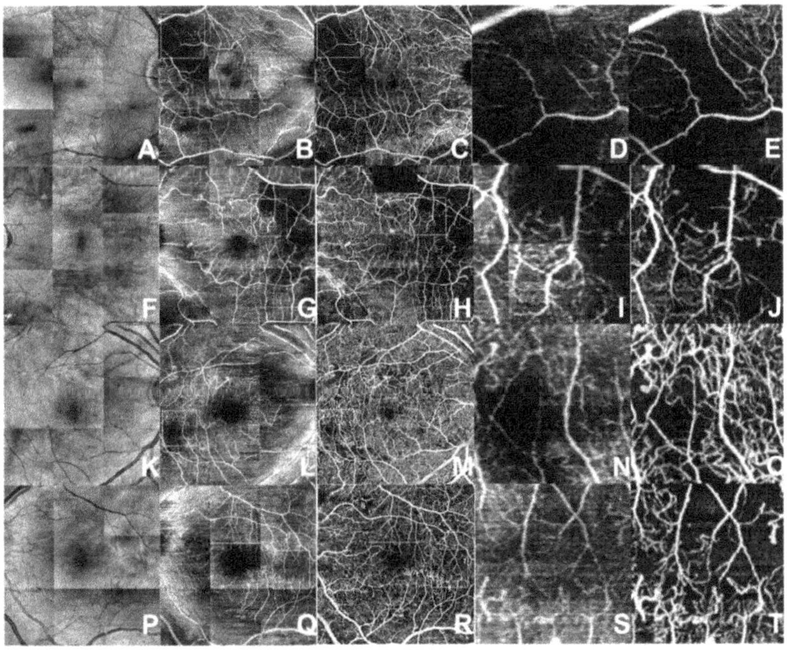

Figure 2.18. En-face projection map of retina from OCT volumes (A,F,K,P), AI generated retinal flow maps (B,G,L,Q) and OCT-A flow images (C,H,M,R). Magnified views of the AI-generated (D,I,N,S) and OCT-A (E,J,O,T) images. [Reprinted with permission from Springer Nature from [70] under the Creative Commons Attribution 4.0 International License (http://creativecommons.org/licenses/by/4.0/).]

Thus deep learning techniques can help in both image processing of OCT-A images to generate real time diagnostic report, enhance image quality and generate OCT-A like images from existing OCT images. Table 2.4 lists the comparison of sensitivity and specificity achieved by various diagnostic techniques. Ophthalmoscopy technique has a low sensitivity in detecting lesions that develop in the earliest stage during diabetic retinopathy. Although fundus photography facilitates storage, retrieval, remote transmission of digital images and grading through telemedicine, it is considered to be less effective compared to FA in identifying closure and leakages in capillaries, as well as has a low sensitivity in detecting DME [10]. Albeit FA is capable of detecting closure in the capillary, formation of microaneurysms, occurrence of DME and DMI, at an early stage, its sensitivity and specificity are low when compared to OCT technique. FA also fails to repeat measurements immediately as it necessitates dye injection. Recently, SD-OCT technique based on computer-aided diagnostic (CAD) system provided higher sensitivity and specificity of 92.5% and 95%, respectively, compared to the conventional techniques. An enhanced sensitivity and specificity of 94.3% and 97.9%, respectively, have been achieved using CAD based OCT-A technique, which implies the importance and reliability of utilising deep learning methods combined with OCT-A technique for providing automated detection, classification and grading of diabetic retinopathy [18, 19].

Table 2.4. Comparison of sensitivity and specificity for various diagnostic techniques.

S. No.	Diagnostic techniques	Sensitivity	Specificity	References
1	Ophthalmoscopy	63%	66%	[10]
2	Fundus photography	77.5%–88.5%	78%–88%	[71]
3	Fluorescein angiography	87%	87%	[72]
4	SD-OCT	87.3%	96%	[12]
5	CAD based SD-OCT	92.5%	95%	[18]
6	CAD based OCT-A	94.3%	97.9%	[19]

2.6 Conclusion

OCT technology, since 1991, has taken rapid strides in developing into a versatile diagnostic tool in the field of ophthalmology, especially in the diagnosis of complications related to diabetic retinopathy. Although conventional Fluorescein angiography has been considered to be a gold standard in imaging retinal vasculature, there are several limitations such as its invasive approach requiring dye injection and lack of providing depth-resolved details of retinal layers. Alternatively, OCT technique overcomes these limitations by assisting in non-invasive detection of diabetic retinopathy by providing high-resolution structural 3D images. OCT Angiography, a functional extension of OCT is an advanced promising technique based on either SD-OCT or SS-OCT for visualisation of retinal and choroidal microvasculature. Several research works described in this chapter have shown the potential of this technique in the diagnosis and assessment of diabetic retinopathy. OCT-A provides layer-wise imaging thereby covering superficial and deep retinal vascular plexus which is not available in Fluorescein angiography imaging. The fastest scanning rate available commercially is 100 000 A-scans per second covering a field of view of 23×20 mm. The application of various deep learning techniques has further improved the quality and speed of OCT-A imaging. The future scope lies in the improved speeds of SS-OCT imaging for enhanced wide field imaging and towards real time diagnosis and automatic grading of retinopathies with the help of deep learning techniques.

References

[1] Louise Bye N M 2013 *Basic Sciences for Ophthalmology* (Oxford Specialty Training: Basic Science) (Oxford: Oxford University Press)

[2] Ulbig M W and Kollias A N 2010 Diabetische Retinopathie: Frühzeitige Diagnostik und Effiziente Therapie *Dtsch. Arztebl.* **107** 75–84

[3] Yau J W Y, Rogers S L and Kawasaki R *et al* 2012 Global prevalence and major risk factors of diabetic retinopathy *Diabetes Care* **35** 556–64

[4] Fong D S, Aiello L and Gardner T W *et al* 2004 Retinopathy in diabetes *Diabetes Care* **27 Suppl. 1** S84–7

[5] Schimel A M, Fisher Y L and Flynn H W 2011 Optical coherence tomography in the diagnosis and management of diabetic macular edema: time-domain versus spectral-domain *Ophthalmic Surg. Lasers Imaging* **42 Suppl.** S41–55

[6] Garcia J M B, de B, Isaac D L C and Avila M 2017 Diabetic retinopathy and OCT angiography: clinical findings and future perspectives *Int. J. Retina Vitr.* **3** 14

[7] Garg S and Davis R M 2009 Diabetic Retinopathy Screening Update *Clin. Diabetes* **27** 140–5

[8] Flammer J, Mozaffarieh M and Bebie H 2013 *Basic Sciences in Ophthalmology: Physics and Chemistry* (Cham: Springer)

[9] Atchison D A and Smith G 2000 *Optics of the Human Eye* (Amsterdam: Butterworth-Heinemann/Elsevier) p iv

[10] Ciulla T A, Amador A G and Zinman B 2003 Diabetic retinopathy and diabetic macular edema: pathophysiology, screening, and novel therapies *Diabetes Care* **26** 2653–64

[11] Sinthanayothin C 1999 Image analysis for automatic diagnosis of diabetic retinopathy *PhD Thesis* (King's College of London)

[12] Sanborn G E and Wroblewski J J 2018 Evaluation of a combination digital retinal camera with spectral-domain optical coherence tomography (SD-OCT) that might be used for the screening of diabetic retinopathy with telemedicine: a pilot study *J. Diabetes Complications* **32** 1046–50

[13] Tey K Y, Teo K and Tan A C S *et al* 2019 Optical coherence tomography angiography in diabetic retinopathy: a review of current applications *Eye Vis* **6** 1–10

[14] Kwiterovich K A, Maguire M G and Murphy R P *et al* 1991 Frequency of adverse systemic reactions after fluorescein angiography: results of a prospective study *Ophthalmology* **98** 1139–42

[15] Goebel W and Franke R 2006 Retinal thickness in diabetic retinopathy: comparison of optical coherence tomography, the retinal thickness analyzer, and fundus photography *Retina* **26** 49–57

[16] Couturier A, Mané V and Bonnin S *et al* 2015 Capillary plexus anomalies in diabetic retinopathy on optical coherence tomography angiography *Retina* **35** 2384–91

[17] Khadamy J, Aghdam K and Falavarjani K 2018 An update on optical coherence tomography angiography in diabetic retinopathy *J. Ophthalmic Vis. Res* **13** 487

[18] Sandhu H S, Eltanboly A and Shalaby A *et al* 2018 Automated diagnosis and grading of diabetic retinopathy using optical coherence tomography *Investig. Ophthalmol. Vis. Sci.* **59** 3155–60

[19] Sandhu H S, Eladawi N and Elmogy M *et al* 2018 Automated diabetic retinopathy detection using optical coherence tomography angiography: a pilot study *Br. J. Ophthalmol.* **102** 1564–9

[20] Braaf B, Gräfe M G O and Uribe-Patarroyo N *et al* 2019 *High Resolution Imaging in Microscopy and Ophthalmology* (Cham: Springer International Publishing)

[21] John P, Vasa N J and Sujatha N 2019 Glucose sensing in the anterior chamber of the human eye model using supercontinuum source based dual wavelength low coherence interferometry *Sens. Bio-Sensing Res* **23** 100277

[22] Shu X, Beckmann L and Zhang H F 2017 Visible-light optical coherence tomography: a review *J. Biomed. Opt.* **22** 1

[23] Yi J, Chen S and Shu X *et al* 2015 Human retinal imaging using visible-light optical coherence tomography guided by scanning laser ophthalmoscopy *Biomed. Opt. Express* **6** 3701

[24] Hee M R, Izatt J A and Swanson E A *et al* 1995 Optical coherence tomography of the human retina *Arch. Ophthalmol.* **113** 325–32

[25] Brezinski M E 2006 *Optical Coherence Tomography. Principles and Applications,* (Amsterdam: Elsevier) pp 97–140

[26] Fujimoto J, Pitris C, Boppart S and Brezinski M 2000 Optical coherence tomography: an emerging technology for biomedical imaging and optical biopsy *Neoplasia (New York)* **2** 9–25

[27] Fercher A F 2010 Optische Kohärenz-Tomographie—Entwicklung, Grundlagen, Anwendungen *Z. Med. Phys.* **20** 251–76

[28] Chen S, Shu X and Yi J *et al* 2016 Dual-band optical coherence tomography using a single supercontinuum laser source *J. Biomed. Opt.* **21** 066013

[29] Chong S P, Bernucci M, Radhakrishnan H and Srinivasan V J 2017 Structural and functional human retinal imaging with a fiber-based visible light OCT ophthalmoscope *Biomed. Opt. Express* **8** 323

[30] Podoleanu A G 2012 Optical coherence tomography *J. Microsc.* **247** 209–19

[31] Damodaran V, Rao S R and Vasa N J 2016 Optical coherence tomography based imaging of dental demineralisation and cavity restoration in 840 nm and 1310 nm wavelength regions *Opt. Lasers Eng.* **83** 59–65

[32] Fasihinia M, Khalesi H and Gholami M 2011 Dental caries diagnostic methods *J. Dent. Res.* **2** 1–12

[33] Drexler W, Liu M and Kumar A *et al* 2014 Optical coherence tomography today: speed, contrast, and multimodality *J. Biomed. Opt.* **19** 071412

[34] Yaqoob Z, Wu J and Yang C 2005 Spectral domain optical coherence tomography: a better OCT imaging strategy *Biotechniques* **39** S6–13

[35] Fujimoto J and Swanson E 2016 The development, commercialization, and impact of optical coherence tomography *Investig. Ophthalmol. Vis. Sci.* **57** OCT1–OCT13

[36] Gao S S, Jia Y and Zhang M *et al* 2016 Optical coherence tomography angiography *Investig. Ophthalmol. Vis. Sci.* **57** OCT27–36

[37] Porta M and Bandello F 2002 Diabetic retinopathy: a clinical update *Diabetologia* **45** 1617–34

[38] Potsaid B, Gorczynska I and Srinivasan V J *et al* 2008 Ultrahigh speed spectral/Fourier domain OCT ophthalmic imaging at 70,000 to 312,500 axial scans per second *Opt. Express* **16** 15149

[39] Grulkowski I, Liu J J and Potsaid B *et al* 2012 Retinal, anterior segment and full eye imaging using ultrahigh speed swept source OCT with vertical-cavity surface emitting lasers *Biomed. Opt. Express* **3** 2733

[40] Miller A R, Roisman L and Zhang Q *et al* 2017 Comparison between spectral-domain and swept-source optical coherence tomography angiographic imaging of choroidal neovascularization *Investig. Ophthalmol. Vis. Sci.* **58** 1499–505

[41] Kolb J P, Pfeiffer T and Eibl M *et al* 2018 High-resolution retinal swept source optical coherence tomography with an ultra-wideband Fourier-domain mode-locked laser at MHz A-scan rates *Biomed. Opt. Express* **9** 120

[42] Ishibazawa A, Nagaoka T and Takahashi A *et al* 2015 Optical coherence tomography angiography in diabetic retinopathy: a prospective pilot study *Am. J. Ophthalmol.* **160** 35–44
[43] Kashani A H, Chen C L and Gahm J K *et al* 2017 Optical coherence tomography angiography: a comprehensive review of current methods and clinical applications *Prog. Retin. Eye Res.* **60** 66–100
[44] Matsunaga D R, Yi J J and De Koo L O *et al* 2015 Optical coherence tomography angiography of diabetic retinopathy in human subjects *Ophthalmic Surg. Lasers Imaging Retin* **46** 796–805
[45] Nesper P L, Roberts P K and Onishi A C *et al* 2017 Quantifying microvascular abnormalities with increasing severity of diabetic retinopathy using optical coherence tomography angiography *Invest. Ophthalmol. Vis. Sci.* **58** BIO307–15
[46] Kim A Y, Chu Z and Shahidzadeh A *et al* 2016 Quantifying microvascular density and morphology in diabetic retinopathy using spectral-domain optical coherence tomography angiography. Investig *Ophthalmol. Vis. Sci.* **57** OCT362–70
[47] Durbin M K, An L and Shemonski N D *et al* 2017 Quantification of retinal microvascular density in optical coherence tomographic angiography images in diabetic retinopathy *JAMA Ophthalmol.* **135** 370–6
[48] Toto L, D'Aloisio R and Nicola M D *et al* 2017 Qualitative and quantitative assessment of vascular changes in diabetic macular edema after dexamethasone implant using optical coherence tomography angiography *Int. J. Mol. Sci.* **18** 1–12
[49] Freiberg F J, Pfau M and Wons J *et al* 2016 Optical coherence tomography angiography of the foveal avascular zone in diabetic retinopathy *Graefe's Arch. Clin. Exp. Ophthalmol.* **254** 1051–8
[50] Bhanushali D, Anegondi N and Gadde S G K *et al* 2016 Linking retinal microvasculature features with severity of diabetic retinopathy using optical coherence tomography angiography *Investig. Ophthalmol. Vis. Sci.* **57** 519–25
[51] Alam M, Zhang Y and Lim J I *et al* 2020 Quantitative optical coherence tomography angiography features for objective classification and staging of diabetic retinopathy *Retina* **40** 322–32
[52] Sasongko M B, Wong T Y and Nguyen T T *et al* 2011 Retinal vascular tortuosity in persons with diabetes and diabetic retinopathy *Diabetologia* **54** 2409–16
[53] Huang F, Dashtbozorg B and Zhang J *et al* 2016 Reliability of using retinal vascular fractal dimension as a biomarker in the diabetic retinopathy detection *J. Ophthalmol.* **2016** 6259047
[54] Salz D A, De Carlo T E and Adhi M *et al* 2016 Select features of diabetic retinopathy on swept-source optical coherence tomographic angiography compared with fluorescein angiography and normal eyes *JAMA Ophthalmol.* **134** 644–50
[55] Eladawi N, Elmogy M and Khalifa F *et al* 2018 Early diabetic retinopathy diagnosis based on local retinal blood vessel analysis in optical coherence tomography angiography (OCTA) images *Med. Phys.* **45** 4582–99
[56] de Carlo T E, Romano A, Waheed N K and Duker J S 2015 A review of optical coherence tomography angiography (OCTA) *Int. J. Retin. Vitr.* **1** 5
[57] Rocholz R, Teussink M M, Dolz-Marco R, Holzhey C, Dechent J F, Tafreshi A and Schulz S *et al* 2018 SPECTRALIS optical coherence tomography angiography (OCTA): principles and clinical applications *Heidelb. Eng. Acad.* 1–12

[58] Abdelsalam M M 2020 Effective blood vessels reconstruction methodology for early detection and classification of diabetic retinopathy using OCTA images by artificial neural network *Informatics Med. Unlocked* **20** 100390
[59] Pellegrini M, Cozzi M, Staurenghi G and Corvi F 2019 Comparison of wide field optical coherence tomography angiography with extended field imaging and fluorescein angiography in retinal vascular disorders *PLoS One* **14** 8–12
[60] Gendelman I, Alibhai A Y and Moult E M *et al* 2020 Topographic analysis of macular choriocapillaris flow deficits in diabetic retinopathy using swept-source optical coherence tomography angiography *Int. J. Retin. Vitr.* **6** 1–8
[61] Parrulli S, Corvi F and Cozzi M *et al* 2020 Microaneurysms visualisation using five different optical coherence tomography angiography devices compared to fluorescein angiography *Br. J. Ophthalmol.* **105** 526–30
[62] Chalam K V and Sambhav K 2016 Optical coherence tomography angiography in retinal diseases *J. Ophthalmic Vis. Res* **11** 84–92
[63] Zhang M, Hwang T S and Dongye C *et al* 2016 Automated quantification of nonperfusion in three retinal plexuses using projection-resolved optical coherence tomography angiography in diabetic retinopathy *Investig. Ophthalmol. Vis. Sci.* **57** 5101–6
[64] Lauermann J L, Treder M and Heiduschka P *et al* 2017 Impact of eye-tracking technology on OCT-angiography imaging quality in age-related macular degeneration *Graefe's Arch. Clin. Exp. Ophthalmol.* **255** 1535–42
[65] Zhang M, Hwang T S and Campbell J P *et al* 2016 Projection-resolved optical coherence tomographic angiography *Biomed. Opt. Express* **7** 816
[66] Wang J, Hormel T T, Gao L, Zang P, Guo Y, Wang X, Bailey S T and Jia Y 2020 Automated diagnosis and segmentation of choroidal neovascularization in OCT angiography using deep learning *Biomed. Opt. Express* **11** 927–44
[67] Schottenhamml J, Moult E M and Ploner S *et al* 2016 An automatic, intercapillary area-based algorithm for quantifying diabetes-related capillary dropout using optical coherence tomography angiography *Retina* **36** S93–S101
[68] Alam , Le and Lim *et al* 2019 Supervised machine learning based multi-task artificial intelligence classification of retinopathies *J. Clin. Med.* **8** 872
[69] Kadomoto S, Uji A and Muraoka Y *et al* 2020 Enhanced visualization of retinal microvasculature in optical coherence tomography angiography imaging via deep learning *J. Clin. Med.* **9** 1322
[70] Lee C S, Tyring A J and Wu Y *et al* 2019 Generating retinal flow maps from structural optical coherence tomography with artificial intelligence *Sci. Rep.* **9** 1–11
[71] Sinthanayothin C, Boyce J F and Williamson T H *et al* 2002 Automated detection of diabetic retinopathy on digital fundus images *Diabet. Med* **19** 105–12
[72] Newsom R, Moate B and Casswell T 2000 Screening for diabetic retinopathy using digital colour photography and oral fluorescein angiography *Eye* **14** 579–82

IOP Publishing

Photo Acoustic and Optical Coherence Tomography Imaging, Volume 1

Diabetic retinopathy

Ayman El-Baz and Jasjit S Suri

Chapter 3

Multimodal photoacoustic microscopy, optical coherence tomography, and fluorescence microscopy molecular retinal imaging in health and disease

Van Phuc Nguyen, Wei Zhang, Xueding Wang and Yannis M Paulus

Photoacoustic microscopy (PAM) is an emerging non-ionizing and non-invasive imaging technique that has been developed as an advanced molecular imaging platform to evaluate retinal pathologies including diabetes. PAM can provide high resolution and high image contrast for visualization of both structural and functional information of the retina. PAM has a strong capability to precisely identify the location of ocular diseases and distinguish abnormal vasculature from the normal vasculature in the local area in the eye at an earlier stage in three dimensions, allowing for improved diagnosis, understanding of pathophysiology, and treatment outcomes to prevent vision impairment. Advanced imaging technology for precise visualization of retinal tissues such as the retinal pigment epithelium, retinal vessels, choroidal vessels, choroidal neovascularization, and retinal neovascularization can combine PAM with optical coherence tomography (OCT), scanning laser ophthalmoscopy (SLO), and fluorescence microscopy. This allows for improved diagnosis and monitoring the progress of diseases by co-registration of the images acquired from a single imaging modality on the same orthogonal imaging plane. In this chapter, the potential application of multimodal molecular photoacoustic microscopy, optical coherence tomography, and fluorescence microscopy retinal imaging is described as a novel advanced imaging tool for evaluation of different eye disorders. The principle and major requirements, safety evaluation, and the applications in eye disorders like diabetic retinopathy retinal neovascularization (RNV), retinal vein occlusion (RVO), and choroidal neovascularization (CNV) are discussed.

doi:10.1088/978-0-7503-2052-8ch3

3.1 Introduction

Vision impairment is a major public health problem and increasing in both the United States and worldwide. An estimated 36 million people worldwide are blind [1], and 5.4 million people will be affected in the USA by 2030 [2]. Much of the vision impairment is caused by various eye disorders such as diabetic retinopathy, age-related macular degeneration (AMD), and glaucoma [3]. Once the function of various ocular components, such as photoreceptors and retinal pigment epithelium (RPE), are degenerated, or retinal blood oxygen saturation (SO_2) levels are reduced, vision can be compromised. For example, the loss of RPE and the accumulation of lipofuscin in the subretinal space has been associated to the progression of AMD [4–6]. Abnormal SO_2 levels in the retinal vessels are related to retinal venous occlusion, diabetic retinopathy, and glaucoma [7–11]. Therefore, early detection of these abnormal pathologies can improve diagnosis of eye diseases, resulting in reduced vision loss and enhanced treatment outcomes.

Several imaging modalities, including color fundus photography, fluorescein angiography (FA), indocyanine green angiography (ICGA) [12–14], optical coherence tomography (OCT), scanning laser ophthalmoscopy (SLO), fundus autofluorescence (FAF), fluorescent lifetime imaging ophthalmoscopy (FLIO), and optical coherence tomography angiography (OCTA) [12, 15, 16] have greatly facilitated diagnosis and evaluation of patients in clinics. With high-quality image contrast and high resolution obtained from the emitted light from an exogenous contrast agent (ICGA, FA), endogenous autofluorescence (FAF), or backscattering light from the retinal tissues (OCT, and OCTA), these imaging modalities have been used to assess anatomical and functional information of various retinal tissues. However, each modality has its own limitation. For example, ICGA, FA, and FAF are clinically limited by their shallow imaging depth in tissues, and they cannot provide 3D volumetric images. OCT and OCTA could not visualize the deep choroidal vascular network because the light is quickly degraded by strong scattering effects when it travels into the retinal tissues and particularly the RPE. Therefore, a clinical need exists to develop an advanced optical imaging technology with the ability to achieve deep penetration with high resolution.

Compared to conventional fluorescence and optical imaging, photoacoustic imaging (PAI) provides deeper imaging depth with high resolution and high image contrast by combining high contrast of the optical absorption and low scattering of the acoustic wave in tissues [17–22]. PAI has several benefits, such as achieving imaging depth up to ~7 cm with a spatial resolution of a few hundred micrometers [23–25]; ability to image many different optical absorbers including exogenous (ICG, astaxanthin, Prussian blue) [21, 26–28] and endogenous (hemoglobin, melanin, lipid) [29–31]; and distinguishing features between normal and abnormal tissues using multiple-wavelength imaging [22, 32–35]. With such unique advantages, PAI has a great potential for application in ophthalmological applications. In 2010, de la Zerda *et al* first described PAI as a novel imaging technology for evaluating both the anterior and posterior segments of the ocular tissues [36]. However, the resolution of this system was too low (i.e., 200 μm) to visualize single capillaries. PA imaging systems with higher resolution are more suitable to quantify microvasculature in the eye. Photoacoustic microscopy (PAM) is a form of PAI that provides high spatial resolution from several micrometers to tens of micrometers and relatively shallow penetration depth from one to a few millimeters [24]. With this high resolution and high sensitivity, PAM has been reported as a promising imaging system for quantification of both structural information (e.g., size, shape), and functional information (e.g., oxygenated and deoxygenated hemoglobin and oxygen saturation) of the ocular tissues [31]. Jeon *et al* and Kelly-Goss *et al* have described the feasibilty of PAM for detection of corneal neovascularization in mice with high resolution and image contrast [37, 38]. SO_2 and blood velocity were also determined using multiple-wavelength PAM imaging [37–39]. To better visualize the ocular tissues, PAM has been combined with other optical imaging techniques such as OCT, SLO, and fluorescence microscopy to perform multimodal imaging. Multimodal imaging has been applied to assess retinal vessels (RVs), choroidal vessels (CVs), retinal pigment epithelium (RPE) cells [40, 41], melanin, corneal neovascularization, retinal neovascularization (RNV), retinal vein occlusion, and choroidal neovascularization (CNV) in rodents (e.g., mice and rats) and large animal eyes (e.g., rabbits) [22, 29, 42–44]. In this chapter, we will summarize the state of the art of multimodal PAM, OCT, and FM imaging for the early-stage detection of ocular diseases. We first present the basic principle of FM and its applications. Then, we will introduce the operation as well as application of OCT. Finally, we will discuss multimodal PAM and OCT in the eye, including the principle of PAM, system requirements for the eye, the main applications of a multimodality system for the study of retinal diseases, and the application of contrast agents to enhance PAM and OCT imaging.

3.2 Fluorescence imaging

Currently, fundus angiography with intravenous administration of fluorescein or fluorescence dye is widely used in ophthalmology [45]. For some situations, it is the gold standard for ophthalmologists to diagnosis ocular diseases. Typically, it is divided into two methods with different dyes: fluorescein angiography and indocyanine green angiography.

3.2.1 Fluorescein angiography

As a technique to identify the circulation of the retina and choroid, fluorescein angiography (FA), or fundus fluorescein angiography (FFA), has been widely used in ophthalmology since 1961 [46]. Fluorescein sodium is intravenously administered to the systematic circulation, typically in the antecubital fossa [47]. With the retina illumined by a blue light, the emission light of the dye will be acquired by a fundus camera with a green light filter. Fluorescence images will be repeatedly captured starting from the time when the dye is administered to track the circulation of the dye in the retinal and choroidal blood vessel. The progress of the angiogram is divided in to five different phases, including choroidal flush, arterial, arteriovenous, venous, and late phase [48].

3.2.2 Indocyanine green angiography

Indocyanine green angiography (ICGA) use ICG as the fluorescent dye to investigate the circulation of the choroidal blood vessels. ICG is highly protein bound, and thus is larger and does not readily escape from the choriocapillaris [49]. Different from the FA, a near-infrared light will be used to excite the fluorescence signal, and the image will be captured with a near-infrared light filter. From a clinical standpoint, the progress of ICGA can be grossly divided into early, middle, and late phases [50].

3.2.3 Advantages and limitations

The use of an exogenous dye gives advantages, and fluorescence-based angiography can provide information that cannot be evaluated by label-free imaging. Besides circulation information, FA and ICG can also show the hyperfluorescence and hypofluorescence of the injected dye to provide additional information for diagnosis. The types of hyperfluorecence on angiography include leakage (e.g., neovascularization), pooling (e.g., serous pigment epithelial detachment), staining (e.g., drusen), and window defect (e.g., geographic atrophy). Each type can distinguish different pathologies. There exist two types of hypofluorescence: blocking (e.g., hemorrhage) and filling defect (e.g., nonperfusion) [47]. Each type can distinguish various causative pathologies.

Although FA can visualize retinal circulation in detail, it has limited ability to evaluate choroidal circulation with the diffuse leakage of fluorescein from the choroid resulting in a choroidal flush. Using a larger fluorescent molecule and near-infrared light, ICGA can evaluate choroidal circulation, but artifacts and pitfalls still exist with ICGA interpretation. In addition, both FA and ICGA are invasive techniques and require intravenous administration of exogenous dyes and a relative long imaging time [51]. Some cases of complications have been reported with these two imaging methods, including nausea, emesis, anaphylactic reactions, and even death [52, 53].

3.3 Optical coherence tomography (OCT)

Since it was developed by Dr Fujimoto's group around 30 years ago, optical coherence tomography (OCT) has been applied and has been highly beneficial to the field of ophthalmology [54]. By providing high resolution, three-dimensional (3D) images, angiography information of the retina and anterior segment at a penetration depth up to a few millimeters, OCT has become the standard of care for the assessment of many retinal conditions.

3.3.1 Basic principle of OCT

As a non-invasive imaging technique, OCT employs light from a low-coherence light source, which is split into a reference arm and a sample arm. The interference pattern of these two arms is then evaluated with an interferometer [54]. Typically, it will use a near-infrared light to achieve better penetration. However, some studies use visible light for imaging. The backscattered light from the detected sample will interfere with the reference beam to generate an interference pattern, which is recorded by the detector. The reflection index based light echoes versus the depth profiles can be decoded from the recorded interference pattern. Here we will briefly discuss the most widely used OCT technologies.

3.3.1.1 Spectral domain OCT

The schematic of SD-OCT system is shown in figure 3.1(A). A broadband, low-coherence light source, such as a super luminescent diode (SLD) or mode-locked laser, is employed in a SD-OCT system [55]. With a beam splitter or fiber coupler, the light will be divided into a reference arm and sample arm, while the echo light will form an interference pattern in the spectrometer arm. In the spectrometer arm, the interference pattern is split by a grating into a function of wavelength, which is simultaneously detected by a charge-coupled device (CCD). The frequency component of the detected signal will be extracted by performing a Fast Fourier Transform (FFT). Each frequency detected corresponds to a certain depth of the sample, which allows all the points to form a A-scan image [56].

3.3.1.2 Swept source OCT

Figure 3.1(B) presents a schematic of an SS-OCT system [55]. Compared with SD-OCT, SS-OCT has the same imaging protocol except the source operating and detection mechanism. SS-OCT uses a tunable narrowband laser that dispatches a particular wavelength/wavenumber at a particular time into the sample. Unlike the grating-based spectrometer, the interference pattern will be detected by a photodetector with respect to time of sweep, later rescaled into wavenumber space. After performing the Fourier transform of the detected signal, the depth information of the sample will be resolved [57].

3.3.1.3 Optical coherence tomography angiography (OCTA)

OCT angiography (OCTA) is a functional extension of OCT, which provides high-resolution volumetric blood flow information in microvasculature by using the light reflectance of the moving red blood cells [58]. To generate angiographic images, the blood flow-induced change in the reflectance signal is detected. Multiple approaches have been used to extract differences in the backscattered OCT signal between sequential OCT b-scans, including intensity, phase, or both [59].

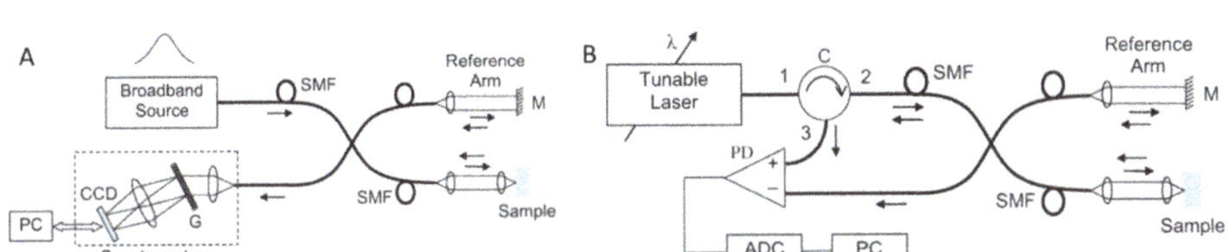

Figure 3.1. Schematic of an SD-OCT system (A) and SS-OCT system (B) [55].

3.3.1.4 Visible-light OCT

Currently, most of the OCT technologies use near-infrared light for its deeper penetration. Benefiting from use of a supercontinuum light source, visible-light OCT had attracted the attention of researchers interested in optimizing imaging resolution and the unique scattering and absorption properties of tissue in the visible wavelengths [60]. Sharing the same imaging protocol with near-infrared light OCT, visible-light OCT follows the same mechanism as SD-OCT and SS-OCT. The biggest advantage of visible-light OCT is the unique ability for improved imaging resolution and characterization of retinal layers for more functional assessment. As demonstrated by several groups, time-frequency analysis is able to be performed in visible-light OCT to reveal the optical properties of biological tissue, including scattering properties and absorption properties. Therefore, various chromophore concentrations can be extracted from the OCT signal [61].

3.3.2 Resolution and image dimensions

Unlike other optical imaging methods, which have a compromise between the axial resolution and lateral resolution, the axial and lateral resolutions of OCT are independent from each other [62, 63]. The lateral resolution of an OCT imaging system is related to the diameter of the spot size of the scanning laser beam. Given a Gaussian beam at the focal spot, the lateral resolution of the OCT system is defined by the beam diameter at half maximum (FWHM), which is also related to the optical limit of the focal system. The lateral resolution is defined as below.

$$\delta x = \sqrt{2\ln 2}\,\frac{\lambda_0}{\pi NA} \tag{3.1}$$

where λ_0 is the wavelength used in imaging, NA is the numerical aperture, which is related to the objective lens.

Another important factor to determine the focal volume is the axial range, which will be determined by the depth of focus b.

$$b = \frac{n \cdot \lambda_0}{2\pi \cdot NA^2} \tag{3.2}$$

where n is the refractive index of the media.

As an interferometer-based imaging technique, the coherence properties of the light source and the sampling rate of the signal at the detector define the axial resolution of the OCT system. The axial resolution is defined as:

$$\delta z = \frac{2\ln 2}{\pi} \cdot \frac{\lambda_0{}^2}{\Delta\lambda_{\mathrm{FWHM}}} \tag{3.3}$$

where $\Delta\lambda_{\mathrm{FWHM}}$ is the spectral bandwidth of the light source.

The axial range of image is defined by imaging depth, which is related to the maximum frequency of the interference spectrum. The imaging depth is defined below.

$$z_{\max} = \frac{1}{4} \cdot \frac{\lambda_0{}^2}{\Delta\lambda} N \tag{3.4}$$

where N is the number of sample points.

3.3.3 Advantages and limitations

By providing anatomic information with high resolution and excellent penetration depth, OCT has unique advantages to evaluate the multi-layer retinal structures. It has been widely used in the diagnosis of retinal anatomic structure-related disease, including macular hole, macular pucker, macular edema, age-related macular degeneration, central serous chorioretinopathy, diabetic retinopathy, and vitreo-macular traction [64]. With the extension of OCT to evaluate function and vasculature, OCTA can evaluate microvasculature-related retinal disease, such as age-related macular degeneration, glaucoma, diabetes, and glaucoma [65]. In addition, OCTA can also evaluate the choriocapillaris and choroidal blood vessels, which are blocked by the melanin in the RPE layer and cannot be accessed by some of the imaging methods. Currently, the newly developed visible-light OCT could provide the functional imaging by investigating the hemodynamics of the retinal blood vessel but has limited penetration beyond the RPE [66].

Despite these advantages, OCT still has limitations. Due to the strong absorption of melanin, OCT has limited penetration to the structures below the RPE layer. The situation is worse for visible-light OCT. OCTA can evaluate the choriocapillaris and upper layer choroidal blood vessels, but it does not demonstrate leakage, provides limited visualization of microaneurysms, and has a restricted field of view, often with artifacts [58].

3.4 Photoacoustic imaging of the eye

3.4.1 Principle of photoacoustic microscopy

Photoacoustic microscopy (PAM) is based on the photoacoustic (PA) effect where optical energy is converted into acoustic energy [18, 24]. When short-pulsed laser light is used to illuminate biological tissues, photons travel into these tissues and some of them are absorbed by molecules, and the absorbed energy is partially converted to induce vibration of the tissue. Due to a rapid localized temperature increase, initial pressure is generated and propagated as acoustic waves, which are termed photoacoustic waves. The initial PA pressure generated, referred to as PA signal amplitudes, can be determined using the formula [67]:

$$p_0 = \Gamma \mu_a \Phi \tag{3.5}$$

where μ_a (cm^{-1}) is the absorption coefficient of the chromophores (i.e., lipid, melanin, hemoglobin, etc), Φ denotes the excitation laser fluence (J cm^{-2}), and Γ represents the Grüneisen coefficient. An ultrasound transducer is used to detect the acoustic wave. The ultrasound transducer is placed in contact with the conjunctiva of the eyes and aligned coaxially and confocally. At a fixed position, a single laser pulse excitation creates a one-dimensional (1D) depth-resolved PA image along the

Z-axis, which is referred to as an A-line without mechanical scanning. By scanning along horizontal direction, a two-dimensional (2D) depth-sensitive PA image (B-scan) is obtained. A 2D raster scanning along the X and Y-directions yields complete 3D volumetric information. To induce PA signal, the excitation laser fluence on the surface of the tissue should be kept within the American National Standards Institute (ANSI) safety limit [29, 68]. The PA signal amplitude is dominated by the optical absorption of the melanin in the retinal pigment endothelium (RPE) or hemoglobin in blood as shown in figure 3.2. Typically, PAM aims to image *in vivo* microvasculature with micron-scale spatial resolution and millimeter-scale depth. In addition, PAM can provide both structural and functional information such as the concentration of oxygenated and deoxygenated hemoglobin in blood vessels, variations in retinal blood oxygen saturation (SO_2), and metabolism by performing spectroscopic PAM imaging using different excitation wavelengths [69–71]. PAM can also be utilized to perform molecular imaging by using exogenous contrast agents to distinguish between normal and abnormal tissues [33, 35, 72, 73]. These characteristics are very critical for ophthalmologic applications since the morphological and functional abnormalities of the retinal vessels can be monitored and treated at an early stage before they cause severe eye disorders and permanent vision loss.

When exogenous contrast agents are used for molecular imaging, the energy deposited in the blood vessels will be absorbed by both the exogenous absorbing agent and hemoglobin (Hb). Therefore, the PA signal amplitude is proportional to the absorption properties of Hb and the absorber. Thus, equation (3.1) can be rewritten as follows:

$$p_0 = \Gamma \mu_t \Phi \tag{3.6}$$

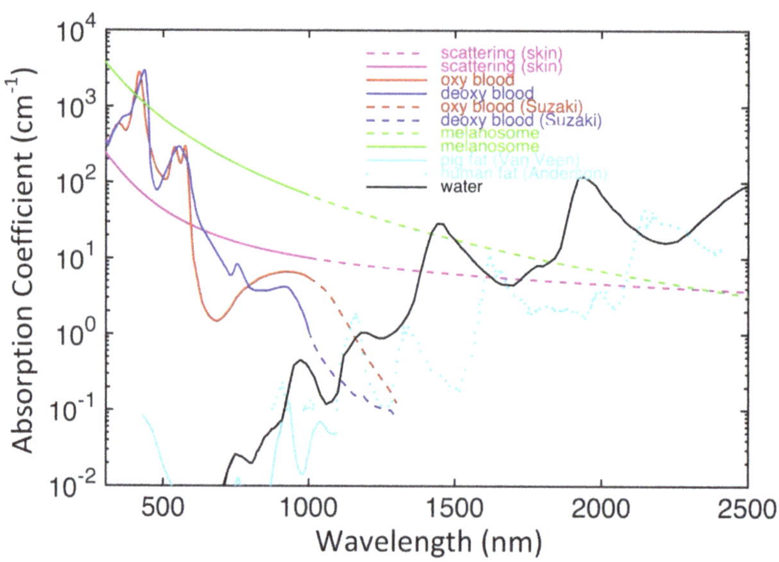

Figure 3.2. Optical absorption coefficient spectra of endogenous biomolecules in tissues (i.e., water, oxygenated hemoglobin, deoxygenated hemoglobin, melanin, and fat). Reprinted with permission from [103].

where:

$$\mu_t = \mu_a + \mu_{Hb} \tag{3.7}$$

where μ_t (cm^{-1}) is the total optical absorption coefficient of tissue contained contrast agents, μ_a (cm^{-1}) is the absorption coefficient of photoabsorber, and μ_{Hb} (cm^{-1}) is the optical absorption coefficient of hemoglobin.

The oxygen saturation (SO_2) of hemoglobin can be obtained as follows:

$$SO_2 = \frac{C_{HhO2}}{C_{HhO2} + C_{Hb}} \tag{3.8}$$

where C represents the concentration of oxygenated hemoglobin (HbO_2) and deoxygenated hemoglobin (Hb).

The concentration of hemoglobin is determined by the following formulas [74]:

$$\mu_a(\lambda_1) = 2.303 \times (E_{Hh}(\lambda_1)C_{Hb} + E_{HbO2}(\lambda_1)C_{HhO2}) \tag{3.9}$$

$$\mu_a(\lambda_2) = 2.303 \times (E_{Hh}(\lambda_2)C_{Hb} + E_{HbO2}(\lambda_2)C_{HhO2}) \tag{3.10}$$

where $\mu_a(\lambda_1)$ and $\mu_a(\lambda_2)$ are the absorption coefficients of the blood at selected wavelengths λ_1 and λ_2. E represents the molar extinction coefficients.

PAM systems have been developed with different designs by several groups for evaluation of different retinal tissues such as melanin, retinal vessels, choroidal vessels, RPE, corneal neovascularization, retinal neovascularization, choroidal neovascularization, retinal vein occlusion, and choroidal vessel occlusion [29, 36, 75]. PAM system can be classified into two subsystems: optical resolution PAM (OR-PAM) and acoustic resolution PAM (AR-PAM) [37, 71, 76]. Figure 3.3 shows schematics of AR-PAM and OR-PAM systems. In OR-PAM, both the excitation laser light and ultrasound transducer are tightly focused onto the ocular tissue and co-registered in the confocal plane. The optical focal beam that is used to generate the PA signal determines the lateral resolution. The lateral resolution of OR-PAM is determined by the following formulas [77]:

$$LR_{OR-PAM} = \frac{0.51\lambda_o}{NA_o} \tag{3.11}$$

where LR_{OR_PAM} is the lateral resolution of OR-PAM, λ_o is the excitation wavelength, and NA_o is the numerical aperture (NA) of the optical objective. In AR-PAM, the tissue is excited by a diffuse laser light, and the laser-induced PA signals are detected by a focused ultrasound transducer. The focus spot size of the transducer determines the lateral resolution of AR-PAM, whereas the axial resolution is determined by the ultrasound frequency and bandwidth [67, 78–80]. The lateral resolution of AR-PAM is determined using the following formula [77]:

$$LR_{AR-PAM} = \frac{0.71\lambda}{NA_a} \tag{3.12}$$

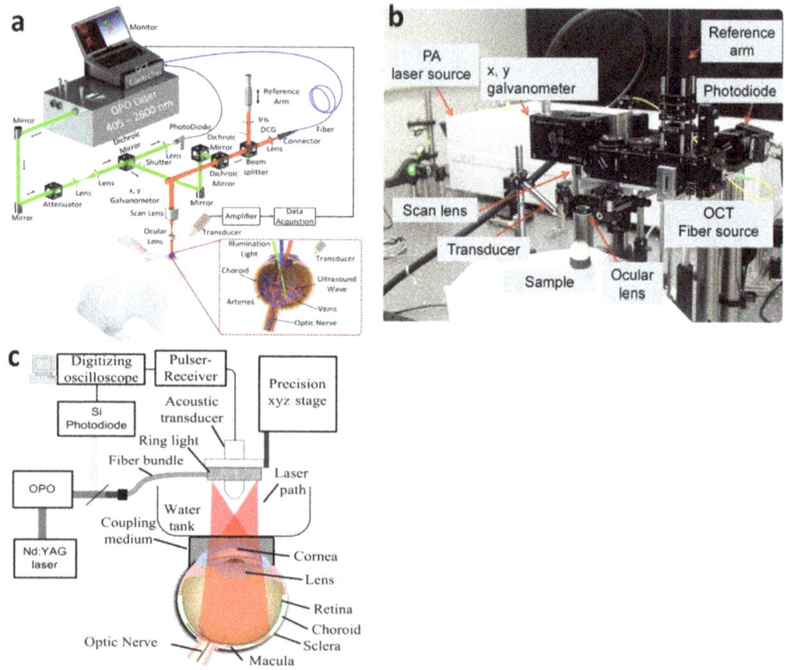

Figure 3.3. Schematics of the typical photoacoustic microscopy of the eye. (a) Reflection-mode OR-PAM setup [35]. (b) Corresponding physical setup. (c) Reflection-mode AR-PAM. Figures reprinted with permission from [35, 36].

Table 3.1. Lateral and axial resolution of OR-PAM.

Lateral resolution (μm)	Axial resolution (μm)	Frequency (MHz)	Bandwidth (MHz)	References
5	15	75	100	Hao *et al* [78]
3		50	50	Jeon *et al* [37]
4.1	37	27	16	Tian *et al* [75]
3.6	27.7	50	50	Kim *et al* [106]

where LR_{AR_PAM} is the lateral resolution of AR-PAM, λ is the acoustic wavelength, and NA_a is the numerical aperture of the focused ultrasound transducer.

OR-PAM provides about 10 times higher lateral resolution than that of AR-PAM. Table 3.1 shows a summary of lateral and axial resolution of OR-PAM. The axial resolutions of both OR-PAM and AR-PAM are similar and mainly determined by the bandwidth of the ultrasound transducer. The axial resolutions are estimated using the following formula [77]:

$$AR = \frac{0.88\, \upsilon_s}{\Delta f} \tag{3.13}$$

where AR represents the axial resolution (μm), υ_s is the speed of sound in medium ($c = 1.54$ mm μs^{-1}), and Δf is the bandwidth of the ultrasound transducer. The PA signal has a widely bandwidth ranging from 1 to 100 MHz [81]. This range is much wider than the detectable bandwidth of the ultrasound transducer. Thus, the bandwidth of the received PA signals is considered to be equal to the bandwidth of the transducer.

3.4.2 Requirement for PAM Imaging in the retina

Ocular tissues are very fragile and extremely sensitive to the light that illuminates the eye. Thus, the optical fluence needs to be safely delivered into the eye without damaging sensitive neural tissue. Due to this major concern, the excitation laser light that is used for ocular imaging should be below the American National Standards Institute (ANSI) safety limit standard [29, 68]. Typically, the pulse energy is less than 1 μJ for OR-PAM and 1 mJ for AR-PAM due to the difference laser spot size to excite the sample [25]. In contrast, the illumination intensity is much lower for ophthalmic imaging safety. High exposure to light energy may cause several types of damage to the retina, including thermal damage, thermoacoustic damage, and photochemical damage [82, 83]. The optical safety exposure dose depends on the optical wavelength, exposure duration, optical spot size, pulse repetition rate, and the average focal length of the eye. In accordance with the ANSI safety limit, the optical fluence is limited by three rules: single-pulse limit, the average power limit, and the repetitive pulse limit to regulate the maximum permissible exposure (MPE) [29, 84]. The ANSI maximum permissible exposure (MPE) for retinal exposure to single nanosecond pulses in the 400 to 700 nm spectral range is:

$$MPE_{sp} = 5.0 \times 10^{-7}\ J.cm^{-2} \tag{3.14}$$

For the repetitive pulse limit:

$$MPE_{rp} = n_{total}^{-0.25}\ MPE_{sp} \tag{3.15}$$

Within a laser spot of 20 μm, there are at most two overlapping laser pulses ($n_{total} = 2$). Thus,

$$MPE_{rp} = n_{total}^{-0.25} MPE_{sp} = 2^{-0.25} MPE_{sp} = 4.2 \times 10^{-7}\ J.cm^{-2} \tag{3.16}$$

Fluence on the cornea:

$$\Gamma = \frac{E}{A} = \frac{E_p}{\pi\left(\frac{D}{2}\right)^2} = 0.026\, E_p \tag{3.17}$$

where E_p is the maximum permissible single laser pulse energy on the retina (J).

For a single-pulse MPE quantification, assuming an incident laser spot size matched to a fully dilated pupil (7 mm). The MPE energy on the retina is

$$\Gamma < MPE_{rp} = 4.2 \times 10^{-7}\,\text{J.cm}^{-2} \tag{3.18}$$

Form (3.16), (3.17) and (3.18), the max optical energy for single-pulse exposure is

$$E = \frac{\Gamma}{0.026} = \frac{4.2 \times 10^{-7}}{0.026} = 160\,\text{nJ} \tag{3.19}$$

Another challenge in ophthalmic imaging is achieving PAM images with high resolution rapidly within the eye fixation time. Eyes are often moving and have a very short fixation time of approximately 500 ms [85]. Therefore, fast acquisition time is essential to avoid any potential motion artifacts. This unexpected motion artifact can cause image blurring or image disruption. For example, figure 3.4(a) shows an example of the PAM image associated with the movement of the rabbits during the experiment. Table 3.2 shows a summary of acquisition time in systems.

Non-invasive or minimally invasive imaging techniques are highly desirable to reduce systemic risk and side effects such as nausea, vomiting, allergic reactions, and patient discomfort by the administration of exogenous contrast agents such as occurs with fluorescein and ICG.

3.4.3 Multimodal PAM, OCT, and FM imaging

Both AR-PAM and OR-PAM have been developed for label-free imaging of different eye tissues from anterior segment structures such as the cornea, iris vasculature, and corneal neovascularization to the posterior segment structures, such as the retinal and choroidal blood vessels, RPE and melanin [40, 41, 76, 86, 87]. Figure 3.4(a) shows the AR-PAM image of the iris, cornea, lens, retina, choroid, sclera, and blood vessels obtained in living rabbit eyes [36]. However, retinal microvasculature is hard to visualize on the maximum intensity projection (MIP) PA image due to limited ultrasound spatial resolution (approximately 200 μm) and

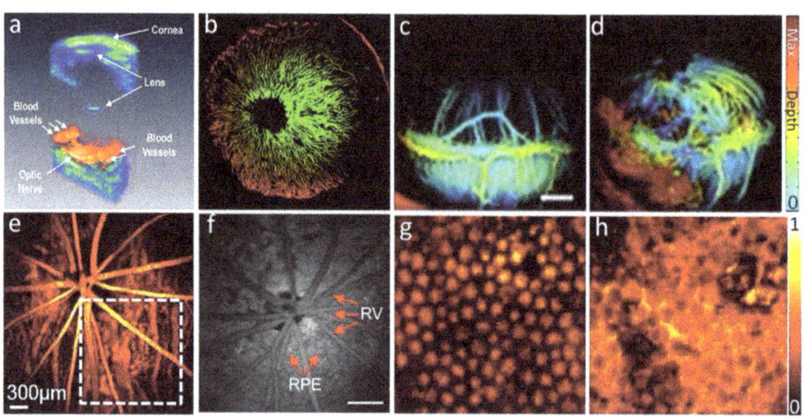

Figure 3.4. Photoacoustic microscopy of ocular tissues: (a) AR-PAM image of rabbit retinal vessels, cornea, and optic nerve [36]. (b–d) OR-PAM image of a rat iris (b), normal corneal vessels (c), and corneal neovascularization (d). (e–h) OR-PAM image of posterior segment structures including retinal and choroidal vessels obtained from albino rat (e), RPE obtained from pigmented rats (f), and melanin (g–h) [88, 90–92, 94].

Table 3.2. Summary of scanning time of the photoacoustic microscopy.

Scanning method	Acquisition time	Imaging size	Wavelength	Energy	Application	References
Mechanical scanning	90 min	12×8 mm^2	740 nm	0.5 mJ cm^{-2}	Eye tissues	de la Zerda *et al* [36]
Mechanical scanning	120 min	2×2 mm^2	570 and 578 nm	40 nJ	Iris microvasculature	Hu *et al* [71]
Mechanical scanning	20 min	3×3 mm^2	532 nm	80 nJ	Corneal neovascularization	Liu *et al* [107]
Mechanical scanning	6.5 min	2×2 mm^2	532 nm	500 nJ	Iris microvasculature	Wu *et al* [108]
Optical scanning	2.7 s	2×2 mm^2	532 nm	40 nJ	Retinal blood vessels, and RPE	Jiao *et al* [41]
Optical scanning	2.7 s	2×2 mm^2	570, 578, and 588 nm	40 nJ	SO_2, retinal and choroidal vessels	Song *et al* [94]
Optical scanning	65 s	3×3 mm^2	570 nm	80 nJ	Retinal and choroidal blood vessels in rabbit	Chao *et al* [29, 75]

the acquisition time is too long to achieve a volumetric PA image (approximately 90 min). Compared to AR-PAM, OR-PAM can provide better resolution, thus, the morphology of microvasculature can be clearly visualized. Figures 3.4(b)–(c) exhibit the anterior segment (iris vasculature and corneal neovasculature) [88, 89] and posterior segment (retinal vessels, RPE, and melanin) achieved by an OR-PAM imaging system [figures 3.4(d)–(h)] [88, 90–93].

Due to the transparent nature of the eye, the imaging of different ocular tissues has been obtained by several imaging modalities such as fluorescence angiography, indocyanine green angiography, optical coherence tomography, OCTA, and SLO [12–15]. Importantly, these imaging modalities are compatible with some PAM systems. Thus, to precisely evaluate the dynamic changes of the retinal microvasculature, PAM can be integrated with OCT, SLO, or fluorescence microscopy (FM) to achieve a multimodality imaging platform. Most multimodality imaging can provide a high scanning speed by using optical scanning (galvanometer scanning or MEMs) and high laser repetition rate (~1 kHz to 30 kHz) [40, 41, 76]. By using a multimodal PAM, OCT, and FM imaging system, the image of retinal tissues can be achieved from each modality and then the acquired images can be overlaid on the same orthogonal imaging plane. The OCT system provides additional information such as location, diameter, and thickness of the retina, choroid, and sclera. Importantly, this information can be monitored with real-time OCT. During an *in vivo* experiment, OCT can be utilized to guide the subretinal injection of agents (such as vascular endothelial growth factor, or VEGF) into the subretinal space or performed as an alignment tool to guide PAM. Thus, the structure and location of the tissues can be visualized.

3.4.4 Evaluation of retinal vasculature network and layers using multimodal imaging

Ophthalmic PA imaging of the anterior and posterior segments has been investigated to visualize normal and abnormal structures of retinal microvasculature as well as functional information such as measurement of SO_2 and retinal oxygen metabolic rate. In this section, the major applications of multimodal PAM, OCT, and FM imaging are described including label-free retinal and choroidal vessels, retinal neovascularization, retinal vein occlusion, and choroidal neovascularization.

3.4.1.1 Label-free evaluation of retinal & choroidal vessels, retinal vein occlusion, and choroidal vein occlusion

The retinal microvasculature plays an important role in providing oxygen and nutrients and removing metabolic waste in the eye. Visualization of retinal microvasculature can provide physiological and pathophysiological information for early diagnosis of many eye disorders such as diabetic retinopathy, glaucoma, and age-related macular degeneration (AMD). With high spatial resolution and the capability to distinguish single microvessels ranging from 10 to 300 μm, multimodal PAM and OCT has been used as an advanced imaging tool for visualization and characterization of the microvasculature in the retina in living animal models [29]. Both mice and rabbits with and without disease models such as retinal vein occlusion, laser-induced choroidal neovascularization (CNV), and choroidal vein occlusion have been widely used. Several quantitative parameters such as vessel diameter, vessel thickness, retinal thickness, vessel perfusion, blood flow speed, oxygen saturation (SO_2), retinal metabolic rate of oxygen ($rMRO_2$), total hemoglobin concentration (HbT), melanin concentration, newly formed blood vessels, and safety evaluation post-PAM imaging have been monitored and described [94, 95]. Those parameters were first examined in rats [96]. Figure 3.5 shows PAM and OCT images of ocular microvasculature and functional information. The location and

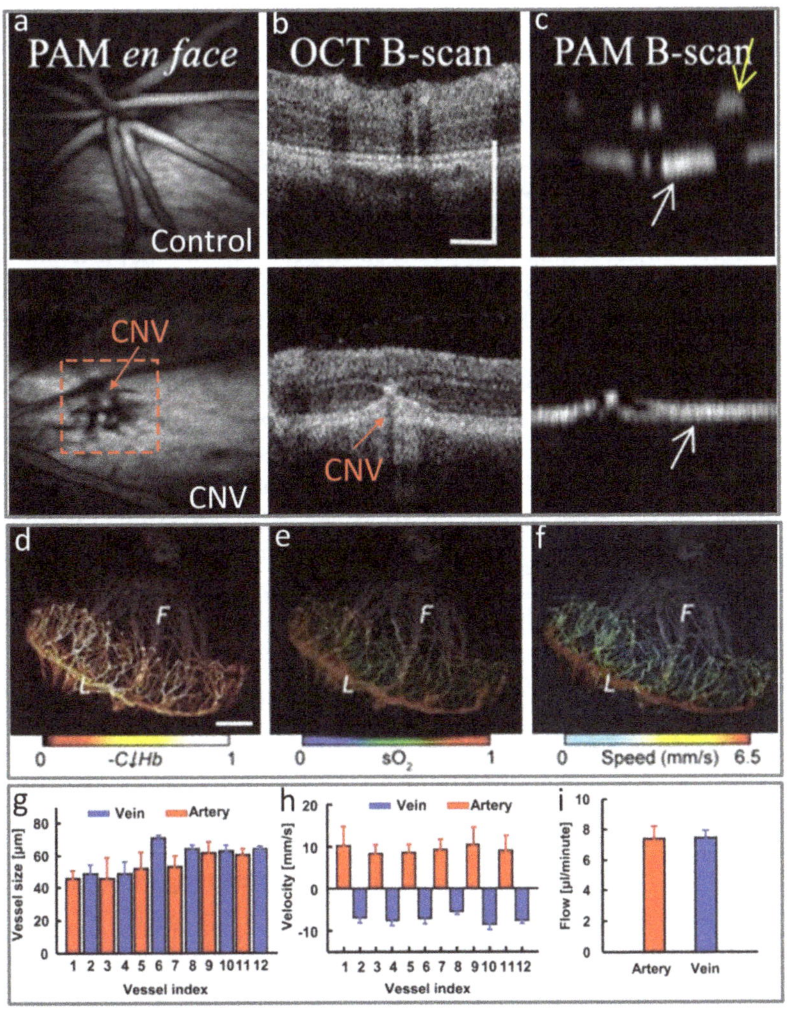

Figure 3.5. Multimodal PAM and OCT imaging of choroidal neovascularization and corneal neovascularization in mice: choroidal and multimodal PAM and OCT: (a) Maximum intensity projection (MIP) PAM image of the laser-induced CNV. Red arrow indicates the location and margin of CNV. (b) and (c) B-Scan OCT and PAM, respectively. (d–f) PAM maps of hemoglobin concentration (d), oxygen saturation (e), blood velocity (f). (g–k) Quantitative measurements of vessels diameter and flow speed in artery and vein. CNV: choroidal neovascularization Bar: 100 μm. Reproduced with permission from reference [89, 93, 105].

morphology of retinal vessels and CNV are identified. Capillaries and newly developed CNV can be resolved with high resolution and high imaging contrast without loss of lateral resolution. SO_2 and $rMRO_2$ were determined using different wavelengths of 570, 578 and 588 nm [96]. As shown in figure 3.5, the SO_2 level was significantly different between artery and vein (i.e., $SO_2 = 93.0 \pm 3.5\%$ in artery versus $77.3 \pm 9.1\%$ in vein). The retinal blood flow was measured to be 7.43 ± 0.51 μl min^{-1} within the vein and 7.38 ± 0.78 μl min^{-1} within the artery [97]. The $rMRO_2$ is approximately 297.86 ± 70.23 nl min^{-1} and reperfusion rate were also changed at the location of ischemia [96]. Another significant piece of information that needs to be considered is the change in blood velocity (figure 3.5). The blood flow rate must be kept constant to provide the same flux of oxygen and nutrients for the retina. When the vessel diameter is decreased or occluded, the blood flow velocity in the artery and vein increased or reduced accordingly. These results provide significant information for the early diagnosis of retinal pathophysiology. However, a major challenge for clinical translation of this murine data is that the axial length of mouse eye is significantly smaller than that of the human eye (i.e., ~3 mm for mice versus ~23 mm for human), resulting in difficulty in clinical translation, particularly for the acoustic portion of PAM [42, 98–100]. Recently, a multimodality PAM and OCT system for visualization of different retinal tissues in larger animals like rabbits was developed and reported by Tian *et al*, Zhang *et al* and Nguyen *et al* [22, 29, 87]. The normal and abnormal microvasculature were successfully visualized in living New Zealand White and Dutch Belted Pigmented rabbits as shown in figure 3.6. Individual healthy retinal vessels, choroidal vessels, and capillaries were clearly identified (figures 3.6(a) and (b)). The thickness of the retina is around 200–300 μm and the major retinal vessels diameter is approximately 70–100 μm [29]. The location of retinal vein occlusion and choroidal vein occlusion can also be identified and distinguished on PAM and OCT images (figures 3.6(c) and (d)). It was noted that the PAM image contrast was reduced at the location of the occlusion site due to the reduction of vessel diameter and concentration of hemoglobin within the vessels. The SO_2 level in the choroid and retina was also revealed at the ischemia by using two different wavelengths of 750 and 850 nm [101]. The SO_2 measurement panel shows that blood oxygen was obviously reduced and exhibited a 6-fold lower signal at the ischemic site when compared to the control areas (conjunctiva). The SO_2 was only changed at the ischemic site in the choroid and retina, but no significant changes were observed on the other parts of the eye. Longitudinal observation in rabbits having retinal vein occlusion and choroidal vein occlusion shows that newly developed blood vessels, named choroidal neovascularization (CNV), were established and observed on the PAM image acquired at day 28 post-ischemia (figure 3.6). The CNV morphology and location were segmented and distinguished from the surrounding vessels using post-image processing (figure 3.6).

3.4.1.2 Retinal neovascularization imaging in vivo *using triple PAM, OCT, and FM*

Although dual modality PAM and OCT has been used to image both normal and abnormal microvasculature in rodents and large animals with promising results

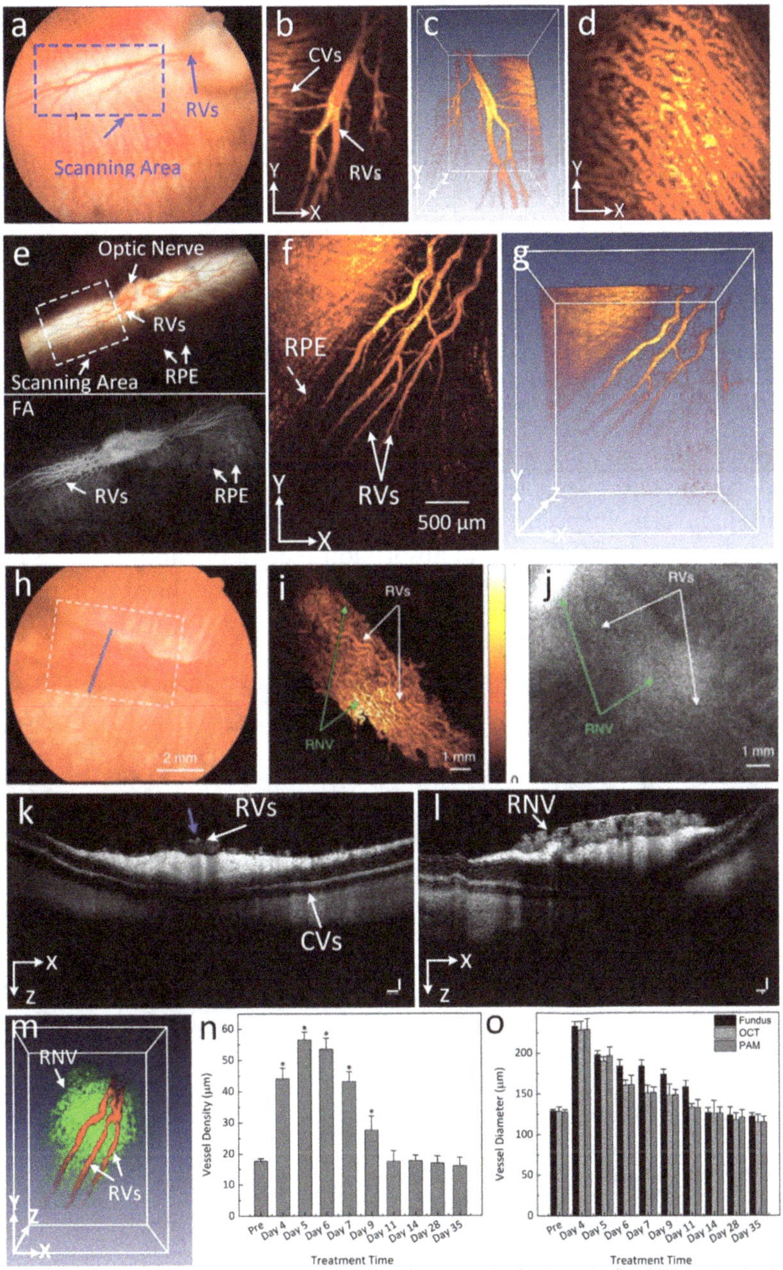

Figure 3.6. Multimodal PAM, OCT, and FM imaging in living albino and pigmented rabbits: (a) Color fundus photography. (b–c) 2D and 3D PAM image of retinal vessels (RV). (d) PAM image of choroidal vessels (CVs) in White New Zealand rabbits. (e) Color fundus photography (top) and fluorescein angiography (FA; bottom) images of Dutch Belted rabbits. (f–g) 2D and 3D PAM images acquired at 578 nm. (h) Color fundus photograph of retinal neovascularization (RNV). (i) corresponding PAM image acquired along the selected area shown in i. (j) Fluorescence microscopy (FM). (k) B-scan OCT of normal retina. (l) B-scan OCT of rabbit retina with having RNV models. (m) Image segmentation. (n) Quantitative vessels density. (o) Measurement of vessel diameters. Figures reproduced with the permission from reference [84, 102, 104].

since it was described, additional improvements to better visualize and assist with the precise diagnosis of different eye disorders have been described. Very recently, a triple PAM, OCT and FM imaging has been reported by Zhang *et al* for detection of retinal neovascularization (RNV) [102]. The imaging was implemented in living rabbits with RNV models and the results are summarized in figure 3.7. It is clear that both native retinal vessels and RNV were clearly visualized and the diameter of native retinal artery and vein were significantly changed over time (figure 3.7(a)), which has been explained by the inflammation of retinal vessels after administration of vascular endothelial growth factor (VEGF). Overlying 3D volumetric images acquired from each imaging modality helped to precisely classify RNV from the retinal vessels. Image segmentation was also used to distinguish between RNV and healthy retinal vessels (figure 3.7(b)).

3.4.1.3 Molecular imaging with contrast agents

Most endogenous chromophores (i.e., melanin and hemoglobin) in the retinal tissues can absorb light at their specific wavelengths, and thus they can produce a strong PA signal without labelling. The optical absorption spectra of chromophores are shown in figure 3.2. In accordance with the absorption spectra, hemoglobin has a very

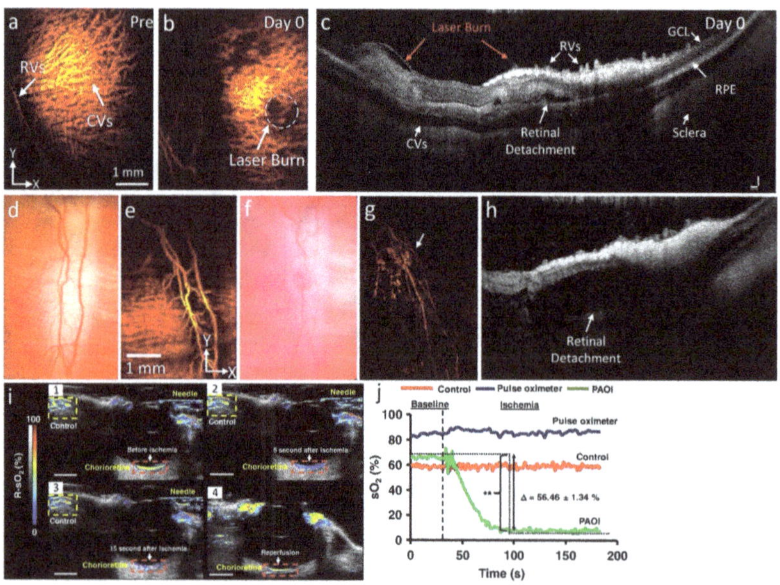

Figure 3.7. Multimodal PAM and OCT of retinal ischemia: (a) MIP PAM images pre- and post-laser-induced choroidal vein occlusion (CVO). White dotted circles show the detected location of CVO. (c) B-scan OCT showing different retinal layers such as retina, choroid, sclera, and ganglion cell layer (GCL). Red arrow shows the treated site. (d) and (f) Color fundus images before and after laser-induced retinal vein occlusion. (e) and (g) corresponding PAM images. The location of vein occlusion was clearly detected (white arrow). (h) B-scan OCT image. (i) Oxygen saturation (SO_2) maps before and after ischemia obtained at different time points. (j) Quantitative SO_2 profile as a function of treatment time. Figures reproduced with permision from reference [32, 34, 101].

strong optical absorption in the visible window ranging from 532 to 580 nm and very low absorption properties in NIR window (i.e., 650 to 1100 nm) (for example, μa = 236.93 cm^{-1} at 532 nm versus μa = 1.98 cm^{-1} at 650 nm) [103]. Therefore, the spectroscopic PAM can be implemented by acquiring the PAM image at multiple excitation wavelengths and the PA signal obtained by these excitation wavelengths can be used to evaluate and identify the location of abnormal vessels. However, the intrinsic PA signal generated in the NIR window is low, resulting in a difficulty to visualize the targeted tissues. To boost the sensitivity of PAM, several exogenous contrast agents have been evaluated as contrast agents for PAM and OCT from small-molecule organic dyes (indocyanine green) [27] to inorganic contrast agents (i.e., gold nanospheres, gold nanorods, gold nanostars, and chain-like clusters gold nanoparticles) (figure 3.7) [24, 33, 104]. The development of PA contrast agents in ophthalmology can open a great opportunity to advance the molecular imaging for detecting the early stages of retinal diseases such as neovascularization or assisting with the development of cell-based therapy. For molecular PAM imaging, exogenous contrast agents with high optical absorption can be conjugated with specific ligands such as arginine–glycine–aspartate (RGD). When the contrast agents are administrated into the biological tissue through intravenous injection or intravitreal injection, these agents such as RGD bind to the integrin $\alpha_v\beta_3$, which is overexpressed in neovascularization. The accumulation of these agents at the specific location of the disease can provide molecular PAM imaging with high spatial resolution, resulting in improved distinguishing of the margin of disease from the surrounding normal microvasculature. Up until now, several exogenous contrast agents have been described to target RNV and CNV as shown in figure 3.8. Due to the improved optical absorption and scattering by targeted contrast agents, the location and margin of RNV and CNV can be visualized and distinguished by spectroscopic PAM and OCT with high resolution. This allows one to create a novel technique for studying retinal pathologies along with conventional fluorescein angiography.

Some challenges of exogenous contrast agents are evaluating ones that are biodegradable, biocompatible, and photostable. Although organic exogenous contrast agents like ICG are approved by the US Food and Drug Administration (FDA) for clinical use, some of these molecules are rapidly cleared from the body after administration and photo-bleaching may occur under illumination with laser light. Thus, it can require the injection of additional doses in order to monitor diseases longitudinally. In contrast, gold nanoparticles can provide better photostability and create better PA signals than organic dyes. However, clinical approval of gold nanoparticles contrast agents in ophthalmology remains very limited. Therefore, more research is required to evaluate the biosafety concerns of the exogenous contrast agents.

3.5 Conclusions

In conclusion, multimodal PAM, OCT, and FM has several potential applications for the study of retinal pathologies. In this chapter, we have introduced the fundamental principle, advantages, and major applications of multimodality

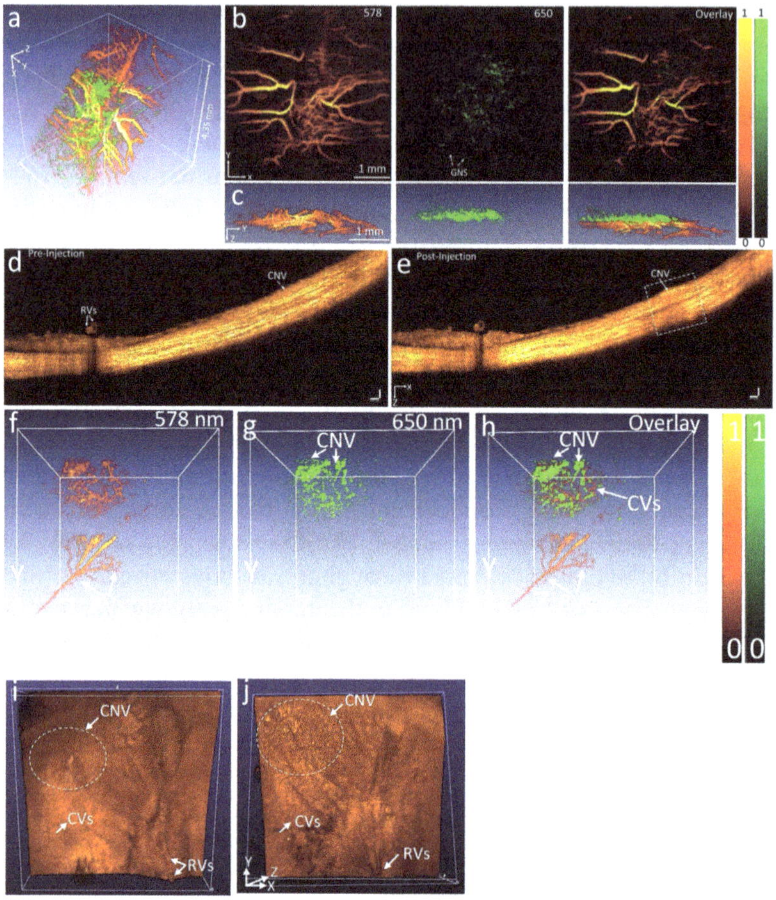

Figure 3.8. *In vivo* multimodal molecular PAM and OCT imaging: (a) 3D volumetric rendering of PAM images of choroidal neovascularization (CNV) at 24 h post-injection of targeting gold nanostars (GNS). (b) Horizontal (x–y) PAM image (post 24 h). (c) Vertical (y–z) PAM image. Pseudo-green color indicates the distribution of GNS. (d-e) B-scan OCT images obtained before and 2 h after IV injection of GNS. The white arrows show the position of CNV. PA detection of CNV in living rabbits using CGNP clusters-RGD. (f) and (g) 3D PAM images of CNV created by laser-induced retinal vein occlusion at day 28 post-laser illumination after the injection of 0.5 ml CGNP clusters-RGD at concentration of 2.5 mg ml^{-1} under nanosecond pulsed laser illumination at wavelength of 578 and 650 nm, respectively. (h) Overlay 3D images showed the distribution of CGNP clusters-RGD accumulated at CNV location in rabbit retina (pseudo-green color). (i) and (j) 3D OCT images before and after the injection of CGNP clusters-RGD. The distribution of CGNP clusters-RGD were clearly observed on 3D OCT image after injection (white dotted circle).[33, 35]

imaging systems. The structural, functional, and molecular imaging of various retinal diseases such as retinal vein occlusion, choroidal vein occlusion, retinal neovascularization, and choroidal neovascularization in both small animals, such as mice and rats, and larger animals, such as rabbits, with great potential for clinical applications have been described. The multimodal PAM, OCT, and FM offers a novel research tool to measure oxygen saturation (SO_2), blood flow, and retinal oxygen metabolic rate ($rMRO_2$). These measurement parameters are very important for accurate diagnosis and treatment of the retinal ischemia and metabolic diseases. Molecular imaging of CNV and RNV is another critical possibility of the multimodal imaging system for early-stage diagnosis and treatment of AMD, and retinopathy. Consequently, multimodal PAM, OCT, and FM holds a great translational tool for studying human ocular diseases.

References

[1] Flaxman S R *et al* 2017 *Lancet Global Health.* **5** e1221–34

[2] Chan T, Friedman D S, Bradley C and Massof R 2018 *JAMA Ophthalmol.* **136** 12–9

[3] Stevens G A, White R A, Flaxman S R, Price H, Jonas J B, Keeffe J, Leasher J, Naidoo K, Pesudovs K and Resnikoff S 2013 *Ophthalmology* **120** 2377–84

[4] Berendschot T T, Goldbohm R A, Klopping W A, van de Kraats J, van Norel J and van Norren D 2000 *Investig. Ophthalmol. Vis. Sci.* **41** 3322–6

[5] Landrum J T, Bone R A and Kilburn M D 1996 *Adv. Pharmacol.* **38** 537–56

[6] Hyman L G, Lilienfeld A M, FERRIS F L and Fine S L 1983 *Am. J. Epidemiol.* **118** 213–27

[7] Hardarson S H and Stefánsson E 2010 *Am. J. Ophthalmol.* **150** 871–5

[8] Hammer M, Vilser W, Riemer T, Mandecka A, Schweitzer D, Kühn U, Dawczynski J, Liemt F and Strobel J 2009 *Graefe's Arch. Clin. Exp. Ophthalmol.* **247** 1025–30

[9] Hardarson S H and Stefánsson E 2012 *Br. J. Ophthalmol.* **96** 560–3

[10] Khoobehi B, Firn K, Thompson H, Reinoso M and Beach J 2013 *Investig. Ophthalmol. Vis. Sci.* **54** 7103–6

[11] Vandewalle E, Abegao Pinto L, Olafsdottir O B, De Clerck E, Stalmans P, Van Calster J, Zeyen T, Stefánsson E and Stalmans I 2014 *Acta Ophthalmol.* **92** 105–10

[12] Coscas F, Glacet-Bernard A, Miere A, Caillaux V, Uzzan J, Lupidi M, Coscas G and Souied E H 2016 *Am. J. Ophthalmol.* **161** e162

[13] Prasad P S, Oliver S C, Coffee R E, Hubschman J-P and Schwartz S D 2010 *Ophthalmology* **117** 780–4

[14] McAllister I L, Yu D-Y, Vijayasekaran S, Barry C and Constable I 1992 *Br. J. Ophthalmol.* **76** 615–20

[15] Soetikno B T, Shu X, Liu Q, Liu W, Chen S, Beckmann L, Fawzi A A and Zhang H F 2017 *Biomed. Opt. Express* **8** 3571–82

[16] Dysli C, Wolf S, Berezin M Y, Sauer L, Hammer M and Zinkernagel M S 2017 *Prog. Retina Eye Res.* **60** 120–43

[17] Xu M and Wang L V 2006 *Rev. Sci. Instrum.* **77** 041101

[18] Beard P 2011 *Interface Focus* **1** 602–31

[19] Cox B T, Laufer J G, Beard P C and Arridge S R 2012 *J. Biomed. Opt.* **17** 061202

[20] Weber J, Beard P C and Bohndiek S E 2016 *Nat. Methods* **13** 639–50

[21] Nguyen V P, Park S, Oh J and Wook Kang H 2017 *J. Biophotonics* **10** 1053–61

[22] Nguyen V P, Li Y, Zhang W, Wang X and Paulus Y M 2018 *Biomed. Opt. Express* **9** 5915–38

[23] Yang J-M, Favazza C, Chen R, Yao J, Cai X, Maslov K, Zhou Q, Shung K K and Wang L V 2012 *Nat. Med.* **18** 1297–302

[24] Wang L V 2009 *Nat. Photonics* **3** 503–9

[25] Wang L V and Yao J 2016 *Nat. Methods* **13** 627

[26] Kim T, Lemaster J E, Chen F, Li J and Jokerst J V 2017 *ACS Nano* **11** 9022–32
[27] Nguyen V P, Li Y, Folz J, Henry J, Aaberg M, Zhang W, Wang X and Paulus Y M 2019 *Front. Opt.* **2** FM5F. 5
[28] Nguyen V P, Oh Y, Ha K, Oh J and Kang H W 2015 *Jpn. J. Appl. Phys.* **54** 07HF04
[29] Tian C, Zhang W, Mordovanakis A, Wang X and Paulus Y M 2017 *Opt. Express* **25** 15947–55
[30] Wang H-W, Chai N, Wang P, Hu S, Dou W, Umulis D, Wang L V, Sturek M, Lucht R and Cheng J-X 2011 *Phys. Rev. Lett.* **106** 238106
[31] Nguyen V P and Paulus Y M 2018 *J. Imaging* **4** 149
[32] Nguyen V P, Li Y, Zhang W, Wang X and Paulus Y M 2019 *Sci. Rep.* **9** 1–14
[33] Nguyen V-P, Li Y, Henry J, Zhang W, Aaberg M, Jones S, Qian T, Wang X and Paulus Y M 2020 *ACS Sens.* **5** 3070–81
[34] Nguyen V P, Li Y, Henry J, Zhang W, Wang X and Paulus Y M 2020 *Int. J. Mol. Sci.* **21** 6508
[35] Nguyen V P, Qian W, Li Y, Liu B, Aaberg M, Henry J, Zhang W, Wang X and Paulus Y M *Nat. Commun.* **12** 1–14
[36] de la Zerda A, Paulus Y M, Teed R, Bodapati S, Dollberg Y, Khuri-Yakub B T, Blumenkranz M S, Moshfeghi D M and Gambhir S S 2010 *Opt. Lett.* **35** 270–2
[37] Jeon S, Song H B, Kim J, Lee B J, Managuli R, Kim J H, Kim J H and Kim C 2017 *Sci. Rep.* **7** 4318
[38] Kelly-Goss M R, Ning B, Bruce A C, Tavakol D N, Yi D, Hu S, Yates P A and Peirce S M 2017 *Sci. Rep.* **7** 9049
[39] Liu W and Zhang H F 2016 *Photoacoustics* **4** 112–23
[40] Song W, Wei Q, Jiao S and Zhang H F 2013 *J. Vis. Exp.* **40** e4390
[41] Jiao S, Jiang M, Hu J, Fawzi A, Zhou Q, Shung K K, Puliafito C A and Zhang H F 2010 *Opt Epxress* **18** 3967–72
[42] Hughes A 1972 *Vis. Res.* **12** 123-IN126
[43] Zhang W, Li Y, Nguyen V P, Huang Z, Liu Z, Wang X and Paulus Y M 2018 *Light: Sci. Appl.* **7** 103
[44] Nguyen V P, Li Y, Zhang W, Wang X and Paulus Y M 2019 *Sci. Rep.* **9** 10560
[45] Brancato R and Trabucchi G 2009 *Semin. Ophthalmol.* **13** 189–98
[46] Novotny H R and Alvis D L 1961 *Circulation.* **24** 82–6
[47] Rabb M F, Burton T C, Schatz H and Yannuzzi L A 1978 *Surv. Ophthalmol.* **22** 387–403
[48] David N J, Norton E W, Gass J D and Beauchamp J 1967 *Arch. Ophthalmol.* **77** 619–29
[49] Stanga P E, Lim J I and Hamilton P 2003 *Ophthalmology* **110** 15–21
[50] Regillo C D, Benson W E, Maguire J I and Annesley W H 1994 *Ophthalmology* **101** 280–8
[51] Hurley B R and Regillo C D 2009 Fluorescein Angiography: General Principles and Interpretation *Retinal Angiography and Optical Coherence Tomography* (Berlin: Springer) pp 27–42
[52] Kornblau I S and El-Annan J F 2019 *Surv. Ophthalmol.* **64** 679–93
[53] Hope-Ross M, Yannuzzi L A, Gragoudas E S, Guyer D R, Slakter J S, Sorenson J A, Krupsky S, Orlock D A and Puliafito C A 1994 *Ophthalmology* **101** 529–33
[54] Huang D, Swanson E A, Lin C P, Schuman J S, Stinson W G, Chang W, Hee M R, Flotte T, Gregory K and Puliafito C A 1991 *Science* **254** 1178–81
[55] Yaqoob Z, Wu J and Yang C 2005 *Biotechniques* **39** S6–S13
[56] Schuman J S 2008 *Trans. Am. Ophthalmol. Soc.* **106** 426

[57] Reddikumar M, Bose K and Poddar R 2016 *Optik* **127** 1656–9
[58] De Carlo T E, Romano A, Waheed N K and Duker J S 2015 *Int. J. Retina Vit.* **1** 5
[59] Hagag A M, Gao S S, Jia Y and Huang D 2017 *Taiwan J. Ophthalmol.* **7** 115
[60] Shu X, Beckmann L J and Zhang H F 2017 *J. Biomed. Opt.* **22** 121707
[61] Fleming C P, Eckert J, Halpern E F, Gardecki J A and Tearney G J 2013 *Biomed. Opt. Express* **4** 1269–84
[62] Aumann S, Donner S, Fischer J and Müller F 2019 *Optical Coherence Tomography (OCT): Principle and Technical Realization* (Berlin: Springer) pp 59–85
[63] Fercher A F, Drexler W, Hitzenberger C K and Lasser T 2003 *Rep. Prog. Phys.* **66** 239
[64] Bhende M, Shetty S, Parthasarathy M K and Ramya S 2018 *Indian J. Ophthalmol.* **66** 20
[65] Wang J C and Miller J B 2019 *Semin. Ophthalmol.* **34** pp 211–7
[66] Soetikno B T, Yi J, Shah R, Liu W, Purta P, Zhang H F and Fawzi A A 2015 *Sci. Rep.* **5** 16752
[67] Nguyen V P, Kim J, Ha K-l, Oh J and Kang H W 2014 *J. Biomed. Opt.* **19** 105010–0
[68] The Laser Institute 2007 https://lia.org/store/product/ansi-z1361–2014-safe-use-lasers-electronic-version
[69] Song W, Wei Q, Liu W, Liu T, Yi J, Sheibani N, Fawzi A A, Linsenmeier R A, Jiao S and Zhang H F 2014 *Sci. Rep.* **4** 6525
[70] Yao J, Maslov K I, Zhang Y, Xia Y and Wang L V 2011 *J. Biomed. Opt.* **16** 076003
[71] Hu S, Rao B, Maslov K and Wang L V 2010 *Opt. Lett.* **35** 1–3
[72] Nguyen V P, Oh J, Park S and Wook Kang H 2016 *J. Biophotonics* **10** 1053–61
[73] Sun Y and O'Neill B 2013 *Appl. Opt.* **52** 1764–70
[74] Hennen S N, Xing W, Shui Y-B, Zhou Y, Kalishman J, Andrews-Kaminsky L B, Kass M A, Beebe D C, Maslov K I and Wang L V 2015 *Exp Eye Res.* **138** 153–8
[75] Tian C, Zhang W, Nguyen V P, Wang X and Paulus Y M 2018 *J. Vis. Exp.* **132** e57135
[76] Liu X, Liu T, Wen R, Li Y, Puliafito C A, Zhang H F and Jiao S 2015 *Opt. Lett.* **40** 1370–3
[77] Li C and Wang L V 2009 *Phys. Med. Biol.* **54** R59
[78] Zhang H F, Puliafito C A and Jiao S 2011 *Ophthalmic Surg., Lasers Imaging Retina.* **42** S106–15
[79] Maslov K, Stoica G and Wang L V 2005 *Opt. Lett.* **30** 625–7
[80] Xing W, Wang L, Maslov K and Wang L V 2013 *Opt. Lett.* **38** 52–4
[81] Vallet M, Varray F, Kalkhoran M A, Vray D and Boutet J 2014 Enhancement of photoacoustic imaging quality by using CMUT technology: Experimental study (Piscataway) (NJ: IEEE) pp 1296–9
[82] Kuo T-R, Hovhannisyan V A, Chao Y-C, Chao S-L, Chiang S-J, Lin S-J, Dong C-Y and Chen C-C 2010 *J. Am. Chem. Soc.* **132** 14163–71
[83] Organisciak D T and Vaughan D K 2010 *Prog. Retin. Eye Res* **29** 113–34
[84] Nguyen V P, Li Y, Aaberg M, Zhang W, Wang X and Paulus Y M 2018 *J. Imaging* **4** 150
[85] Robinson D 1964 *J. Physiol.* **174** 245–64
[86] Zhang W, Li Y, Nguyen V P, Derouin K, Xia X, Paulus Y M and Wang X 2020 *J. Biomed. Opt.* **25** 066003
[87] Zhang W, Li Y, Yu Y, Derouin K, Qin Y, Nguyen V P, Xia X, Wang X and Paulus Y M 2020 *Photoacoustics* **20** 100194
[88] Jeon S, Song H B, Kim J, Lee B J, Managuli R, Kim J H, Kim J H and Kim C 2017 *Sci. Rep.* **7** 1–9

[89] Kelly-Goss M R, Ning B, Bruce A C, Tavakol D N, Yi D, Hu S, Yates P A and Peirce S M 2017 *Sci. Rep.* **7** 1–12

[90] Zhao H, Wang G, Lin R, Gong X, Song L, Li T, Wang W, Zhang K, Qian X and Zhang H 2018 *J. Biomed. Opt.* **23** 046006

[91] Xie D, Li Q, Gao Q, Song W, Zhang H F and Yuan X 2018 *J. Biophotonics* **11** e201700360

[92] Shu X, Li H, Dong B, Sun C and Zhang H F 2017 *Biomed. Opt. Express* **8** 2851–65

[93] Song W, Wei Q, Jiao S and Zhang H F 2013 *JoVE (J. Vis. Exp.)* **71** e4390

[94] Song W, Wei Q, Liu T, Kuai D, Zhang H F, Burke J M and Jiao S 2012 *J. Biomed. Opt.* **17** 061206

[95] Ning B, Kennedy M J, Dixon A J, Sun N, Cao R, Soetikno B T, Chen R, Zhou Q, Shung K K and Hossack J A 2015 *Opt. Lett.* **40** 910–3

[96] Song W, Wei Q, Liu W, Liu T, Yi J, Sheibani N, Fawzi A A, Linsenmeier R A, Jiao S and Zhang H F 2014 *Sci. Rep.* **4** 1–7

[97] Liu W and Zhang H F 2014 Noninvasive *in vivo* imaging of oxygen metabolic rate in the retina (Piscataway, NJ: IEEE) pp 3865–8

[98] Massof R W and Chang F W 1972 *Vis. Res.* **12** 793–6

[99] Hughes A 1979 *Vis. Res.* **19** 569–88

[100] Deering M F 2005 *ACM Tran. Graph. (TOG)* **24** 649–58

[101] Hariri A, Wang J, Kim Y, Jhunjhunwala A, Chao D L and Jokerst J V 2018 *J. Biomed. Opt.* **23** 036005

[102] Zhang W, Li Y, Nguyen V P, Huang Z, Liu Z, Wang X and Paulus Y M 2018 *Light: Sci. Appl.* **7** 1–12

[103] Jacques S L 2015 Generic tissue optical properties https://omlc.org/news/feb15/generic_optics/index.html (accessed 4 November 2023)

[104] Nguyen V P, Li Y, Qian W, Liu B, Tian C, Zhang W, Huang Z, Ponduri A, Tarnowski M and Wang X 2019 *Sci. Rep.* **9** 1–17

[105] Dai C, Li L, Liu W, Wang F and Zhou C 2018 *Int. Soc. Opt. Photon.* **10494** 631–8

[106] Kim J Y, Lee C, Park K, Lim G and Kim C 2015 *Sci. Rep.* **5** 7932

[107] Liu T, Li H, Song W, Jiao S and Zhang H F 2013 *Curr. Eye Res.* **38** 1229–34

[108] Wu N, Ye S, Ren Q and Li C 2014 *Opt. Lett.* **39** 2451–4

IOP Publishing

Photo Acoustic and Optical Coherence Tomography Imaging, Volume 1
Diabetic retinopathy
Ayman El-Baz and Jasjit S Suri

Chapter 4

OCT-Leakage: non-invasive identification and measurement of abnormal retinal fluid

Conceição Lobo, Torcato Santos and José Cunha-Vaz

4.1 Introduction

Diabetes mellitus is one of the most common diseases, particularly in developed countries, with its prevalence and incidence increasing in recent years. Indeed, the International Diabetes Federation reports that the worldwide number of individuals with diabetes will rise from 463 million (in 2019) to 700 million (in 2045), corresponding to a prevalence of 9.3% and 10.9%, respectively [1].

Among several complications associated with this multifactorial disease, diabetic retinopathy (DR) is one of the most frequent and the main cause of vision loss in the active working population in Western countries, being, each year, responsible for 10% of new cases of blindness [1].

This particular complication has been associated with breakdown or failure of the blood-retinal barrier (BRB), which regulates the passage of molecules, solutes, and other compounds, prevents fluid diffusion and contributes to the immune privilege of the retina [2]. In DR, the BRB has been shown to be altered as early as the initial stages [3, 4]. This event results in leakage that promotes abnormal increase of retinal extracellular fluid.

For several years, fluorescein angiography (FA) has been considered the gold-standard imaging technique used to document these BRB alterations associated with DR. This methodology is based on the use of sodium fluorescein dye intravenously administered, which allows the observation of retinal vessels and leakage sites. However, the use of this technique may lead to minor adverse reactions (5% of cases) or some rare severe complications, requiring the presence of a medical doctor for its administration.

More recently, the use of FA has been partially replaced by the use of optical coherence tomography (OCT) microangiography for imaging in the eye [5], which takes advantage of multiple B-scans acquired on structural OCT, to detect motion contrast, allowing the non-invasive identification and quantification of retinal vascular alterations, as capillary dropout [6]. Still, this approach does not have the ability, per se, to identify sites of retinal capillary leakage, corresponding to BRB alterations. Indeed, to our knowledge, thus far, there has been no non-invasive method able to properly evaluate and to image leakage or breakdown of the BRB [7].

In an attempt to respond to this need, we have developed a novel algorithm, OCT-Leakage, a non-invasive imaging technique capable of identifying, locating and measuring increases in the retinal extracellular space, which are surrogate indicators of BRB alterations and therefore complement the conventional OCT microangiography.

4.2 OCT-Leakage: analysis of retinal extracellular fluid

The concept of OCT-Leakage is based on the notion that, when assessed by structural OCT, retinal cell nuclei present higher reflectivity than water-filled areas. Hence, it is considered that areas with low optical reflectivity (LOR) on OCT imagens depict the sites with the presence of retinal fluid, consequently allowing the identification of sites with abnormal accumulation of extracellular fluid, corresponding to damage of the BRB.

This measurement of retinal extracellular fluid distribution, in the whole retina, in a given retinal region, or layer-by-layer can be, therefore, achieved by the measuring of the LOR area ratio, which represents the number of A-scans obtained in structural OCT with reflectivity values below the threshold divided by the total number of A-scans within the considered area.

The establishment of a threshold value for optical reflectivity is an important step when aiming for the identification of these abnormal increases in extracellular space. It is obtained from A-scans from a population of healthy control eyes, in this case using the HD-OCT Cirrus 5000 from Carl Zeiss Meditec (Dublin, CA). The selection of the threshold value must always be specified for the OCT equipment used (figure 4.1). To select the threshold value, a full retinal scan should be considered, and the value is set below the mean plus 2 standard deviations of the value obtained in the healthy control eyes. It is also important to have in consideration the quality of the images and therefore only scans with signal strength of 7 and above can be considered.

In our studies [8], using HD-OCT Cirrus 5000, the LOR area ratios defined in the retinal central subfield of the healthy control population were: retinal nerve fiber layer (RNFL): 0.17±0.05; ganglion cell and inner plexiform layers (GCL+IPL): 0.42 ±0.10; inner nuclear layer (INL): 0.31±0.08; outer plexiform layer (OPL): 0.37 ±0.10; outer nuclear layer and inner segment (ONL+IS): 0.96±0.03; outer segment (OS): 0.22±0.09; retinal pigment epithelium (RPE): 0.02±0.01. In normal eyes, the highest value of extracellular space is located on the ONL+IS.

As seen in figure 4.2, when applying the OCT-Leakage algorithm, sites of LOR are depicted in two-dimensional images for the different retinal layers, designating a

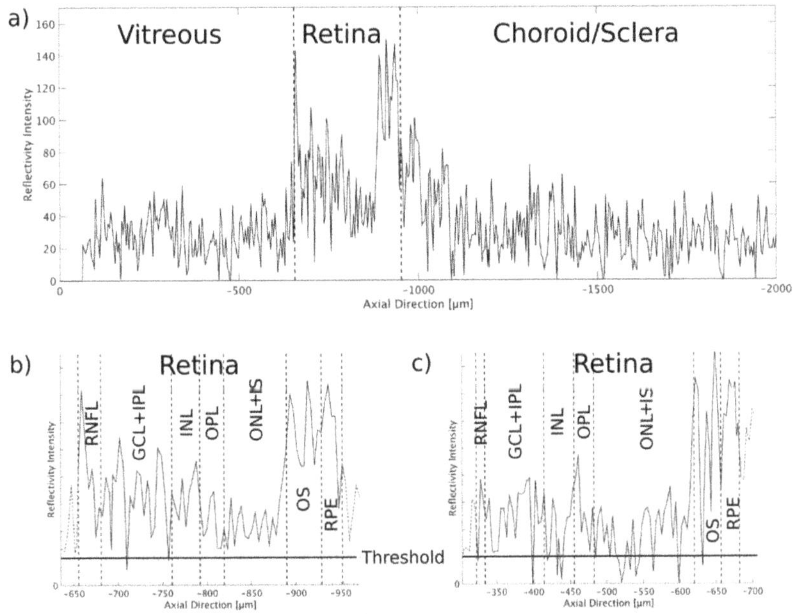

Figure 4.1. Cirrus SD-OCT A-scan optical reflectivity profiles. (a) Full-length SD-OCT A-scan from a healthy subject. (b), (c) Detail of retinal SD-OCT A-scan in a healthy subject (bottom left) and in an individual with non-proliferative diabetic retinopathy clinical macular edema (bottom right). Vertical dashed lines differentiate the optical reflectivity of each segmented layer. The optical reflectivity threshold is depicted as a horizontal line.

representative value for each A-scan evaluated. These LOR maps reflect the locations of A-scans having reflectivity values below the defined threshold (white areas), while the black areas represent a reflectivity above the predefined threshold.

Individuals with clinical macular edema present decreased reflectivity, particularly observed in the GCL+IPL, INL and in the ONL+IS (see example in figures 4.2(m)–(o)).

To perform a more detailed analysis of the abnormal extracellular fluid accumulation in DR, we have assessed the central subfield values of retinal thickness (RT) in a group of healthy control individuals (n = 25) and individuals with non-proliferative DR using structural OCT and performed segmentation of the distinct retinal layers (RNFL, GCL+IPL, INL, OPL, ONL+IS, OS and RPE). This primary evaluation showed that there is a selective involvement of specific retinal layers in the observed structural alterations, namely between diabetic individuals with normal RT (n = 10), individual with subclinical macular edema (n = 30; defined as eyes with non-proliferative diabetic retinopathy (NPDR) and 1SD increase in central RT) and clinical macular edema (n = 8; defined as eyes with NPDR and 2SD increase in central RT) [9], (table 4.1).

Although the overall RT shows an apparent increase in diabetic eyes with subclinical macular edema in the GCL+IPL, INL, OPL and ONL+IS layers, the RNFL and GCL+IPL revealed thinning in the NPDR eyes without edema.

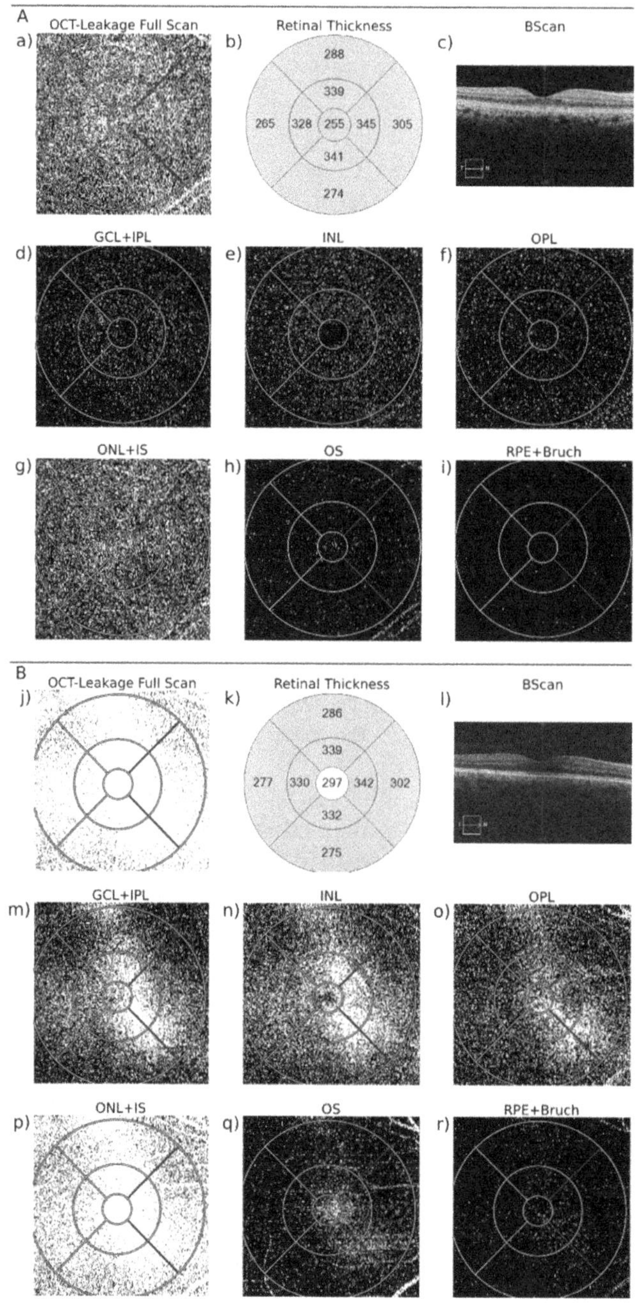

Figure 4.2. OCT-Leakage LOR maps in a healthy subject (A) and in a diabetic individual (B) with subclinical macular edema. (a), (j) Full retinal scan LOR maps; (b), (k) retinal thickness ETDRS grid map; (c), (l) B-scan centered on the fovea; (d)–(i), (m)–(r) LOR maps layer by layer.

Table 4.1. Retinal thickness values for the central subfield obtained from the different segmented retinal layers (layer thickness, μm).

Layer		Layer thickness (μm) Central subfield						
		Healthy (n = 25)	NPDR (n = 10)	Healthy vs NPDR (p-value)	SME (n = 30)	Healthy vs SME (p-value)	CME (n = 8)	Healthy vs CME (p-value)
RNFL	**Mean**	7.59	4.64	0.005[a]	6.34	0.112	9.00	0.268
	SD	2.62	2.50		3.01		4.34	
GCL+ IPL	**Mean**	44.79	40.15	0.095	48.42	0.066	58.60	<0.001[a]
	SD	7.37	6.75		6.99		7.70	
INL	**Mean**	17.85	19.90	0.263	24.74	<0.001[a]	33.78	<0.001[a]
	SD	3.67	7.00		4.45		5.49	
OPL	**Mean**	24.34	23.69	0.738	29.01	0.002[a]	33.07	<0.001[a]
	SD	4.69	6.25		5.56		6.14	
ONL+IS	**Mean**	109.19	110.25	0.763	121.78	<0.001[a]	124.58	0.001[a]
	SD	9.17	9.86		8.99		12.06	
OS	**Mean**	44.86	43.89	0.327	43.89	0.400	44.52	0.732
	SD	2.56	2.67		5.19		1.67	
RPE	**Mean**	24.27	25.33	0.354	25.08	0.308	24.56	0.814
	SD	3.18	2.52		2.66		2.44	

Notes

[a] Statistically significant using Bonferroni adjustment. NPDR: Non-proliferative diabetic retinopathy; SME: Subclinical macular edema; CME: Clinical macular edema.

The INL has showed the higher thickness increases in the central subfield, both in subclinical and clinical macular edema, being of 49.0% and 104.7%, respectively. This increase was also observed in the inner and outer ring areas. Likewise, the OPL revealed a marked increase of 29.2% in individuals with subclinical macular edema and 47.3% in clinical macular edema.

When applying the OCT-Leakage algorithm to the same group of eyes, there were significant LOR ratio increases in the INL in individuals with subclinical and clinical macular edema and in the GCL+IPL in individuals presenting clinical macular edema (table 4.2). Importantly, a positive correlation was observed with RT increases when performing an automated analysis of the retinal extracellular space, particularly strong in the INL, when assessed between the percentage of increase of the LOR area and percentage of increase of INL thickness ($r = 0.71$, $p<0.001$).

Other studies have previously reported the use of optical reflectivity only where evaluating cystoid spaces in different retinal diseases, demonstrating that its assessment could show heterogeneity of reflectivity, especially in diabetic macular edema (DME) [10]. The method described here allows the detection of increases in fluid accumulation in eyes with increased RT in both types of macular edema.

Automated analysis of the retinal extracellular space appears, therefore, to be a useful tool to evaluate fluid accumulation in retinal edema and offers a demonstration of the location and severity of the breakdown of BRB that is responsible for the increase in abnormal extracellular fluid.

These observations highlight the relevance of this innovative approach that allows identification and mapping of areas with abnormal fluid accumulation in the retina, and their correlation with changes in RT. Furthermore, as discussed in the following section, OCT-Leakage was found to identify abnormal retinal fluid with better sensitivity and specificity than FA [11]. Importantly, unlike the situation with FA, the amount of fluid can be quantified.

4.3 Agreement between OCT-Leakage and fluorescein angiography to identify sites of alteration of blood-retinal barrier

The use of FA, by intravenous administration of fluorescein, serving as a fluorescent tracer, has been well-demonstrated to identify sites of vascular leakage in the retinal tissue. This observed leakage enables us to observe the extravasation of molecules with molecular weight similar to fluorescein. However, when the BRB is damaged (i.e., breakdown of BRB), smaller or larger molecules can leak, as well as protein-free fluid, even before the fluorescein leakage can be detected by FA. Therefore, to better monitor disease progression, it is of high relevance the use of sensitive imaging methods with less visualization limitations, such as OCT-Leakage, capable of detecting accumulation of abnormal retinal fluid.

In a validation study for the use of OCT-Leakage, 52 eyes with NPDR from 28 individuals with type 2 diabetes, with a mean age 64.9±7.7 were examined and imaged with OCT angiography using both our OCT-Leakage software and standard FA. Raw structural data from the OCT acquisitions were exported and processed using the OCT-Leakage algorithm.

Table 4.2. Low optical reflectivity area ratio values for the central subfield obtained from the different segmented retinal layers (LOR ratios, arbitrary units).

Layer		Healthy ($n = 25$)	NPDR ($n = 10$)	Healthy vs NPDR (p-value)	SME ($n = 30$)	Healthy vs SME (p-value)	CME ($n = 8$)	Healthy vs CME (p-value)
RNFL	**Mean**	0.17	0.14	0.180	0.14	0.074	0.18	0.809
	SD	0.05	0.05		0.06		0.04	
GCL+ IPL	**Mean**	0.42	0.43	0.720	0.43	0.709	0.55	0.003[a]
	SD	0.10	0.07		0.12		0.07	
INL	**Mean**	0.31	0.34	0.330	0.39	0.002[a]	0.51	<0.001[a]
	SD	0.08	0.11		0.11		0.05	
OPL	**Mean**	0.37	0.38	0.858	0.40	0.328	0.46	0.029
	SD	0.10	0.09		0.10		0.10	
ONL +IS	**Mean**	0.96	0.96	0.799	0.97	0.406	0.97	0.337
	SD	0.03	0.03		0.03		0.02	
OS	**Mean**	0.22	0.19	0.234	0.18	0.075	0.23	0.880
	SD	0.09	0.06		0.08		0.13	
RPE	**Mean**	0.02	0.02	0.893	0.02	0.337	0.03	0.208
	SD	0.01	0.01		0.01		0.01	

Notes

[a] Statistically significant using Bonferroni adjustment. NPDR: Non-proliferative diabetic retinopathy; SME: Subclinical macular edema; CME Clinical macular edema

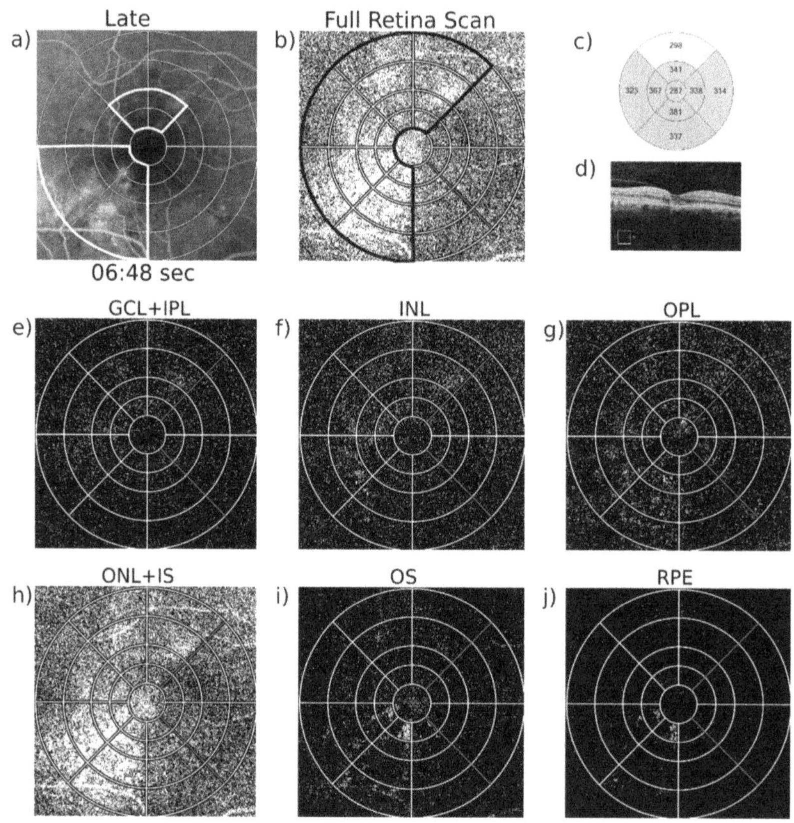

Figure 4.3. Low optical reflectivity (LOR) maps for the right eye of a subject with well-defined leakage on fluorescein angiography (FA). (a) Late FA image, with leakage areas identified by graders, outlined in white. (b) Full retina scan LOR map with the identified area of LOR outlined in black. (c) ETDRS grid of the retinal thickness. (d) B-scan centred on the fovea. (e)–(j) The LOR maps layer by layer for the ganglion cell and inner plexiform layers (GCL+IPL), inner nuclear layer (INL), outer plexiform layer (OPL), outer nuclear layer and inner segment (ONL+IS), outer segment (OS), and retinal pigment epithelium (RPE), respectively. Locations of LOR are identified in white.

Maps of the LOR sites were obtained for the full retina and layer by layer, as en-face images (figure 4.3). To validate the obtained data, intraday reproducibility and interoperator reproducibility for the LOR area ratio was assessed and found to be 0.97 with a 95% confidence interval (CI) of 0.92 to 0.99 and 0.94 with a 95% CI of 0.84 to 0.98, respectively [8].

Quantification of fluorescein leakage in FA imaging was performed by two experienced graders, and graded for the different levels of leakage presence, being 0 = not present; 1 = questionable; 2 = well-defined leakage sites; 3 = diffuse leakage without a well-defined leaking site and grade corresponding to 8 'not classifiable'. When grade 2 was chosen, the graders proceeded to draw the outline of the areas of leakage around leakage sites and to annotate the location of these sites. To discriminate vessel location from the remaining background, two-dimensional en-face reference images were calculated, taking in account OCT shadowing from superficial vessels and OCT high-reflectivity at the RPE layer.

Vessel bifurcations in both OCT en-face image and FA images were matched to calculate an affine transform, in order to allow those areas of leakage to be mapped and compared in both modalities.

The standard 9-area Early Treatment Diabetic Retinopathy Study (ETDRS) grid was divided into smaller sections to provide a finer evaluation of the location of leakage on FA and OCT-Leakage images. Each area of the inner and outer ring was divided into four smaller ones by angle aperture (central cross) and radius (two thinner rings), making each ring composed of 16 smaller sub-sections. This modified ETDRS grid was superimposed on the late FA images and on the OCT-Leakage maps.

For classification of OCT-Leakage data, both in full retina and layer-by-layer, retinal fluid increased was considered as presented when the LOR ratio was higher than the mean plus two standard deviations relative to the healthy population. These sections were thereafter used for comparison with the corresponding sections where fluorescein leakage was identified in the FA images.

In this validation study [11], the study population included 13 eyes with ETDRS 10–20, 28 eyes with ETDRS grade 35, 4 eyes with ETDRS grade 43, 4 eyes with ETDRS grade 47, and 3 eyes with ETDRS grade 53. Of these, 8 presented center-involved macular edema and 26 subclinical macular edema, as defined by DRCR. net [9]. In this population, LOR was linearly correlated with increased fluid accumulation and increased RT, with correlation between the LOR ratios and the increases in RT being highly significant at the OPL, ONL+IS and INL.

Regarding FA, 19 eyes were classified without presence of leakage, 10 eyes with grade 1 (questionable leakage), 11 with grade 2, 11 eyes with grade 3 (well-defined and diffuse leakage, respectively), and 1 eye was considered not classifiable (grade 8).

This co-registration applied in this study allowed to map fluorescein leakage sections identified in the FA image onto the structural OCT data, so locations with lower-than-normal optical reflectivity and leakage could be compared.

Importantly, in 19 eyes graded as absence of leakage by FA (grade 0), the OCT-Leakage was able to identify abnormal fluid accumulation in 10 individuals (52.6%). In eyes with FA grade 2 (well-defined leakage) OCT-Leakage detected the location of the sites of FA leakage and, in general, showed larger areas of abnormal fluid accumulation (figure 4.3).

There is, indeed, very good correspondence between the location of increased LOR area ratios and sites of fluorescein leakage in FA. The changes in extracellular space, represented by the LOR area ratio agreed well with the sites of leakage on the FA examinations.

When assessing the sensitivity and specificity for OCT-Leakage, compared with FA, as the standard method for leakage assessment the achieved accuracy of OCT-Leakage was 79.6%, with 95.9% (95% CI, 81.9 to 100.0) and 75.4% (95% CI, 61.7 to 89.2) for sensitivity and specificity, respectively. The same parameters assessed for the accuracy when evaluating individuals graded with diffuse leakage without a well-defined leaking site (grade 3), OCT-Leakage showed a sensitivity of 93.2% (95% CI,

85.9 to 100.0) and a specificity of 58.9% (95% CI, 43.9 to 74.0) regarding agreement between the sections showing FA leakage and the sections with increased LOR ratios.

Overall, whenever a well-defined leakage site was identified in the FA, an agreement with 95.9% sensitivity was found when applying OCT-Leakage. In general, the algorithm identified larger areas of abnormal fluid accumulation in the retina, thus explaining the specificity of 75.4%. In addition, even in eyes with no visible fluorescein leakage classified by FA, OCT-Leakage found evidence of abnormal fluid accumulation, present in 52.6%, demonstrating its increased sensitivity to detect abnormal fluid in the retina when compared with FA. Likewise, OCT-Leakage was able to identify the location and extent of the abnormal fluid accumulation, in each retinal layer, allowing identification of the location of the FA leaking sites.

Of major relevance, when comparing both methods, FA images only depict leakage of fluorescein molecules, whereas OCT-Leakage can detect abnormal increases of extracellular fluid in the retina due to leakage of water and other solutes. Also, the use of OCT-Leakage offers relevant information regarding the preferential involvement of the different retinal layers in a specific eye, which is not possible when using FA. This layer-by-layer analysis is of critical importance to indicate the predominant accumulation of fluid in the inner or outer retina indicates, respectively, preferential breakdown of the inner BRB (vascular endothelium) or the outer BRB (retinal pigment epithelium). Finally, OCT-Leakage is able to quantify the changes in optical reflectivity, which has potential value for follow-up.

In summary, OCT-Leakage has been shown to have high sensitivity in the identification of sites of retinal vascular leakage, in agreement with the method currently used in clinical practice, FA. In addition, it appears to be more sensitive to detect BRB alterations associated with DR and other retinal vascular diseases [8, 12].

The use of OCT-Leakage for identification of location and quantification of sites of extracellular fluid in complement with OCT microangiography can provide relevant information on retinal circulation and vascular integrity, namely vessel closure, vascular morphology and alterations in the BRB, being of major interest to identify disease progression biomarkers, monitor disease progression, as well as monitoring the response to treatment in retina diseases that impact the vascular component.

4.4 Characterization of progression of macular edema

Retinal edema is the most frequent cause of vision loss due to diabetes. Its characterization and understanding of its progression are fundamental for the development of more timely and targeted treatments. Optical coherence tomography allows objective and quantitative evaluation of retinal edema, providing information on the microstructure of the retina and reproducible thickness measurements. OCT-Leakage is capable of performing automated analyses of the abnormal retinal extracellular space by identifying and measuring sites of lower optical reflectivity.

We examined eyes in the initial stages of NPDR, well identified by the ETDRS grading, for a period of three years with annual examinations, to study the rates of occurrence of macular edema, considering the categories of central-involved diabetic macular edema (CI-DME) and subclinical CI-DME with and without visual loss.

The study was designed to analyse eyes/patients with type 2 diabetes (T2D) and with NPDR who have completed a 3-year follow-up with annual examinations. A total of 90 patients with T2D and ETDRS levels between 10 and 47 were included, with a maximum glycated hemoglobin A_{1C} (HbA_{1C}) value of 10%.

At baseline, when considering central retinal thickness (CRT), there were 25 eyes from 74 eyes/patients (34%) with subclinical CI-DME and 8 eyes with CI-DME (11%). The eyes with CI-DME were distributed similarly in the three ETDRS groups, 10–20 (9%), 35 (10%) and 43–47 (15%). At the end of the three-year period of follow-up the prevalence of CI-DME remained stable with 9% (ETDRS 10–20), 10% (ETDRS 35) and 15% (ETDRS 43–47).

When examining the LOR ratio values it was possible to identify an increase in the retinal extracellular fluid accumulation during the three-year period of follow-up. There was an increase in the LOR ratio values in the eyes with subclinical CI-DME from 22% to 50% in grades ETDRS 10–20, from 40% to 70% in grade ETDRS 35 and from 50% to 88% in grades ETDRS 43–47. An increase in the LOR ratio values was also identified during the 3-year period in all eyes with CI-DME from 50% to 100% (ETDRS 10–20), from 33% to 100% (ETDRS 35) and from 33 to 100% (ETDRS 43–47).

All six eyes with CI-DME at the end of the three-year period of follow-up, showed marked increases in extracellular retinal fluid, represented by increased LOR ratio values ⩾1 SD (6/6, 100%).

In the eyes with subclinical CI-DME and CI-DME the LOR values showed marked fluctuations between visits, with presence of these fluctuations being associated with vision loss.

The results reported here confirm that eyes in the initial stages of diabetic retinopathy in T2D individuals show evidence of CI-DME identified by OCT in a relatively small number of eyes (~10%) and that CI-DME show similar prevalence in eyes with different grades of ETDRS severity below level 47. In summary, CI-DME, i.e., ⩾2SD CRT increase, can occur very early in the disease process and is, in general, independent of DR severity.

Our study shows that subclinical CI-DME and CI-DME can occur at any ETDRS level, with similar prevalence, and that subclinical CI-DME is not predictive of evolution to CI-DME in a 3-year period.

It is apparent from this study that subclinical CI-DME does not inexorably progress over timescales of one to three years and, furthermore, a fraction of these eyes spontaneously improved. We observed, however, in the 3-year period of follow-up, that some eyes with increased CRT show a progressive increase in LOR values, indicating a progressive increase in retinal extracellular fluid accumulation when there is a chronic and prolonged situation of increased retinal thickness, representing edema. Furthermore, eyes with CI-DME and visual loss show a clear association with increased retinal fluid accumulation, demonstrated by increased LOR ratio values, whereas eyes with CI-DME and no vision loss did not show increases in extracellular retinal fluid. Abnormal accumulation of retinal extracellular fluid appears to be a good indicator for the need of close monitoring with potential for early treatment.

Recent reports have called attention to the association of vision loss to fluctuations in CRT in eyes with macular edema. We confirmed that eyes with subclinical CI-DME and CI-DME showing marked fluctuations in LOR values, representing retinal extracellular fluid accumulation, are associated with vision loss.

In conclusion, eyes from T2D patients with minimal, mild and moderate NPDR followed for a period of a 3 years, with annual visits, show evidence of subclinical CI-DME and CI-DME with and without visual loss in a relatively small percentage of cases. The changes in LOR ratios do not appear to be directly associated with changes in ischemia or neurodegeneration. Furthermore, subclinical CI-DME did not appear to be a predictor of development of CI-DME in a period of 3 years. There was however a progressive increase in retinal extracellular fluid registered in the eyes with subclinical CI-DME and CI-DME with vision loss, suggesting a relevant role for retinal extracellular fluid accumulation and chronicity in the development of vision loss in the initial stages of macular edema.

4.5 OCT-Leakage to identify and monitor abnormal retinal fluid in treatment of diabetic macular edema

For several years, laser photocoagulation was considered as the standard treatment, shown to be able of stopping disease progression in 50% of the cases [13]. More recently, the use of vascular endothelial growth factor (VEGF) inhibitors, administered by intravitreal (IVT) injection, have become the standard of care in DME, with a growing interest in exploring different regimens and different anti-VEGF drugs to maximize efficacy while minimizing burden on patients and health care systems [14].

Of increased relevance is the fact that the resolution of DME is not always followed by recovery of visual function, and there is no defined anti-VEGF standard regimen, as different patients present different responses, with higher or lower reversibility after treatment, and within different time periods [15]. It is, therefore, of major importance to identify a prognostic biomarker, able to identify the different visual outcomes observed in different patients after the anti-VEGF injections, and to predict poor response to treatment.

To try to fulfil this gap, we have explored the possibility of using OCT-Leakage in DME naive patients which were under clinical recommendation for anti-VEGF treatment. Our goal was to quantify and identify the location of abnormal extracellular fluid accumulation before and after treatment [16], therefore correlating retinal fluid changes with best corrected visual acuity (BCVA) outcome, particularly focusing on the response to initial stages of anti-VEGF therapy. Furthermore, we wanted to evaluate the predictive value of OCT-Leakage as a prognostic biomarker in the management of DME treatment with anti-VEGF. The 21 eyes included in the study presented distinct patterns of response to treatment, according to visual acuity changes, as 38% were classified as good responders (⩾8 ETDRS letters gained), 24% as moderate responders (⩾5 and <8 ETDRS letters gained), and the other 38% as poor responders (<5 ETDRS letters gain or loss of visual acuity).

Regarding LOR ratio evaluation, in the full retina, after 1 week of treatment (figure 4.4), larger LOR ratio decreases were observed in the good responders' group (−21%) in comparison with moderate responders (−20%) and poor responders (+1%) groups. A more detailed analysis, performed layer-by-layer, revealed that the higher LOR ration decreases were observed in OS and OPL, with a statistically significant association between LOR decrease and BCVA changes in response to treatment being achieved in the OS (good responders: −49%, responders: 18%, and poor responders: 5% [p = 0.026]) and in OPL (good responders: −53%, responders: −12%, and poor responders: 7% [p = 0.010]).

Indeed, LOR ratio decreases after 1 week of anti-VEGF treatment were shown to have a moderate to strong correlation with visual acuity decreases assessed 1 month after initiation of anti-VEGF treatment, a correlation that was borderline significant in the full retina ($r_s = -0.42$ [95% confidence interval: −0.72 to 0.01]; $p = 0.060$) and achieved statistical significance when focusing on the INL, OPL, and OS.

In fact, assessment of LOR ratios were shown to be the best predictor of good BCVA response to anti-VEGF treatment, when compared with retinal structural parameters, such as central retinal thickness, disorganization of retinal inner layers (DRIL), and ellipsoid zone (EZ) disruption [15, 16].

The layer-by-layer analysis of LOR ratios is able to identify the presence of fluid in OPL with higher ability for discriminating the effects of the treatment response (receiver operating characteristic area under the curve = 0.90, sensitivity 85%, and specificity 80%), with OS showing almost similar predictive capabilities (receiver operating characteristic area under the curve = 0.83, sensitivity 62%, and specificity 80%).

In conclusion, the OCT-Leakage results show that the location of abnormal retinal fluid in DME and the degree of its elimination after anti-VEGF treatment are more robust biomarkers of BCVA response to treatment than the degree of central RT reduction or presence of DRIL or EZ disruption [16]. Diabetic macular edema represents, mainly, increased accumulation of fluid in the retina [17, 18]. Therefore, the availability and capability of OCT-Leakage—this new algorithm based on OCT—to identify and quantify abnormal retinal fluid, offers the possibility of examining eyes with DME before and after anti-VEGF treatment.

The degree of retinal fluid elimination and drying of the retina, particularly in the OPL and OS retina layers, resulting from IVT administration of an anti-VEGF drug, and the degree of its decrease appear to determine the degree of BCVA recovery when treating DME with IVT anti-VEGF injections.

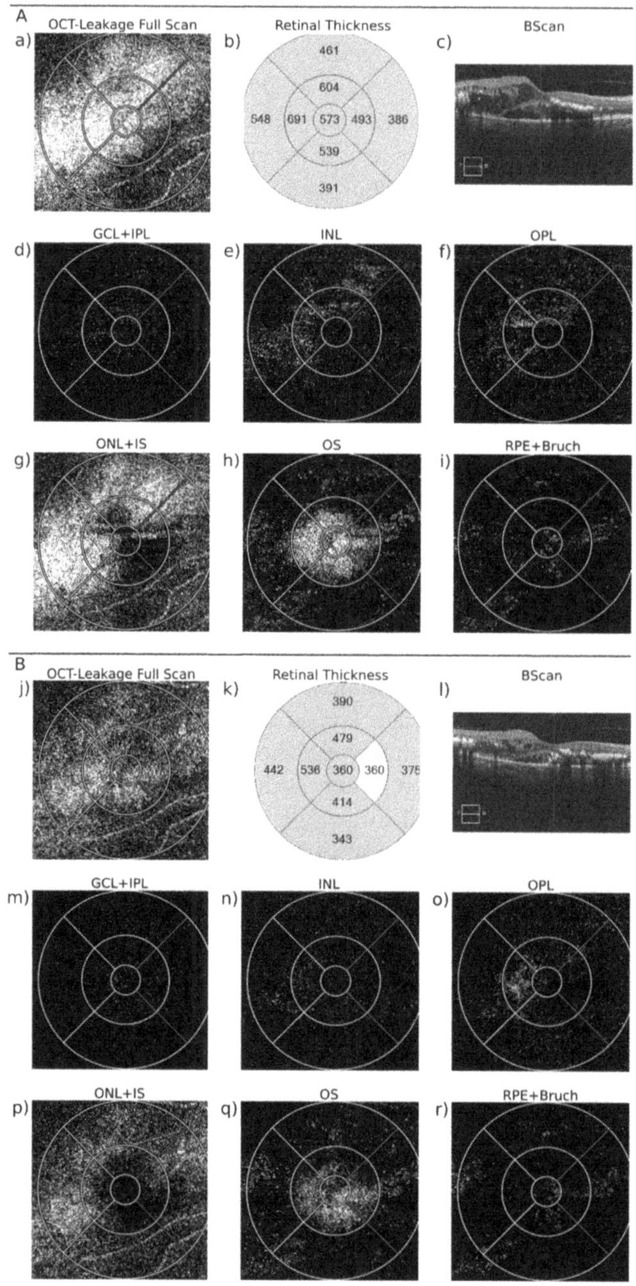

Figure 4.4. Optical coherence tomography leakage maps of a good responder patient before treatment (A) and after treatment (B). (a),(j) Full retina scan LOR map; (b),(k) retinal thickness ETDRS grid map; (c),(l) B-scan centered on the fovea; (d)–(i), (m)–(*r*) LOR maps layer by layer. Decrease of LOR is observed in all layers, especially in the inner nuclear layer (INL), outer nuclear layer and inner segments (ONL+IS) and outer segments (OS).

4.6 Conclusion

Developed in 2016, OCT-Leakage is able to identify and to display sites of low optical reflectivity (LOR), representing retinal extracellular space. The software was developed to provide en-face fundus images, full-scan and layer-by-layer, in which areas of LOR are represented in white and considered to correspond to areas of extracellular fluid.

OCT-Leakage can locate and quantify the alteration of the BRB that occurs in DR and DME. Furthermore, the work developed so far has shown that the retinal fluid distribution and accumulation measured by OCT-Leakage is a better predictor of response to treatment than previously proposed biomarkers, with improved sensitivity and specificity over other imaging techniques used in clinical practice. As it is a non-invasive approach it is expected to replace FA.

References

[1] International Diabetes Federation 2019 *IDF Diabetes Atlas* (International Diabetes Federation) 9th edn

[2] Cunha-Vaz J, Bernardes R and Lobo C 2011 Blood-retinal barrier *Eur. J. Ophthalmol.* **21 (Suppl.)** 3–9

[3] Cunha-Vaz J G, Goldberg M F, Vygantas C and Noth J 1979 Early detection of retinal involvement in diabetes by vitreous fluorophotometry *Ophthalmology* **86** 264–75

[4] Cunha-Vaz J, Faria de Abreu J R and Campos A J 1975 Early breakdown of the blood-retinal barrier in diabetes *Br. J. Ophthalmol* **59** 649–56

[5] Jia Y, Tan O and Tokayer J *et al* 2012 Split-spectrum amplitude-decorrelation angiography with optical coherence tomography *Opt. Express* **20** 4710

[6] Kuehlewein L, Tepelus T C, An L, Durbin M K, Srinivas S and Sadda S 2015 Noninvasive visualization and analysis of the human parafoveal capillary network using swept source OCT optical microangiography *Investig. Ophthalmol. Vis. Sci.* **56** 3984–8

[7] Jiaa Y, Baileya S T and Hwanga T S *et al* 2015 Quantitative optical coherence tomography angiography of vascular abnormalities in the living human eye *Proc. Natl Acad. Sci. USA* **112** E2395–402

[8] Cunha-Vaz J, Santos T, Ribeiro L, Alves D, Marques I and Goldberg M 2016 OCT-leakage: a new method to identify and locate abnormal fluid accumulation in diabetic retinal edema *Investig. Ophthalmol. Vis. Sci.* **57** 6776–83

[9] Chalam K V, Bressler S B and Edwards A R *et al* 2012 Retinal thickness in people with diabetes and minimal or no diabetic retinopathy: Heidelberg spectralis optical coherence tomography *Investig. Ophthalmol. Vis. Sci.* **53** 8154–61

[10] Horii T, Murakami T and Nishijima K *et al* 2012 Relationship between fluorescein pooling and optical coherence tomographic reflectivity of cystoid spaces in diabetic macular edema *Ophthalmology* **119** P1047–55

[12] Farinha C, Santos T and Marques I P *et al* 2017 OCT-leakage mapping: a new automated method of OCT data analysis to identify and locate abnormal fluid in retinal edema *Ophthalmol. Retina* **1** 486–96

[11] Cunha-Vaz J, Santos T and Alves D *et al* 2017 Agreement between OCT leakage and fluorescein angiography to identify sites of alteration of the blood–retinal barrier in diabetes *Ophthalmol. Retina* **1** 395–403

[13] Early Treatment Diabetic Retinopathy Study Research Group 1987 Treatment techniques and clinical guidelines for photocoagulation of diabetic macular edema: early treatment diabetic retinopathy study report number 2 *Ophthalmology* **94** 761–74

[14] Dugel P U, Hillenkamp J and Sivaprasad S *et al* 2016 Baseline visual acuity strongly predicts visual acuity gain in patients with diabetic macular edema following anti-vascular endothelial growth factor treatment across trials *Clin. Ophthalmol.* **2016** 1103–10

[15] Santos A R, Costa M and Schwartz C *et al* 2018 Optical coherence tomography baseline predictors for initial best-corrected visual acuity response to intravitreal anti-vascular endothelial growth factor treatment in eyes with diabetic macular edema *Retina* **38** 1110–9

[16] Santos A R, Alves D, Santos T, Figueira J, Silva R and Cunha-Vaz J G 2019 Measurements of retinal fluid by optical coherence tomography leakage in diabetic macular edema: a biomarker of visual acuity response to treatment *Retina* **39** 52–60

[17] Cunha-Vaz J 2017 The blood-retinal barrier in the management of retinal disease: EURETINA award lecture *Ophthalmologica* **237** 1–10

[18] Marmor M F 1999 Mechanisms of fluid accumulation in retinal edema *Doc. Ophthalmol.* **97** 239–49

IOP Publishing

Photo Acoustic and Optical Coherence Tomography Imaging, Volume 1
Diabetic retinopathy
Ayman El-Baz and Jasjit S Suri

Chapter 5

Comparison of ocular ultrasound with optical coherence tomography in the evaluation of diabetic retinopathy

João Heitor Marques and Bernardete Pessoa

5.1 Introduction

The prevalence of visual impairment and blindness caused by diabetic retinopathy increased substantially between 1990 and 2015 according to the latest report of the Vision Loss Expert Group of the Global Burden of Disease Study [1], and diabetic retinopathy remains the leading cause of vision loss and preventable blindness in adults aged 20–74 years [2]. Vision loss in ocular diabetes develops from the sequelae of maculopathy (macular edema and ischemia) or neovascularization of the retina (vitreous hemorrhage and tractional retinal detachment) and of the iris (neovascular glaucoma).

Ocular ultrasound (US) and optical coherence tomography (OCT) are both valuable tools in diagnosis and staging of several ocular disorders, namely diabetic retinopathy, and its complications.

Despite their different principles, both methods rely on the reflection of waves on the structures to be imaged: mechanical waves in US and electromagnetic light waves in OCT [3, 4].

Analogously, both imaging methods may be considered tomographic, as a linear group of depth-resolved scanning points (A-scans) form a two-dimensional cross-sectional B-scan. Successive B-scans form a tridimensional C-scan that may be represented on most OCT interfaces. On US, successive B-scans are not automatic as they are dependent on the movement of the probe by the device operator. [3]

5.2 Technical differences

Given its distinctive principles, each technique measures a different biological material property and has different resolution and penetration that prove advantageous for specific clinical situations (table 5.1).

The much higher spacial resolution of OCT (almost two orders of magnitude) is a clear advantage and allows near cell-level observation of tissues [5].

The main advantage of US in ophthalmology is its ability to scan through opaque light-conducting media, such as the normal iris, corneal opacities, dense cataract, or vitreous hemorrhage.

Newer probes using frequencies higher than conventional US (up to 50 MHz), such as ultrasound biomicroscopy (UBM), allow for great improvement in axial and lateral resolution reaching 30 and 60 μm, respectively [6]. The higher frequency reduces the tissue penetration depth so that only the anterior segment of the eye is reached, including the cornea, iridocorneal angle, anterior chamber, iris, ciliary body, and lens. Although still not comparable with OCT, UBM retains the ability to visualize through typically opaque structures such as the iris and the sclera. Therefore, it remains the best method to study the ciliary body, where OCT signal does not reach [7].

Moreover, US is the only imaging method that, thanks to is large focus depth, can capture the entire eye in a single cross-section [3].

Ultrasound may be performed dynamically in time, as the patients moves the eye. This may help distinguish the vitreous (totally loose and mobile) from a detached retina (loose but attached to the optic nerve) and from solid lesions such as hemorrhage, fibrosis or tumor.

5.3 Clinical applications in diabetic retinopathy

Although ocular US may be considered the gold-standard exam to establish the vitreous status, spectral-domain OCT is the retinal imaging method most used worldwide. OCT is currently the backbone in diagnosis and staging of diabetic macular edema, where US has little to no applicability. However, some other clinical situations pose a challenge for OCT due to its narrower scanning area and difficult penetration in case of optical media opacity. The role of US is therefore emphasized in:

Table 5.1. Technical differences between conventional ocular ultrasound (US) and optical coherence tomography (OCT)—Heidelberg Spectralis® used as reference for OCT and Ellex Eye Cubed™ used as reference for US.

Feature	Conventional ocular ultrasound	Optical coherence tomography
Emission wave	10 MHz soundwave	880 nm light wave
Lateral resolution (μm)	150	6
Axial resolution (μm)	450	4
Depth of penetration (mm)	25	1.9

5.3.1 Posterior vitreous detachment and vitreoschisis

Posterior vitreous detachment (PVD) is typically an age-related process that originates at the perifoveal site, progresses to the optic disk, and terminates at the vitreous base. It occurs due to progressive vitreous liquefaction associated with weakening of the vitreoretinal adhesions. Anomalous PVD occurs when liquefaction is not concurrent with dehiscence at the vitreoretinal interface leading to vitreo-macular traction or vitreo-papillary adhesion. Anomalous PVD is believed to be highly prevalent in diabetic patients [8]. Vitreo-macular traction is an important factor when deciding the treatment for diabetic macular edema and may even motivate vitreoretinal surgery [9]. On the other hand, vitreo-papillary adhesion plays a role in promoting neovascularization and vitreous hemorrhage in proliferative diabetic retinopathy. It is therefore important to properly address the posterior vitreous status in diabetic retinopathy [10–12].

A classification for PVD on OCT has been developed: [13]

- stage 0—absence of PVD;
- stage 1—incomplete PVD that was localized in the perifovea, in 1 to 3 quadrants, with persistent attachment to the fovea, optic nerve head, and midperipheral retina;
- stage 2—perifoveal PVD across all quadrants, with persistent attachment to the fovea, optic nerve head, and midperipheral retina;
- stage 3—detachment of the posterior vitreous face from the fovea, with persistent attachment to the optic nerve head and midperipheral retina;
- stage 4—complete PVD, with biomicroscopically identified detachment of the posterior vitreous face with Weiss ring.

In stage 4 of this classification, OCT may fail to detect a posterior vitreous signal because the distance from the retina may be out of range. Moreover, a vitreous premacular bursa can be misdiagnosed as PVD on standard 20° OCT [14]. Widefield 55° OCT has been shown to have good agreement with US in the detection of complete PVD, unlike regular 20° OCT [15–18] (figure 5.1).

Vitreoschisis may be another form of anomalous PVD. It corresponds to a split in the posterior vitreous cortex, leaving the external layer attached to the macula, while the remainder of the vitreous collapses forward. Vitreoschisis is mainly an ultrasonographic entity [19] and its recognition is important in the surgical management of diabetic retinopathy [20]. Vitreoschisis has also been described in OCT [21, 22] but it is unclear if it corresponds to the same entity on US. In addition, the variability in the wall thickness of the vitreoschisis cavity may, in many cases, be below the level of resolution of the method used to define the vitreous structure with the standard OCT macular. Although, posterior vitreous status in its full length may be better seen in US due to the larger scanning area, its sensitivity is not 100%: Schwartz *et al* identified a higher incidence of posterior vitreoschisis detected during vitrectomy (81%) than by US (17%). [23]

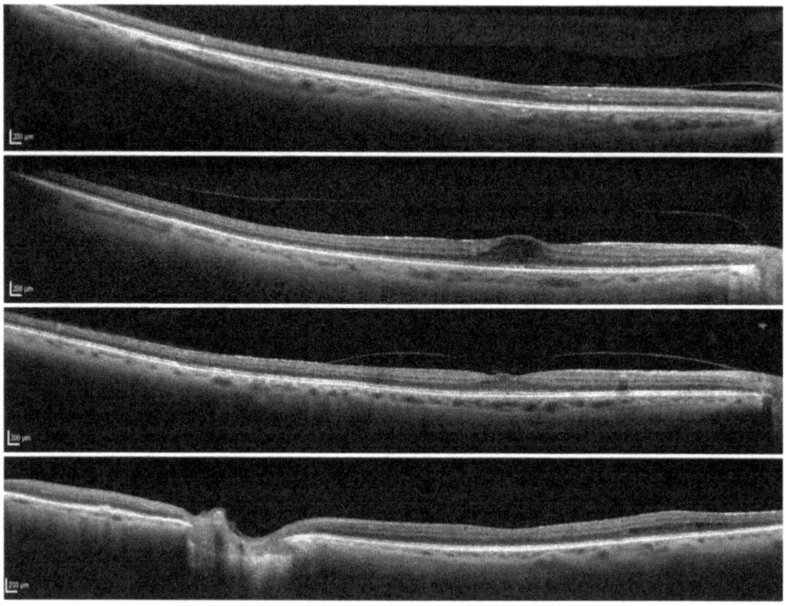

Figure 5.1. Shows different posterior vitreous status on 55° OCT: from top to bottom, complete posterior vitreous attachment; anomalous PVD with papillary adhesion; anomalous PVD with foveal adhesion and complete posterior vitreous detachment.

5.3.2 Tractional retinal detachment

Tractional retinal detachment is a dangerous and advanced form of proliferative diabetic retinopathy (PDR). It develops from neovascular stimuli that cause fibrovascular proliferation along the vitreoretinal interface. Contractile forces then detach the neuroretina from the underlying pigmented epithelium [24]. Its presence may be an indication for surgical intervention in diabetic retinopathy. Vitreous hemorrhage may occur together with tractional retinal detachment because of bleeding from neovessels or from a vessel ruptured by the tractional component. Therefore, direct visualization or OCT scanning of the retina may not be possible. Ocular US is a well-established tool in the diagnosis of retinal detachment, including the tractional type, particularly in the case of opaque media [25–27]. Tractional retinal detachment in diabetic retinopathy has been studied with optical coherence tomography [28] but its importance in clinical practice remains unclear.

5.3.3 Vitreous hemorrhage

Proliferative diabetic retinopathy is the most common cause of vitreous hemorrhage, usually due to bleeding from neovessels [29]. However, even in diabetic patients, other causes for vitreous hemorrhage, such as retinal tear or retinal detachment, must be excluded as their management may be entirely different [30]. Ocular US is the method of choice to approach these patients [31]. Newer OCT devices that reach longer wavelengths may have improved penetration through the hemorrhage [32], but their role in the investigation of vitreous hemorrhage is not established.

5.3.4 Axial length measurement

Diabetes is a well-known risk factor for cataract development [33]. Axial length measurement is key for intraocular lens power calculation in cataract surgery. The most recent biometers use OCT technology for axial length measurement that show high precision and accuracy [34] but fail to acquire measurements in some cases of dense cataract or vitreous hemorrhage [32, 35]. Therefore, an ultrasonic biometer must always be present in an Ophthalmology Department.

5.3.5 Measuring blood perfusion

OCT can localize the presence of blood flow by repeatedly scanning the same region (OCT-angiography). Changes in signal with time detect blood cells' movement and microscopic vessels can be seen with no need for intravenous contrast administration. In diabetic retinopathy, OCT-A has proved useful [36]. OCT angiography is able to image the different retinal capillary plexuses in a 3-dimensional way, as well as to visualize in isolation the individual plexuses. In diabetes, the earliest changes are seen in the deep capillary plexus [37]. Individual visualization of the retinal plexuses allows for better identification of the changes associated with diabetic retinopathy such as capillary dropout, enlargement of the foveal avascular zone and neovascularization. However, commercially available OCT devices cannot quantify blood flow. Ocular US using Doppler studies is capable of measuring blood flow, velocity and resistance in the central retinal artery and detecting its changes with diabetic retinopathy [38]. Its lower resolution impairs the visualization of vessels smaller that the central retinal artery.

5.3.6 Patient collaboration

In clinical practice, an important issue that is seldom discussed is the need for patient collaboration during the ophthalmological examination and while performing complementary imaging exams. Notwithstanding the recent motion artifacts correction algorithms on OCT, this technology requires the patient to fixate on a target. Ultrasound requires less collaboration from the patient, and it may even be performed on the bedside and with the eyelids closed. This may be the only eye imaging method available for critically ill or immobile patients.

5.4 Conclusion

With technological evolution over time, an increased accuracy in detecting the real vitreous status is also expected. Currently the most recent US and widefield OCT have demonstrated equivalent results regarding the definition of the posterior pole vitreous macular status. An accurate knowledge of vitreous macular interface is of an undeniable importance for a more effective approach of the two mainstay causes for DR vision loss, DME and PDR. This is especially important in a background of a disease where a high prevalence of some degree of vitreous adherence to the posterior pole is also expected.

References

[1] Flaxman S R, Bourne R R A and Resnikoff S *et al* 2017 Global causes of blindness and distance vision impairment 1990–2020: a systematic review and meta-analysis *Lancet Glob. Heal.* **5** e1221–34

[2] Vujosevic S, Aldington S J and Silva P *et al* 2020 Screening for diabetic retinopathy: new perspectives and challenges *Lancet Diabetes Endocrinol.* **8** 337–47

[3] E G J S 1979 Ophthalmic ultrasound as a diagnostic tool *J. Am. Optom. Assoc* **50** 73–8 https://pubmed.ncbi.nlm.nih.gov/310824/

[4] P D P, C-S L P and F C *et al* 2011 Optical coherence tomography: fundamental principles, instrumental designs and biomedical applications *Biophys. Rev.* **3** 155–69

[5] Podoleanu A G 2012 Optical coherence tomography *J. Microsc.* **247** 209–19

[6] Silverman R H 2009 High-resolution ultrasound imaging of the eye—a review *Clin. Exp. Ophthalmol. NIH Public Access* **37** 54–67

[7] Abbas R 2021 *Ophthalmic Ultrasonography and Ultrasound Biomicroscopy: A Clinical Guide* (Springer)

[8] Ophir A and Martinez M R 2011 Epiretinal membranes and incomplete posterior vitreous detachment in diabetic macular edema, detected by spectral-domain optical coherence tomography *Invest. Ophthalmol. Vis. Sci.* **52** 6414–20

[9] Johnson M W 2012 Posterior vitreous detachment: evolution and role in macular disease *Retina* **32** S147–78

[10] Sebag J 1987 Age-related changes in human vitreous structure *Graefes Arch. Clin. Exp. Ophthalmol.* **225** 89–93

[11] Sebag J 2008 Vitreoschisis *Graefe's Arch. Clin. Exp. Ophthalmol.* **246** 329–32

[12] Sebag J 2004 Anomalous posterior vitreous detachment: a unifying concept in vitreo-retinal disease *Graefes Arch. Clin. Exp. Ophthalmol.* **242** 690–8

[13] Uchino E, Uemura A and Ohba N 2001 Initial stages of posterior vitreous detachment in healthy eyes of older persons evaluated by optical coherence tomography *Arch. Ophthalmol.* **119** 1475–9

[14] Spaide R F 2014 Visualization of the posterior vitreous with dynamic focusing and windowed averaging swept source optical coherence tomography *Am. J. Ophthalmol.* **158** 1267–74

[15] Pessoa B, Coelho J and Malheiro L *et al* 2020 Comparison of ocular ultrasound versus SD-OCT for imaging of the posterior vitreous status in patients with DME *Ophthalmic Surg. Lasers Imaging Retin.* **51** S50–S53

[16] Wang M D, Zaid C T and Syed M *et al* 2021 Swept source optical coherence tomography compared to ultrasound and biomicroscopy for diagnosis of posterior vitreous detachment *Clin. Ophthalmol.* **2021** 507–12

[17] Bertelmann T, Goos C and Sekundo W *et al* 2016 Is optical coherence tomography a useful tool to objectively detect actual posterior vitreous adhesion status? *Case Rep. Ophthalmol. Med.* **2016** 1–5

[18] Pang C E, Freund K B and Engelbert M 2014 Enhanced vitreous imaging technique with spectral-domain optical coherence tomography for evaluation of posterior vitreous detachment *JAMA Ophthalmol. Am. Med. Assoc.* **132** 1148–50

[19] Chu T G, Lopez P F and Cano M R *et al* 1996 Posterior vitreoschisis: an echographic finding in proliferative diabetic retinopathy *Ophthalmology* **103** 315–22

[20] Schwartz S D, Alexander R and Hiscott P *et al* 1996 Recognition of vitreoschisis in proliferative diabetic retinopathy: a useful landmark in vitrectomy for diabetic traction retinal detachment *Ophthalmology* **103** 323–8

[21] Sebag J, Gupta P and Rosen R R *et al* 2007 Macular holes and macular pucker: The role of vitreoschisis as imaged by optical coherence tomography/scanning laser ophthalmoscopy *Trans. Am. Ophthalmol. Soc.* **105** 121–9

[22] Muqit M M K and Stanga P E 2014 Swept-source optical coherence tomography imaging of the cortical vitreous and the vitreoretinal interface in proliferative diabetic retinopathy: assessment of vitreoschisis, neovascularisation and the internal limiting membrane *Br. J. Ophthalmol.* **98** 994–7

[23] Schwartz S D, Alexander R, Hiscott P and Gregor Z J *et al* 1996 Recognition of vitreoschisis in proliferative diabetic retinopathy. A useful landmark in vitrectomy for diabetic traction retinal detachment *Ophthalmology* **103** 323–8

[24] McMeel J W 1971 Diabetic retinopathy: fibrotic proliferation and retinal detachment *Trans. Am. Ophthalmol. Soc.* **69** 440–93

[25] Blumenkranz M S and Byrne S F 1982 Standardized echography (ultrasonography) for the detection and characterization of retinal detachment *Ophthalmology* **89** 821–31

[26] Rabinowitz R, Yagev R and Shoham A *et al* 2004 Comparison between clinical and ultrasound findings in patients with viterous hemorrhage *Eye* **18** 253–6

[27] Dibernardo C, Blodi B and Byrne S F 1992 Echographic evaluation of retinal tears in patients with spontaneous vitreous hemorrhage *Arch. Ophthalmol.* **110** 511–4

[28] Kim Y C and Shin J P 2016 Spectral-domain optical coherence tomography findings of tractional retinal elevation in patients with diabetic retinopathy *Graefe's Arch. Clin. Exp. Ophthalmol.* **254** 1481–7

[29] Spraul C W and Grossniklaus H E 1997 Vitreous hemorrhage *Surv. Ophthalmol.* **42** 3–39

[30] El Annan J and Carvounis P E 2014 Current management of vitreous hemorrhage due to proliferative diabetic retinopathy *Int. Ophthalmol. Clin.* **54** 141–53

[31] Parchand S, Singh R and Bhalekar S 2014 Reliability of ocular ultrasonography findings for pre-surgical evaluation in various vitreo-retinal disorders *Semin. Ophthalmol.* **29** 236–41

[32] Wang Q, Huang Y and Gao R *et al* 2020 Axial length measurement and detection rates using a swept-source optical coherence tomography-based biometer in the presence of a dense vitreous hemorrhage *J. Cataract Refract. Surg.* **46** 360–4

[33] Drinkwater J J, Davis W A and Davis T M E 2019 A systematic review of risk factors for cataract in type 2 diabetes *Diabetes. Metab. Res. Rev.* **35** e3073

[34] Moshirfar M, Buckner B and Ronquillo Y C *et al* 2019 Biometry in cataract surgery: a review of the current literature *Curr. Opin. Ophthalmol.* **30** 9–12

[35] Huang J, Chen H and Li Y *et al* 2019 Comprehensive comparison of axial length measurement with three swept-source OCT-based biometers and partial coherence interferometry *J. Refract. Surg* **35** 115–20

[36] Choi W, Waheed N K and Moult E M *et al* 2017 Ultrahigh speed swept source optical coherence tomography angiography of retinal and choriocapillaris alterations in diabetic patients with and without retinopathy *Retina* **37** 11–21

[37] Moore J, Bagley S and Ireland G *et al* 1999 Three dimensional analysis of microaneurysms in the human diabetic retina *J. Anat* **194** 89–100

[38] Päivänsalo M, Pelkonen O and Rajala U *et al* 2004 Diabetic retinopathy: sonographically measured hemodynamic alterations in ocular, carotid, and vertebral arteries *Acta Radiol.* **45** 404–10

IOP Publishing

Photo Acoustic and Optical Coherence Tomography Imaging, Volume 1

Diabetic retinopathy

Ayman El-Baz and Jasjit S Suri

Chapter 6

Optical coherence tomography biomarkers in diabetic macular edema

Pablo Carnota-Méndez, Carlos Méndez-Vázquez, Nuria Olivier-Pascual, Carlos Torres-Borrego, Daniel Velázquez-Villoria, María Gil-Martínez and Sara Rubio-Cid

6.1 Introduction

Diabetic retinopathy (DR) is an ophthalmologic complication in patients with diabetes mellitus (DM), and it is the leading cause of blindness in people under 75 years of age in developed countries [1, 2]. Diabetic macular edema (DME) is a chronic, multifactorial, sight-threatening condition that critically impacts on the patient's quality of life [3]. DME can occur at any stage of DR, being the major cause of central vision loss in diabetic patients and the leading cause of vision loss in working age patients in developed countries [3, 4].

Optical coherence tomography (OCT) is a fast, non-invasive method of imaging the retinal microstructure. In cross-sectional images, it allows to assess the vitreous, all the layers of the retina (from inner limiting membrane (ILM) to retinal pigment epithelium (RPE)), the choriocapillaris and the choroid and, with techniques such as Enhanced-Depth-Imaging (EDI-OCT) or Swept-Source (SS-OCT), the sclero-choroid junction and the sclera. OCT has become an essential tool for retina specialists in daily practice, and today is the gold standard for the diagnosis of DME.

In 1998, the National Institutes of Health Biomarkers Definitions Working Group established a definition for biomarker as 'a characteristic that is objectively measured and evaluated as an indicator of normal biological processes, pathogenic processes, or pharmacologic responses to a therapeutic intervention' [5]. Thus, biomarker is a general term used to find a relationship between measurable biological processes and a clinical outcome.

Over the last years, several OCT biomarkers have been described in DME [6, 7]. However, not all of them provide the same type of information. First, there are

doi:10.1088/978-0-7503-2052-8ch6

quantitative biomarkers such as central retinal thickness or macular volume that are indicators of disease activity. In this chapter we try to address the complex question of whether there is a correlation between these biomarkers and visual acuity (VA) in DME. The other type of biomarkers is qualitative. Among them, some correlate with poor functional outcome, independently of treatment of choice, such as disorganization of inner retinal layers (DRIL), disruption of external limiting membrane (ELM), and disruption of ellipsoid zone (EZ). Others suggest a more inflammatory component of DME, such as intraretinal fluid (IRF), subfoveal neurosensory detachment (SND) and hyperreflective foci (HRF). Finally, abnormalities in the vitreomacular interface (VMI), like epiretinal membrane (ERM) or vitreomacular traction (VMT), can be detected by OCT and corrected surgically.

In 2020, an international expert panel proposed a grading protocol termed 'TCED-HFV' (To Classify Edema Discerning Hidden Functional Variables), taking into account 7 tomographic qualitative and quantitative features, which include foveal thickness (T), intraretinal cysts (C), the EZ and/or ELM status (E), presence of DRIL (D), number of HRF (H), subfoveal fluid (F), and vitreoretinal relationship (V) [49]. The aim of this classification was to provide a simple, direct, objective tool to classify diabetic maculopathy [49].

This chapter is intended to be a comprehensive review of the existing OCT biomarkers in DME with a pragmatic approach.

6.2 Retinal thickness and macular volume

Since the development of OCT, retinal thickness has been an important measure in order to detect several pathological conditions in the macula [8, 9]. Retinal thickness is obtained by directly measuring the distance between the ILM (inner boundary) and the RPE (outer boundary) in every scan [9–12].Then, the OCT devices reconstruct the macula as a topographic map, with numerical averages of thickness of each of the nine sectors defined by the ETDRS [10, 13]. The average thickness of the central circle 1mm of diameter is usually referred as central retinal thickness (CRT) [9, 10, 31].

These measurements allow clinicians to measure small changes in macular thickness and quantitatively assess the efficacy of different treatment modalities [10, 12]. Retinal thickness increases in processes that break the internal and external blood-retinal barrier (like DME) and diminishes in macular atrophy [10, 13, 15, 28]. A threshold between normal and increased CRT has been used in many studies. However, CRT presents great variability in healthy individuals and, furthermore, its value depends on the device used [9, 17, 30].

Despite these limitations, CRT has been used extensively as a quantitative biomarker to detect DME, to determine its severity, and to assess the response to treatment [16] (figure 6.1). In fact, a classification of anatomical response to treatment in DME has been proposed according to the change of CRT after treatment [29]:

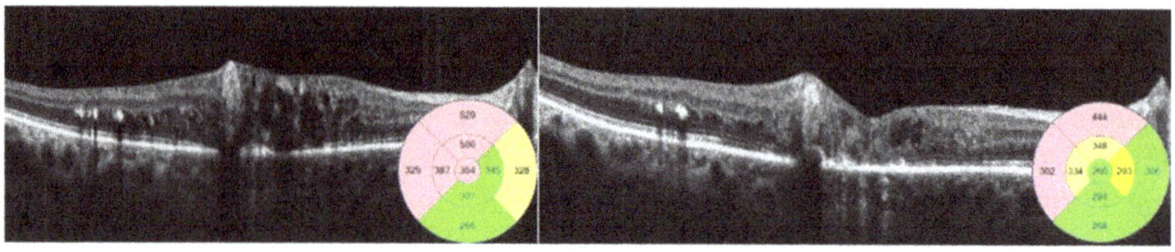

Figure 6.1. 63-year-old male with diabetes of 15 years of evolution. (A) Right eye with naïve DME, BCVA 20/50 and central retinal thickness (CRT) of 354 microns. (B) Two months after an intravitreal dexamethasone implant, BCVA improved to 20/40 and CRT decreased to 260 microns. This response would be classified by OCT as 'improvement'. BCVA is not better probably due to the presence of DRIL and absence of ELM and EZ at the fovea.

- 'Success' as CRT < 250 microns;
- 'Improvement' as a decrease of at least 10% in CRT;
- 'No improvement' as a change of +/−10% in CRT;
- 'Failure' as >250 microns with loss of 10 or more ETDRS letters.

On the other hand, CRT is not an objective indicator of VA and has no either prognostic or predictive ability of the visual outcome because this depends on the integrity of the retinal tissue and not only on the accumulation of fluid [10, 13, 17]. It has been hypothesized that cross-sectional area of retinal tissue between inner and outer plexiform layers (IPL and OPL) could better predict VA than CRT [6, 13, 14]. However, currently this measure is useful only for research since OCT devices do not calculate it automatically.

Another quantitative feature of DME obtained using OCT is the macular volume (MV). MV refers to the total cross-sectional volume of the macula calculating from all points of measurements, usually in a 6 mm×6 mm cube centered in the fovea. Although is not essential to the diagnosis of DME, VM is a useful tool to evaluate the response to treatment in the macula as a whole (and not only in the fovea like CRT) [18]. MV is unlikely to be a prognostic factor but it has been used as an anatomical endpoint in some studies about DME [19, 20].

In conclusion, CRT is a useful tool to diagnose DME and evaluate the response to treatment but, in order to make an overall assessment, clinicians should look for other OCT biomarkers and not rely only in CRT when facing a patient with DME [6, 21].

6.3 Subfoveal choroidal thickness

Subfoveal choroidal thickness (SChT) is the distance from the posterior edge of the RPE to the sclerochoroidal junction at the level of the fovea [22–25]. SChT is measured manually by using a caliper tool in a central OCT B-scan obtained with a software suitable to view the choroid such as Swept-Source (SS-OCT) or Enhanced-Depth-Imaging (EDI-OCT). SChT is highly variable in healthy individuals with values between 225 to 311 microns [24], but in diabetic patients, SChT is slightly decreased compared to healthy controls [22, 23]. This thinning could be the result of

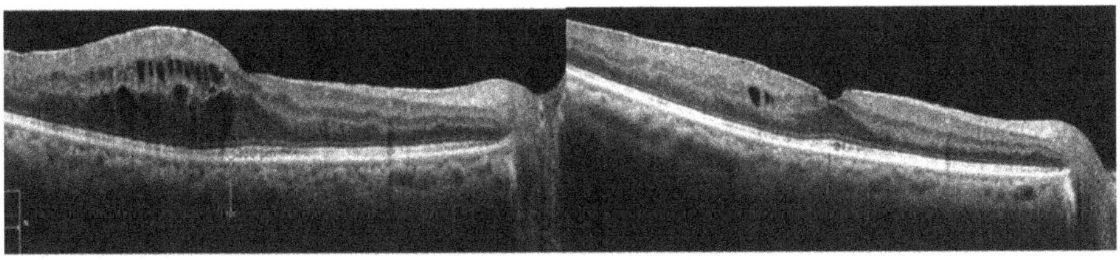

Figure 6.2. (A) EDI-OCT of the right eye of a patient with naïve DME. Subfoveal choroidal thickness (SchT) was 233 microns, BCVA 20/30 and CRT 396 microns. (B) After 3 injections of bevacizumab, SchT was 217 microns, BCVA 20/25 and CRT 335 microns.

contraction or loss of capillaries secondary to a hypoxemic microenvironment associated with early stages of diabetic choroidopathy [22, 23, 27]. However, as the severity of diabetic retinopathy increases, there is an increase in SChT, probably due to an increased vascular endothelial growth factor (VEGF) production [22, 26].

In a retrospective study, Rayess *et al* found an association between a greater SChT and a better anatomical and functional response in patients with DME treated with anti-VEGF [32] (figure 6.2). In patients with greater basal SChT, it is presumed that they would have an intact choriocapillaris with less ischemic repercussion at the level of the outer retina. Consequently, photoreceptors would be better preserved, which could explain the visual results after treatment with anti- VEGF [6].

A new biomarker known as the choroidal vascular index (CVI) allows a quantitative assessment of the vascular and stromal compartments of the choroid. CVI could be used as a more realistic biomarker of the vascular health of the choroid than SChT, but further research is required to confirm this finding [23, 25].

6.4 Vitreomacular interface

The vitreomacular interface (VMI) is the junction of the vitreous and the retina at the posterior pole. The VMI consists of the ILM of the retina, the posterior vitreous cortex, and an intervening extracellular matrix [33]. In 2013, the International Vitreomacular Traction Study Group established a new, OCT-based classification for vitreomacular adhesion (VMA), traction and macular hole. Broad and focal VMA and posterior vitreous detachment (PVD) were considered different stages of normal ageing of the vitreous, while epiretinal membrane (ERM) and vitreomacular traction (VMT) were classified as pathological processes [34]. OCT is more sensitive in detecting VMI abnormalities than standard techniques (slit lamp exam, fluorescein angiography, and color fundus photography) [35], and the setup of VMI has several specific and important implications in eyes with DME.

6.4.1 Vitreomacular adhesion and posterior vitreous detachment

VMA represents a specific stage of vitreous separation wherein partial detachment of the vitreous in the perifoveal area has occurred without retinal abnormalities (figures 6.3(A) and (B)). Gandorfer *et al* found that perifoveal PVD was present in 53% of eyes with DME, compared with 11% of eyes without DME in sex and age

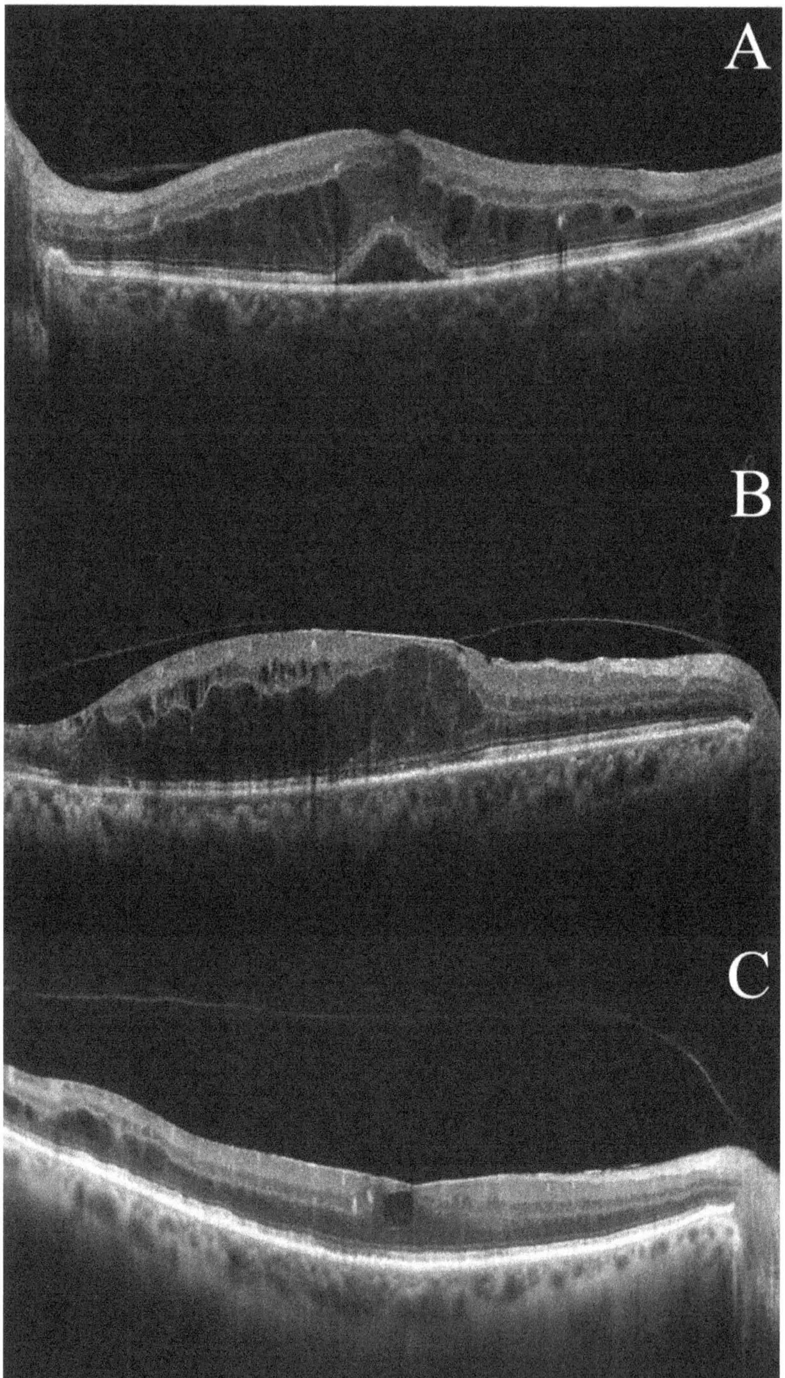

Figure 6.3. (A) Broad vitreomacular adhesion in DME. A slight perifoveal PVD can be seen. (B) Focal vitreomacular adhesion in DME. Perifoveal PVD is greater than in 3A. (C) Mild DME with PVD and non-tractional ERM.

matched diabetic patients [36]. They hypothesized that the vitreous may provide traction on the macula during the perifoveal PVD, promoting the development of edema. In this way, in another study about natural history of DME, the authors found that 55% of eyes with PVD versus 25% of eyes with VMA showed a spontaneous resolution of their macular edema within 6 months of diagnosis [37]. Regarding the relationship between VMA and response to treatment in DME, VMA is related to outcome in patients treated with ranibizumab [38, 39]. The presence of VMA at baseline is associated with better functional and anatomical response to ranibizumab, particularly when VMA release occurs during treatment [39] (figure 6.3(C)). Such VMA release in eyes with DME treated with ranibizumab happens in approximately 1 out of 3 patients [40]. On the other hand, neither VMA nor PVD have been shown to be related to functional or anatomical outcomes in eyes with DME treated with intravitreal dexamethasone implant [8].

6.4.2 Epiretinal membrane and vitreomacular traction

Often, in diabetic eyes, the posterior hyaloid forms a sheet along the posterior pole resulting in tractional force and mechanical retinal distortion. This type of VMT is termed as taut posterior hyaloid membrane (figure 6.4). At other times, PVD triggers the formation of an epiretinal membrane (ERM). VMT is present in 6.2%–28% and ERM in 13.7%–27% of eyes with DME [37, 39–43]. The impact of VMI abnormalities on treatment outcomes in DME is not clear. Some studies found no association between VMI abnormalities and change in BCVA or CRT from baseline to the last follow-up [43], while others showed that patients with significant ERM or VMT at baseline had a lower final vision [39, 44, 45]. On the other hand, vitrectomy in DME with VMT should be assessed cautiously. In protocol D of DRCR.net, VA improved significantly in 38% but deteriorated in 22% after surgery [46], whilst retinal thickening was reduced in most eyes. As for the ERM, surgery should be considered only in tractional cases (figure 6.5). Vitrectomy, membranectomy and ILM peeling does not improve VA in refractory DME with non-tractional ERM [47] (figures 6.3(C) and 6.6(B)). In some cases, different VMI abnormalities can be seen sequentially in the same patient (figure 6.6).

6.5 Disorganization of the inner retinal layers

The retinal dysfunction in diabetes mellitus produces a change in the retinal neurovascular unit. This unit represents the association of Muller cells, astrocytes, ganglion cells, amacrine cells, and retinal vascular endothelial cells [48]. The distortion of this unit traduces a disorganization of the inner retinal layers (DRIL), what is thought to represent an alteration in retinal homeostasis, neuro-transmitter regulation and blood-retina barrier integrity [48]. Therefore, DRIL has been proposed as a potential biomarker in DME.

In OCT, DRIL can be defined as the loss of clear demarcation between the ganglion cell- inner plexiform layer (IPL) complex, the inner nuclear layer (INL), and the outer plexiform layer (OPL) in the central fovea [49]. Therefore, it is considered a qualitative feature, and it has been correlated with macular ischemia [49]. Moreover, it

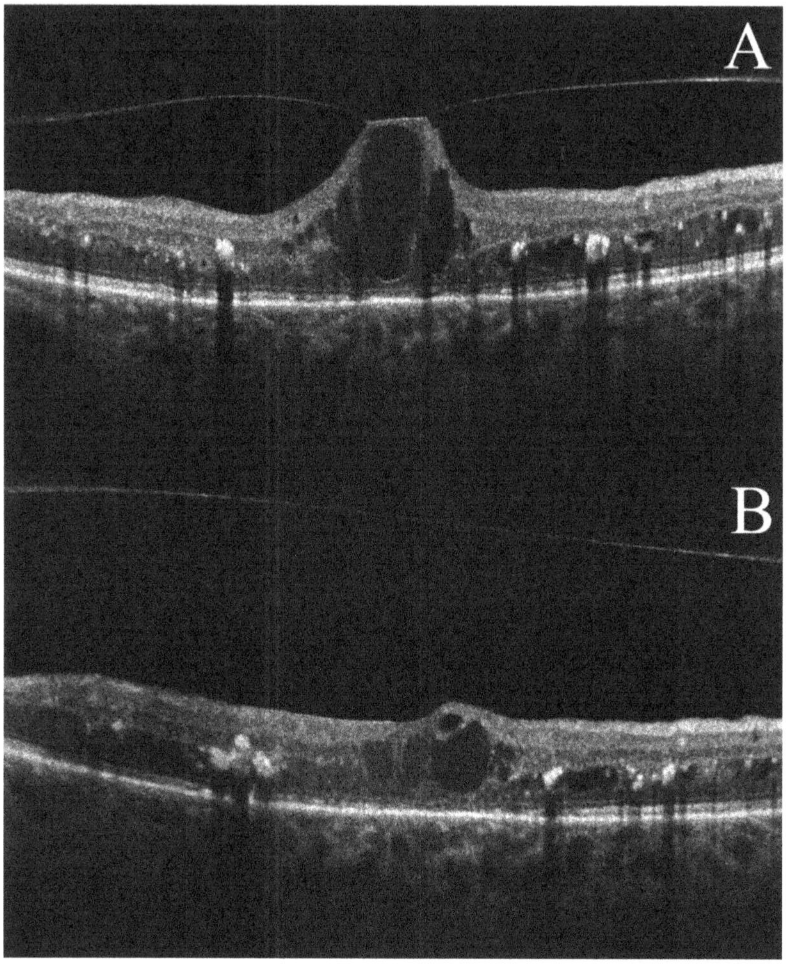

Figure 6.4. (A) Taut posterior hyaloid membrane responsible for the vitreomacular traction seen in a patient with DME. (B) 6 weeks later, spontaneous release of traction could be observed.

has been interpreted as an indirect sign of retinal ischemia and loss of normal vasculature in resolved macular edema [48]. In this line, an association has been found between DRIL and disruption of the outer retina and increasing severity of diabetic retinopathy [50]. Detection of DRIL can be challenging, especially in cases with DME, and it is more easily recognizable after resolution of edema (figures 6.1 and 6.7).

A recent review suggests that eyes with DME and DRIL are at a nearly 8-times greater risk of poor visual recovery with treatment than eyes without DRIL [51]. Specifically, DRIL in the foveal area (1-mm of diameter) is associated with poor VA, and especially, when DRIL is affecting the 50% or more of the central fovea it is associated with worse VA. Also, change in DRIL after treatment predicts future change in VA: resolution of DRIL after treatment is associated with better VA

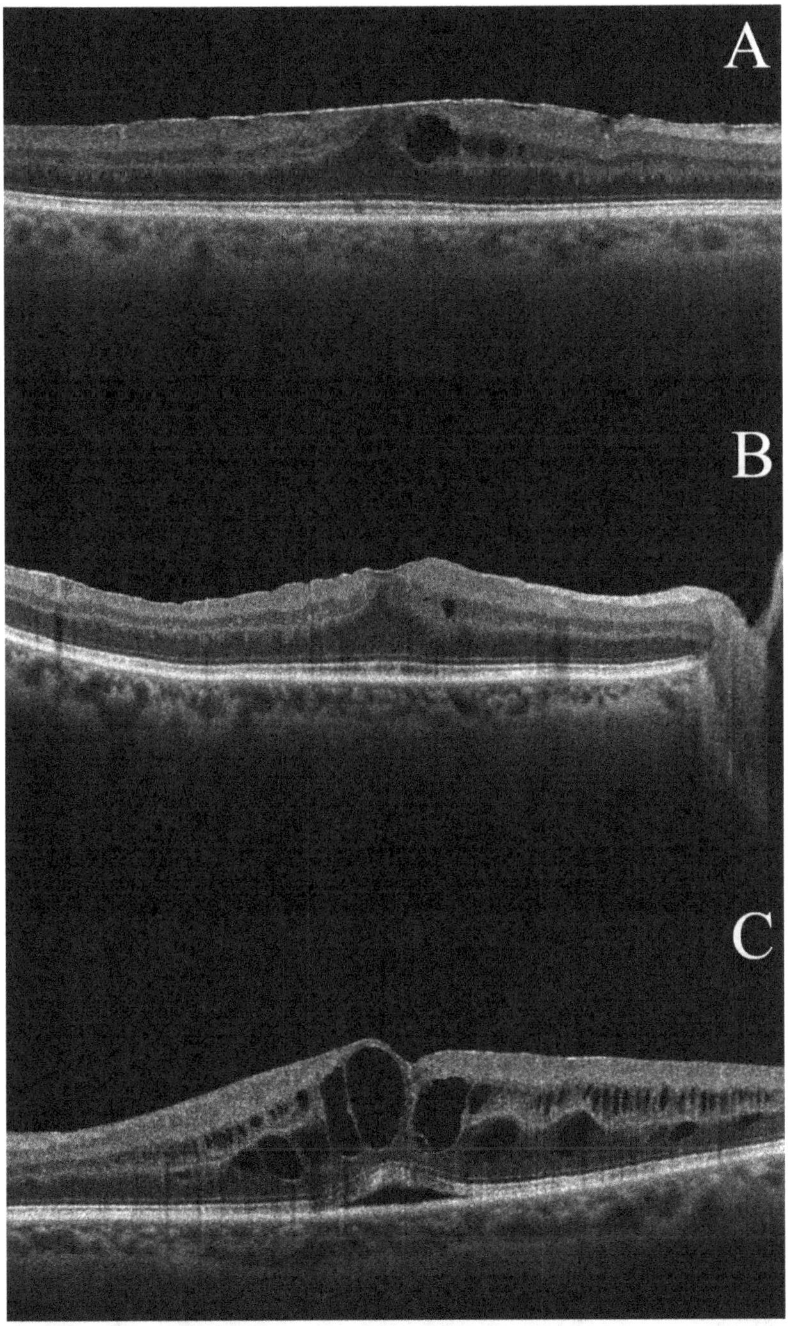

Figure 6.5. (A) Tractional epiretinal membrane in a patient with DME. (B) OCT B-scan in the same location than 5A one year after surgery (vitrectomy and peeling of ERM); no intravitreal treatment was needed. (C) Contralateral eye of the same patient with intraretinal cysts and subfoveal neurosensory detachment. This eye required intravitreal treatment.

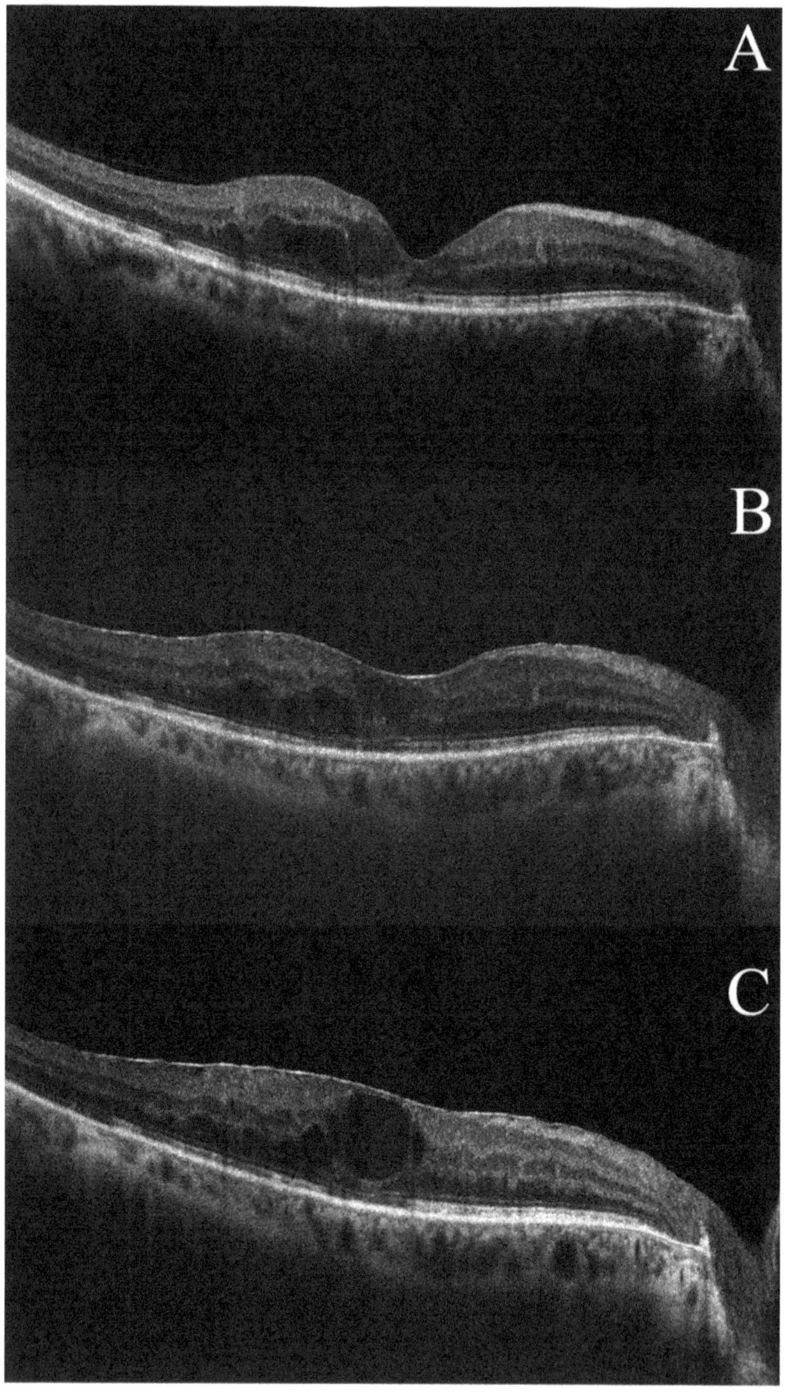

Figure 6.6. (A) OCT B-scan of a patient with DME and vitreomacular adhesion. (B) One year later, posterior vitreous detachment can be observed with development of a non-tractional epiretinal membrane (ERM). (C) Eight months later, the ERM has become tractional and the edema has increased.

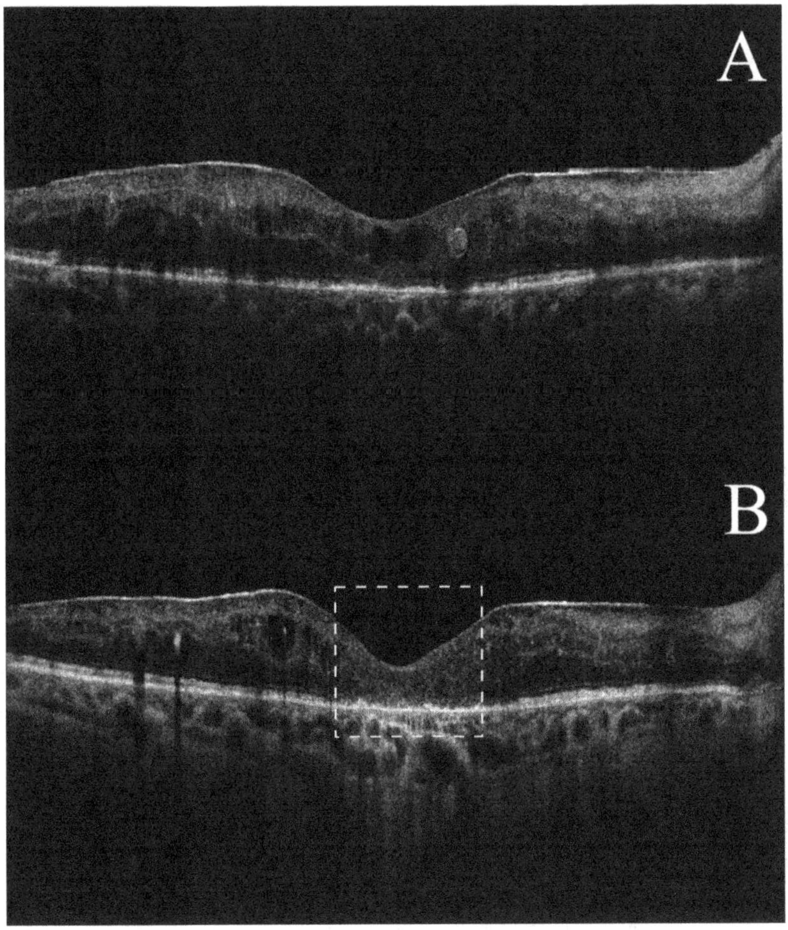

Figure 6.7. (A) OCT B-scan of a patient with DME treated with anti-VEGF and intravitreal dexamethasone implant several times. External limiting membrane (ELM) and ellipsoid zone (EZ) are discontinuous. (B) Two months after a new dexamethasone implant, disorganization of retinal inner layers (DRIL) has become evident. There is no clear demarcation between ganglion cell layer, IPL, INL and OPL inside de dotted box. Disruption of ELM and EZ has also become more evident after resolution of edema.

improvement, and persistent DRIL after resolution of DME is associated with worse visual outcome [52].

The extent of DRIL has been analyzed in several studies, concluding that it is a good predictor of central subfield thickness at 3 months [53] and that is significantly associated with the severity of metamorphopsia secondary to DME [54]. Moreover, early change in DRIL extent could identify eyes with a high likelihood of subsequent visual improvement or decline [55].

6.6 External limiting membrane and ellipsoid zone

The external limiting membrane (ELM) is the first hyper-reflective band in OCT. ELM is a scaffolding structure that maintains photoreceptor integrity and separates

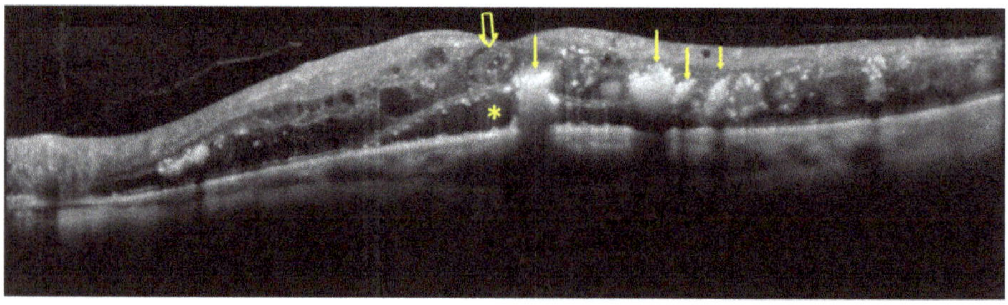

Figure 6.8. OCT B-scan of a patient with DME. Multiple hyperreflective foci (HRF) are seen throughout all the retinal layers, even under the ellipsoid zone (EZ). Some of the HRF are surrounding intraretinal cysts and others are confluent, forming even more hyperreflective structures that could correspond with hard exudates (arrows). These structures disrupt the external limiting membrane (ELM) and the EZ, which, in other parts, are intact. Multiple intraretinal cysts are seen, mainly in inner and outer nuclear layers. One of the cysts shows a hyperreflective content (open arrow). A broad subfoveal neurosensory detachment can be seen (asterisk).

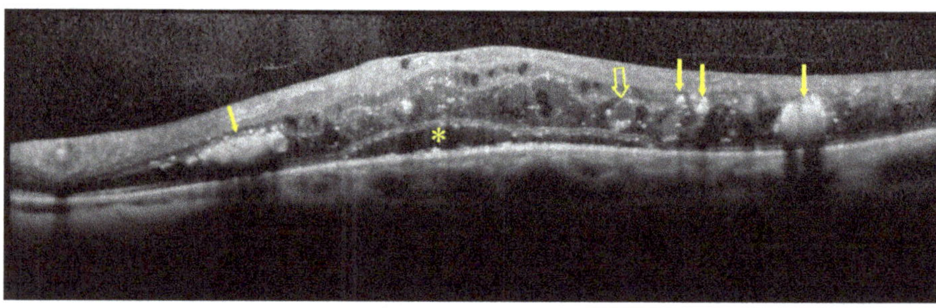

Figure 6.9. OCT B-scan of a patient with DME. Multiple hyperreflective foci (HRF) are seen throughout all the retinal layers. HRF surrounding intraretinal cysts (open arrow) and confluent HRF (arrows) can be identified. ELM and EZ are intact. Multiple intraretinal cysts are seen, mainly in inner and outer nuclear layers. Subfoveal neurosensory detachment can be observed (asterisk).

the inner segments of photoreceptors from the outer nuclear layer, where Müller cells attach to photoreceptor cells. It is believed that ELM is responsible for maintaining a protein balance between the photoreceptor layer and the outer nuclear layer (ONL) [56]. Therefore, it has a critical role in restoration of the photoreceptor microstructures and alignment [56]. In this line, ELM disruption could be a surrogate marker of Müller cell dysfunction [57]. On the other hand, the ellipsoid zone (EZ) is the second hyper-reflective band in OCT [58]. Although there is still a lack of a consensus on the exact anatomical correlate of the EZ, it could correspond to the isthmus between the inner and outer segments of the photoreceptors. It is generally believed that a visible EZ and ELM is a hallmark of the integrity of the photoreceptor layer [48]. In eyes with DME, EZ and/or ELM can be present, discontinuous or absent [49] (figures 6.1, 6.8, 6.9 and 6.10). A literature review has shown a strong correlation between the integrity of ELM and EZ and baseline VA in DME [59–62].

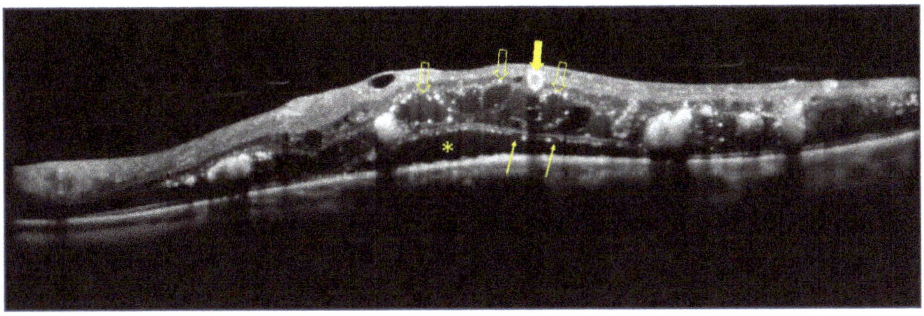

Figure 6.10. OCT B-scan of a patient with DME. Multiple hyperreflective foci surrounding intraretinal cysts (open arrows) can be seen. Areas of focal disruption of EZ with ELM integrity can be identified (arrows). Multiple intraretinal cysts are seen, mainly in inner and outer nuclear layers. Subfoveal neurosensory detachment can be observed (asterisk). A microaneurysm can also be seen (full arrow).

Absence of EZ and/or ELM is defined as a complete loss of foveal reflectivity at that level. These layers are classified as broken if they are not perfectly identifiable [49]. In eyes of diabetic patients, ELM disruption is significantly larger in subjects with clinically significant macular edema than in subjects without DME [56]. In DME patients, the extent of ELM disruption at baseline has been found as a good predictor of best-corrected visual acuity at 3 months [53]. In a study in patients with bilateral DME with asymmetrical anti-VEGF response, extensive ELM disruption was the strongest OCT biomarker to predict anti-VEGF resistance, followed by tractional epiretinal membrane. DRIL and EZ disruption had relatively lower predictive value [63].

EZ integrity has been identified as a potential biomarker for therapy surveillance and outcome prediction of visual acuity [64]. EZ reflectivity ratio correlated with functional outcome in DME patients from baseline to fifth year. EZ reflectivity improved the most in the first year of treatment and declined gradually until year 5 of treatment [64]. It has been shown in a retrospective interventional study that the shorter length of disruption in the EZ correlated with better visual outcome in patients with DME treated with intravitreal triamcinolone [58].

Finally, ELM and EZ are closely related. An increase in VEGF results in sequential ELM and EZ disruption on OCT. In fact, an intact ELM is necessary for an intact EZ in DME. Anti-VEGF therapy could help to restore the barrier effect of ELM, and in some cases, this is followed by EZ restoration [65].

6.7 Hyperreflective foci

The term 'hyperreflective retinal foci' (HRF) was first described by Bolz *et al* as any hyperreflective lesion, focal or dotten in appearance, visualized on OCT in any retinal layer [66]. HRF cannot be identified with clinical examination because of their small size and axial thickness, and appropriate imaging resolution in OCT is necessary for their recognition [67].

There are several hypotheses about the origin of HRF. At first, these hyperreflective lesions were believed to represent extravasated protein and/or lipid

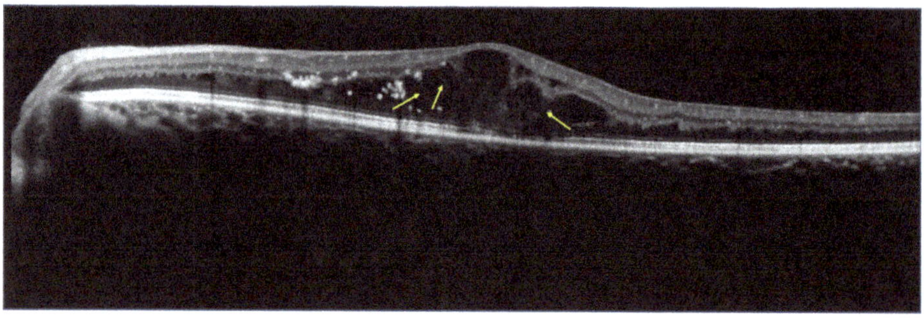

Figure 6.11. OCT B-scan of a patient with DME. Intraretinal cysts in outer nuclear layer are observed, with bridging retinal processes between cysts (arrows). Also, confluent HRF can be seen. In this case, ELM and EZ are continuous and subfoveal neurosensory detachment is absent.

deposits, precursors of hard exudates, that tended to resorb along with intraretinal fluid after laser treatment [66, 68, 69]. Another theory hypothesized that HRF were lipid-laden macrophages migrating into cystoid spaces as a consequence of blood-retinal barrier (BRB) breakdown [70, 71]. However, with the introduction of OCT angiography (OCTA), it was noticed that some HRF presented decorrelation signals, possibly an expression of morphological changes in microglia/macrophages or intracellular organelles containing highly reflective material [72]. In consonance with these findings, Lee *et al* demonstrated that levels in aqueous humor of CD14, a proinflammatory cytokine expressed by microglia, monocytes, and macrophages, correlated with HRF located in the inner retina, as well as with a pattern of diffuse macular edema [73]. It is well known that the role of microglia is essential to maintain retinal homeostasis and to elicit the inflammatory response, and glial cell proliferation represents one of the main alterations in diabetic retinopathy [74]. Vujosevic *et al* have reported that HRF are activated resident microglial cells which are initially present near ganglion cells and other inner retinal layers [75]. With the progression of retinopathy and under the influence of inflammatory mediators which includes VEGF, the inflammatory process spreads to entire retina with outward migration of HRF to outer retinal layers [76].

HRF can be distributed throughout all the retinal layers, either within the septae between cystoid spaces, or as confluent lesions located in the outer retinal layers, or as focal deposits within the vascular wall of microaneurysms (figures 6.8, 6.9, 6.10, 6.11). Combined multimodal analysis has showed that HRF mainly occupy the ONL and OPL, with the smallest foci distributed in the INL and IPL [77]. A recent international consensus has defined the diagnostic criteria for HRF to distinguish them from other subtypes of hyperreflective material (i.e., retinal exudates, hemorrhages or microaneurysms) as follows [78] (figure 6.1):

- Location within the inner retina;
- Size $\leqslant$30 μm;
- Absence of posterior shadowing;
- Reflectivity similar to the retinal nerve fiber layer.

6.7.1 Clinical and prognostic implications of hyperreflective foci in diabetic macular edema

In a study of patients with DME and good visual acuity managed with observation, eyes with DRIL, HRF, and ellipsoid zone disruption at baseline exhibited a high risk of visual loss [79]. Not only the presence but the number of HRF reflects disease severity, exhibiting direct associations with HbA1c values and high levels of total cholesterol, triglycerides, and low-density lipoprotein [76, 80, 81]. The association of HRF with glycometabolic state has been observed even in early stages of diabetic retinopathy without DME, supporting the hypothesis of lipid extravasation conceivable in subjects with poor glycemic control [82, 83]. HRF have been noted in diabetic eyes which did not have diabetic retinopathy which further strengthens the concept of their inflammatory origin [76]. The hypothesis that HRF are of inflammatory origin and actually represent activated microglia is supported by the fact that HRF significantly decrease in number or disappear completely after treatment with anti-VEGF [76].

Hyperreflective foci have also been reported within the choroid of diabetic patients, and this OCT finding has been termed 'hiperreflective choroidal foci' (HCF) [84]. HCF have always been found in conjunction with HRF, suggesting a similar origin [84, 85]. HCF are considered HRF migrated to the choroid [85] (figure 6.2). The finding of HCF denoted worse disease severity and prognosis [84, 86], compared to the presence of HRF alone [84].

HRF is an important biomarker of response to treatment in DME. Presence of HRF has been associated with a poorer visual outcome following treatment both with intravitreal steroid and anti-VEGF agents [87]. Clusters of HRF occupying the central macular area were associated with worse visual acuity than eyes without HRF clusters before any treatment, and this functional difference was maintained following intravitreal ranibizumab and focal laser therapy for up to 5 years [88]. Furthermore, location of HRF has prognostic implications. In eyes with DME, HRF located in the outer retinal layers (ONL, OPL) have been strongly associated with worse visual prognosis, disruption of the ELM, photoreceptor loss, and worse prognosis after vitrectomy [89–91]. The number of HRF in the outer retinal layers as a predictor of poor final visual acuity is independent of the pattern of DME by OCT (diffuse macular edema, cystoid macular edema, and serous retinal detachment) [92]. The prognostic role of HRF has been further corroborated by higher levels of both IL-1β and HRF (>10) found in refractory DME [36]; likewise, a higher HRF number at baseline is predictive of early recurrence of DME and a shorter duration of dexamethasone implant efficacy [93, 94].

A theoretical advantage in favor of a dexamethasone implant as first-line agent over anti-VEGF therapy has been hypothesized for DME with inflammatory biomarkers [78]. In this line, treatment-naïve DME with inflammatory biomarkers (i.e., HRF and SND) showed a superior anatomical response and fewer injections with dexamethasone implant than with intravitreal aflibercept, even if better VA was achieved with intravitreal aflibercept. Lens opacity development explained the lower-than-expected functional outcome in the dexamethasone implant group [95]. While the role of HRF in predicting visual outcome of DME treated with anti-VEGF agents did not reach univocal conclusions [76, 92, 96, 97], final visual gain resulted evident in eyes managed with dexamethasone implant [8, 98]. In fact, treatment with dexamethasone implant significantly reduces the number of HRF with a reduction maintained up to 12 months of follow up [99]. On the other hand, in eyes treated with anti-VEGF, the reduction of the number of HRF located in the outer retina has been associated to an improvement in visual gain [100, 101].

In conclusion, HRF are OCT biomarkers in eyes with DME, believed to be microglial cells which are activated as a result of inflammatory response to retinopathy. HRF are located in the inner retinal layers in the early spectrum of disease in diabetic retinopathy. With the progression of disease under the influence of various inflammatory mediators which includes VEGF, they migrate toward outer retinal layers. The detection of HRF of size $\leqslant$30 μm without posterior shadowing and reflectivity similar to the retinal nerve fiber layer configures the inflammatory phenotype in DME that usually responds better to early intravitreal steroid implant than to anti-VEGF.

6.8 Intraretinal cystoid spaces and bridging retinal processes

Intraretinal cysts (IRC) appear in OCT as rounded or oval hyporreflective spaces separated by hypereflective septae visible in OCT [102]. These cysts combined with retinal thickening lead to formation of macular cystic edema causing significant distortion of the macular anatomy (figures 6.8, 6.9, 6.10, 6.11, 6.12). Like the other patterns, the presence of intraretinal cysts does not constitute a specific finding of DME, since it can appear in other types of edema, including: central/branch retinal vein occlusion, pseudophakia, posterior uveitis and retinitis pigmentosa [103].

The main mechanism that leads to extracellular accumulation of fluid in the retina is an alteration in the permeability of the blood-retina barrier (BRB). The exact pathophysiologic mechanism of DME is not fully understood but it is well known that VEGF plays a key role in its development. Increased VEGF levels in diabetic retinopathy affects the inner BRB leading to increased vascular permeability, resulting in decreased osmotic gradient, extracellular fluid accumulation, and cyst formation, while the outer BRB is not affected significantly [104, 105]. In addition, systemic factors such as arterial hypertension and hypoalbuminemia lead to an increase in vascular permeability and favor the appearance of edema.

In DME, IRC are a valuable prognostic biomarker, so it is important to consider several parameters. On the one hand, the size of the cysts is directly proportional to the degree of final atrophy, so it is essential to classify them according to their size: Small <100 μm, Large 100–200 μm or Giants >200 μm [106]. On the other hand, their location and extension should be assessed, because cysts in the outer nuclear layer (ONL) have shown to damage photoreceptor cells and disrupt inner segment–outer segment (IS/OS) junction causing irreversible damage to visual functions [107]. Likewise, it has been described some hyperreflective signals within the cyst which

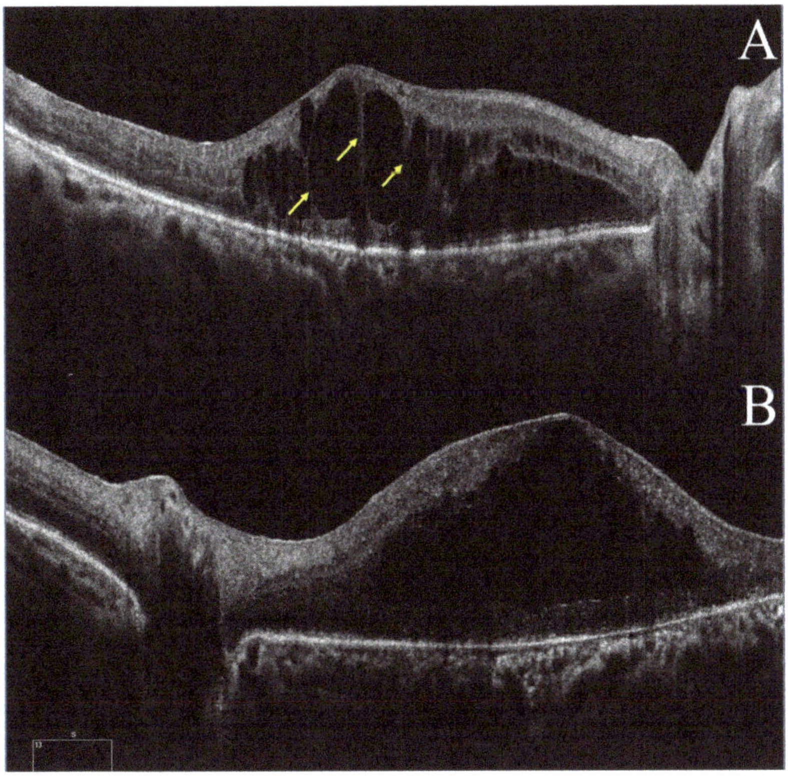

Figure 6.12. (A) Right eye of a patient with DME. Giant cysts (>200 microns of diameter) can be observed, separated by multiple bridging retinal processes (BRP) (arrows). With intravitreal treatment, visual acuity reaches 20/50. (B) Left eye of the same patient with chronic DME. A giant cyst is observed as a result of confluence of several cysts. BRP are scarce. Despite treatment, best-corrected visual acuity is only counting fingers.

are associated with severe disruption of the BRB. The etiology of these hyperreflective cysts is doubtful since there are several hypotheses: some of them suggest that fibrin and other inflammatory by-products may fill these spaces [108] and others bet on as suspended scattering particles in motion (SSPiM) as seen using OCT-angiography [109].

Another OCT sign recently described in DME is the presence of bridging retinal processes (BRP), seen as hyperreflective lines between the cystic cavities (figures 6.11 and 6.12). They represent residual neural elements which connect outer and inner retina, thus helping in transmitting visual impulses from the outer retinal layers to the optic nerve axons. Although the exact nature of these tissues is not known, it is believed that Müller cells and bipolar cells are an important component of BRP. Their presence in eyes with DME has been associated with improved VA after treatment [110]. In contrast to this, absence of these BRP is associated with poor prognosis after treatment. These eyes are unlikely to improve despite resolution of cysts and progress to foveal atrophy and thinning [13].

In the recent years, numerous advances have been made in the treatment of diabetic retinopathy and DME [111, 112]. Anti-VEGF drugs are considered the first-line treatment. However, corticosteroids represent a fundamental alternative for treating these patients due to their strong anti-inflammatory and antiedema properties [113, 114]. It has been proven a greater decrease in CRT and IRC at the level of deep capillary plexus after dexamethasone implant versus intravitreal anti-VEGF treatment [115]. However, no correlation between IRC and VA improvement has been demonstrated [8]. Despite these limitations, persistent IRC after treatment might be a sign of refractory DME, and a switch in the drug used should be considered in these cases.

6.9 Subfoveal neurosensory detachment

Subfoveal neurosensory detachment (SND) is an accumulation of fluid beneath the neurosensory retina. It is visible with OCT as a hyporeflective cavity bordered by two hyperreflective areas: in depth by the RPE and on the surface by the neurosensory retina (figures 6.8–6.10). Together with sponge thickening and the appearance of intraretinal cysts, it constitutes one of the 3 OCT-patterns of fluid accumulation in DME [116] and is included as an essential element in OCT classifications of DME [17], not only in the first ones such as Panozzo's [18], but also in the most recents such as ESASO's [49]. Its prevalence in eyes with DME is estimated to be between 15%–30% of naïve cases [116–119]. However, SND is not a specific finding in DME, since it can also appear in other types of edema and macular inflammatory entities: macular edema secondary to vascular occlusions, pseudophakic macular edema, and uveitic macular edema, among others [120, 121].

The pathophysiology of SND is not completely known. It has been hypothesized that one of the most important factors in the development of SND is the leakage from the retina and choroid to the subretinal space in an amount that exceeds the reabsorption capacity of the RPE [116, 122]. However, other mechanisms have been implicated such as dysfunction of the RPE [123], hypoxia [124], as well as alteration of the ELM [121, 122]. On the other hand, it has been described by different authors an increased level of inflammatory cytokines in aqueous humor and vitreous, mainly interleukin 6 (IL-6), but also IL-8, VEGF, PIGF, ICAM-1, MCP-1, IP-10, and IL-1b, in patients with SND [125–130]. For this reason, SND has been proposed as a non-invasive biomarker of retinal inflammation in diabetic retinopathy and DME [131]. This is why some authors, such as Vujosevic *et al* [132], suggest that SND is a specific subtype of DME; in their study, they found that eyes with SND had more hyperreflective spots, greater central retinal thickness, more extensive damage in the ELM and lower sensitivity determined by microperimetry, compared to eyes without SND. However, other authors think that SND appears at an early stage in the course of the disease and should not be differentiated into a specific group [133]. Moreover,

SND has also been associated with the presence of renal damage and particularly with high serum albumin levels [134].

In patients with DME, SND has been analyzed as a biomarker of both prognosis and response to treatment. As a prognostic biomarker, SND has been extensively studied with contradictory data among different authors: some works suggest worse VA when DNS is present [132, 135, 136], while others find that DNS is a good prognostic factor [125, 137–142]. Another hypothesis is that what is truly relevant for visual prognosis would be the disappearance of SND after the start of treatment and its non-recurrence throughout the course of the disease [119].

Also, SND has its value as a marker of response to treatment. Although SND responds well to both anti-VEGF and steroid treatment, some authors have found a better response with the latter [8, 115, 131, 143–146], suggesting that SND would be a factor to take into account when deciding the treatment of a patient with DME (figure 6.13). This is in line with the proposed inflammatory pathophysiology of SND mentioned above.

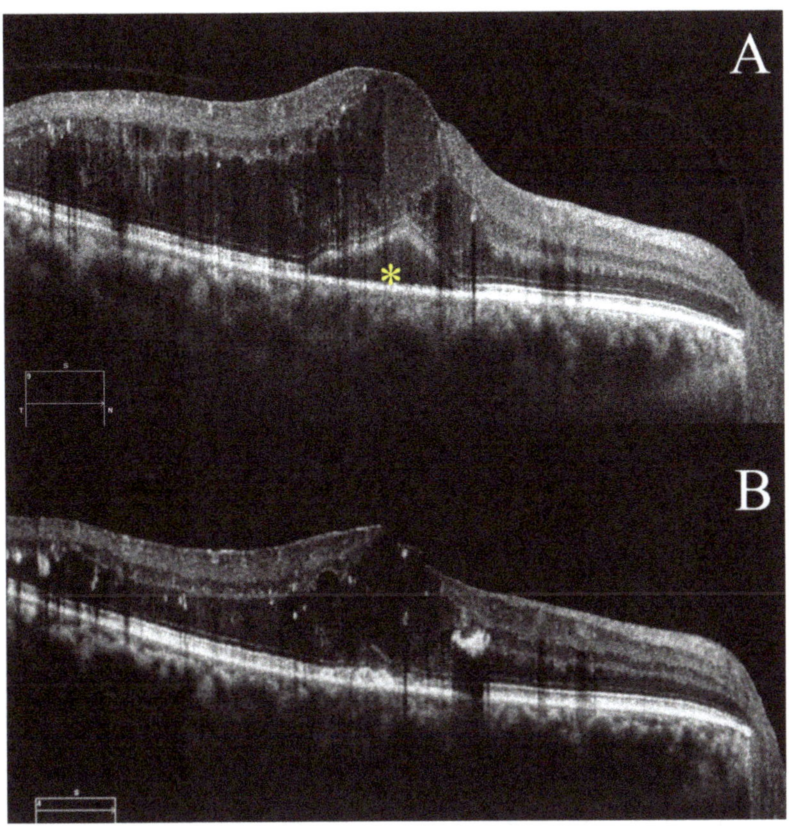

Figure 6.13. (A) OCT B-scan of a patient with naïve DME and subfoveal neurosensory detachment (SND) (asterisk). (B) Two months after an intravitreal dexamethasone implant, the SND has resolved, whilst intraretinal cysts persist.

Finally, in the recent years, it has been possible to detect and quantify SND, as well as other fluid locations, in an automated way with machine learning strategies and neural networks [147, 148]. This opens new avenues for the investigation of its importance and behavior with the evolution of the disease and the response to drugs.

6.10 Conclusion

Several OCT biomarkers have been described in DME. Physicians that face patients with this disease must do a careful review of all the macular OCT B-scans in order to act correctly. VMI abnormalities, location and amount of fluid (IRC, size of cysts, SND), presence of poor prognosis biomarkers (DRIL, disruption of ELM and EZ, absence of BRP) and location and number of HRF must be assessed (table 6.1). This will allow physicians to inform the patient properly about the diagnosis and VA recovery possibilities and to choose the best treatment option.

Table 6.1. OCT biomarkers in diabetic macular edema. CRT: central retinal thickness. MV: macular volume. VMA: vitreomacular adhesion. PVD: posterior vitreous detachment. ERM: epiretinal membrane. VMT: vitreomacular traction. DRIL: disorganization of retinal inner layers. ELM: external limiting membrane. EZ: ellipsoid zone. HRF: hyperreflective foci. IRC: intraretinal cysts. BRP: bridging retinal processes. SND: subfoveal neurosensory detachment.

	Prognostic	Cronicity	Inflammation	Better response to intravitreal dexamethasone than to anti-VEGF
CRT	No	No	No	No
MV	No	No	No	No
VMA	Good (if release)	No	No	No
PVD	No	No	No	No
ERM	Poor (tractional)	No	No	No
VMT	Poor	No	No	No
DRIL	Poor	Yes	No	No
ELM / EZ disruption	Poor	Yes	No	No
HRF	Poor (outer retina)	Yes (outer retina)	Yes	Yes
IRC	No	Yes (giant cysts)	Yes (if hyperreflective)	Yes (if hyperreflective)
BRP	Poor (if absent)	Yes (if absent)	No	No
SND	No	No	Yes	Yes

References

[1] Aiello L P, Gardner T W and Kingetal G L 1998 Diabetic retinopathy *Diabetes Care* **21** 143–56

[2] Klein R, Moss S E, Klein B E, Davis M D and DeMets D L 1989 The Wisconsin epidemiologic study of diabetic retinopathy: XI. The incidence of macular edema *Ophthalmology* **96** 1501–10

[3] Cunha-Vaz J and Coscas G 2010 Diagnosis of macular edema *Ophthalmologica* **224** 2–7

[4] Kocur I and Resnikoff S 2002 Visual impairment and blindness in Europe and their prevention *Br. J. Ophthalmol.* **86** 716–22

[5] Biomarkers Definition Working Group Biomarkers and surrogate endpoints: preferred definitions and conceptual framework 2001 *Clin. Pharmacol. Therap.* **69** 89–95

[6] Markan A, Agarwal A, Arora A, Bazgain K, Rana V and Gupta V 2020 Novel imaging biomarkers in diabetic retinopathy and diabetic macular edema *Ther. Adv. Ophthalmol.* **12** 1–16.

[7] Suciu C I, Suciu V I and Nicoara S D 2020 Dec 31 Optical coherence tomography (angiography) biomarkers in the assessment and monitoring of diabetic macular edema *J. Diabetes Res.* **2020** 6655021

[8] Zur D, Iglicki M, Busch C, Invernizzi A, Mariussi M and Loewenstein A 2018 for the International Retinal Group OCT Biomarkers as functional outcome predictors in diabetic macular edema treated with dexamethasone implant *Ophthalmology* **125** 267–75

[9] You Q S, Tsuboi K, Guo Y, Wang J, Flaxel C J, Bailey S T, Huang D, Jia Y and Hwang T S 2021 Comparison of central macular fluid volume with central subfield thickness in patients with diabetic macular edema using optical coherence tomography *JAMA Ophthalmol.* **139** 734–41

[10] Chan A, Duker J S and Schuman J 2006 Normal macular thickness in healthy eyes using Stratus optical coherence tomography *Arch. Ophtahalmol.* **124** 193–8

[11] Mitsch C, Lammer J, Karst S, Scholda C, Pablik E and Schmidt-Erfurth U M 2020 Systematic ultrastructural comparison of swept-source and full- depth spectral domain optical coherence tomography imaging of diabetic macular edema *Br. J. Ophthalmol.* **104** 88–873.

[12] Virgili G, Menchini F, Murro V, Peluso E, Rosa F and Casazza G 2011 Optical coherence tomography for detection of macular oedema in patients with diabetic retinopathy *Cochrane Database Sys. Rev.* **6** CD008081

[13] Pelossini L, Hull C C, Boyce J F, McHugh D, Stanford M R and Marshall J 2011 Optical coherence tomography in patients with macular edema *Invest. Ophthalmol. Vis. Sci.* **52** 2741–8

[14] Tsuboi K, Sheng You Q, Guo Y, Wang J, Flaxel C, Bailey S T, Huang D, Jia Y and Hwang T S 2022 Association between fluid volume in inner nuclear layer and visual acuity in diabetic macular edema *Am. J. Ophthalmol.* **237** 167–172.

[15] Choi M Y, Jee D and Kwnon J W 2019 Characteristics of diabetic macular edema patients refractory to anti-VEGF treatments and dexamethasone implant *PLoS One* **12;14** e0222364

[16] Bressler S B, Qin H, Beck R W, Chalman K V, Kim J E, Melia M and Well J A 2012 for the Diabetic Retinopathy Clinical Research Network. Factors associated with changes in visual acuity and OCT thickness at 1 year after treatment for diabetic macular edema with ranibizumab *Arch. Ophthalmol.* **130** 1153–61

[17] Hui V W K, Szeto S K H, Tang F, Yang D, Chen H and Lai T Y Y *et al* 2021 Optical coherence tomography classification systems for diabetic macular edema and their associations with visual outcome and treatment responses. An update review *Asia Pac. J. Ophthalmol.* **11** 247–57
[18] Panozzo G, Parolini B and Gusson E *et al* 2004 Diabetic macular edema: an OCT-based classification *Semin. Ophthalmol.* **19** 13–20
[19] Patel J I, Hykin P G and Schadt M *et al* 2006 Pars plana vitrectomy for diabetic macular oedema: OCT and functional correlations *Eye* **20** 674–80
[20] Wielders L H P, Schouten J and Winkens B *et al* 2018 Randomized controlled European multicenter trial on the prevention of cystoid macular edema after cataract surgery in diabetics: ESCRS PREMED Study Report 2 *J. Cataract Refract. Surg.* **44** 836–47
[21] Karla G, Kar S S, Sevgy D D, Madahushi A, Srivastava S K and Ehlers J P 2021 Quantitave imaging biomarkers in age-related macular degeneration and diabetic eye disease: step closer to precision medicine *J. Pers. Med.* **11** 1161
[22] Kim J T, Lee D H, Joe S G, Kim J G and Yoom Y H 2013 Changes in choroidal thickness in relation to the severity of retinopathy and macular edema in type 2 diabetic patients *Invest. Ophthalmol. Vis. Sci.* **54** 3378–84
[23] Dou N, Yu S, Tsui C K, Yang B, Lin J and Lu X *et al* 2021Nov26 Choroidal vascularity index as biomarker for visual response to anti-vascular endothelial growth factor treatment in diabetic macular edema *J. Diabetes Res.* **2021** 3033219
[24] Géhl Z, Kulcsar K, Kiss H J M, Németh J, Maneschg O A and Resch M D 2014 Retinal and choroidal thickness measurements using spectral domain optical coherence tomography in anterior and intermediate uveitis *BMC Ophthalmol.* **14** 103
[25] Kongwattananon W, Kumar A, Oyeniran E, Sen H N and Kodati S 2021 Changes in choroidal vascularity index (CVI) in intermediate uveitis *Transl. Vis. Sci. Technol.* **10** 33
[26] Kupak R, Kumar S, Dhaivat S, Maitreyi C and Sugandha G 2019 Choroidal *Hipereflective foci*: a novel spectral domain optical coherence tomography biomarker in eyes with diabetic macular edema *Asia Pac. J. Ophthalmol.* **8** 314–8
[27] Lai C T, Hsieh T T, Linn C J, Wang J K, Hsia N Y and Bair H *et al* 2021 Age, initial central retinal retinal thickness, and OCT biomarkers have an influence on the outcome of diabetic macular edema treated with ranibizumab- tri-center 12-month treat-and-extend study *Front. Med.* **8** 668107
[28] Abraham J R, Wykoff C C, Arepalli S, Lunasco L, Yu H J and Hu M *et al* 2021 Aqueous cytokine expression and higher order OCT biomarkers: assessment of the anatomic biologic bridge in the imagine DME study *Am. J. Ophthalmol.* **222** 328–39
[29] Parravano M, Costanzo E and Querques G 2020 Profile of non-responder and late responder patients treated for diabetic macular edema: systemic and ocular factors *Acta Diabetol.* **57** 911–21
[30] Cunha-Vaz J, Ribeiro L and Lobo C 2014 Phenotypes and biomarkers of diabetic retinopathy *Prog. Retina Eye Res.* **41** 90–111
[31] Lent-Schochet D, Lo T, Luu K Y, Tran S, Wilson M D and Moshiri A *et al* 2021 Natural history and predictors of vision loss in eyes with diabetic macular edema and good initial visual acuity *Retina* **41** 2132–9
[32] Rayess N, Rahimy E, Ying G-S, Bagheri N, Ho A C and Regillo C D *et al* 2015 Baseline choroidal thickness as a predictor for response to anti-vascular endothelial growth factor therapy in diabetic macular edema *Am. J. Ophthalmol.* **159** 85–91

[33] Sebag J 2015 The vitreoretinal interface and its role in the pathogenesis of vitreomaculopathies *Ophthalmologe* **112** 10–9

[34] Duker J S, Kaiser P K, Binder S, de Smet M D, Gaudric A, Reichel E, Sadda S R, Sebag J, Spaide R F and Stalmans P 2013 The International Vitreomacular Traction Study Group classification of vitreomacular adhesion, traction, and macular hole *Ophthalmology* **120** 2611–9

[35] Ghazi N G, Ciralsky J B, Shah S M, Campochiaro P A and Haller J A 2007 Optical coherence tomography findings in persistent diabetic macular edema: the vitreomacular interface *Am. J. Ophthalmol.* **144** 747–54

[36] Gandorfer A, Haritoglou C and Kampik A 2006 Optical coherence tomography assessment of the vitreoretinal relationship in diabetic macular edema *Am. J. Ophthalmol.* **141** 234–5

[37] Hikichi T, Fujio N, Akiba J, Azuma Y, Takahashi M and Yoshida A 1997 Association between the short-term natural history of diabetic macular edema and the vitreomacular relationship in type II diabetes mellitus *Ophthalmology* **104** 473–8

[38] Sadiq M A, Soliman M K, Sarwar S, Agarwal A, Hanout M and Demirel S *et al* 2016 READ-3 Study Group. Effect of vitreomacular adhesion on treatment outcomes in the Ranibizumab for Edema of the Macula in Diabetes (READ-3) Study *Ophthalmology* **123** 324–9

[39] Wong Y, Steel D H W, Habib M S, Stubbing-Moore A, Bajwa D and Avery P J 2017 Sunderland Eye Infirmary Study Group. Vitreoretinal interface abnormalities in patients treated with ranibizumab for diabetic macular oedema *Graefes Arch. Clin. Exp. Ophthalmol.* **255** 733–42

[40] Veloso C E, Brocchi D N, Singh R P and Nehemy M B 2021 Vitreomacular interface after anti-VEGF injections in diabetic macular edema *Int. J. Retina Vitr.* **7** 23

[41] Jackson T L, Nicod E, Angelis A, Grimaccia F, Prevost A T, Simpson A R H and Kanavos P 2013 Vitreous attachment in age-related macular degeneration, diabetic macular edema, and retinal vein occlusion: a systematic review and metaanalysis *Retina* **33** 1099–108

[42] Kim B Y, Smith S D and Kaiser P K 2006 Optical coherence tomographic patterns of diabetic macular edema *Am. J. Ophthalmol.* **142** 405–12

[43] Mikhail M, Stewart S, Seow F, Hogg R and Lois N 2018 Vitreomacular interface abnormalities in patients with diabetic macular oedema and their implications on the response to anti-VEGF therapy *Graefes Arch. Clin. Exp. Ophthalmol.* **256** 1411–8

[44] Akbar Khan I, Mohamed M D, Mann S S, Hysi P G and Laidlaw D A 2015 Prevalence of vitreomacular interface abnormalities on spectral domain optical coherence tomography of patients undergoing macular photocoagulation for centre involving diabetic macular oedema *Br. J. Ophthalmol.* **99** 1078–81

[45] Agarwal D, Gelman R, Ponce C P, Stevenson W and Christoforidis J B 2015 The vitreomacular interface in diabetic retinopathy *J. Ophthalmol.* **2015** 392983

[46] Diabetic Retinopathy Clinical Research Network Writing Committee *et al* 2010 Vitrectomy outcomes in eyes with diabetic macular edema and vitreomacular traction *Ophthalmology* **117** 1087–93

[47] Ghassemi F, Bazvand F, Roohipoor R, Yaseri M, Hassanpoor N and Zarei M 2016 Outcomes of vitrectomy, membranectomy and internal limiting membrane peeling in patients with refractory diabetic macular edema and non- tractional epiretinal membrane *J. Curr. Ophthalmol.* **28** 199–205

[48] Hui V W K, Szeto S K H, Tang F, Yang D, Chen H and Lai T Y Y *et al* 2021 Optical coherence tomography classification systems for diabetic macular edema and their associations with visual outcome and treatment responses—an updated review *Asia-Pac. J. Ophthalmol.* **11** 247–57
[49] Panozzo G, Cicinelli M V, Augustin A J, Battaglia Parodi M, Cunha-Vaz J and Guarnaccia G *et al* 2020 An optical coherence tomography-based grading of diabetic maculopathy proposed by an international expert panel: The European School for Advanced Studies in Ophthalmology classification *Eur. J. Ophthalmol.* **30** 8–18
[50] Das R, Spence G, Hogg R E, Stevenson M and Chakravarthy U 2018 Disorganization of inner retina and outer retinal morphology in diabetic macular edema *JAMA Ophthalmol.* **136** 202–8
[51] Santos A R, Costa M, Schwartz C, Alves D, Figueira J and Silva R *et al* 2018 Optical coherence tomography baseline predictors for initial best-corrected visual acuity response to intravitreal anti-vascular endothelial growth factor treatment in eyes with diabetic macular edema: the CHARTRES Study *Retina* **38** 1110–9
[52] Radwan S H, Soliman A Z and Tokarev J *et al* 2015 Association of disorganization of inner retinal layers with vision after resolution of center-involved diabetic macular edema *JAMA Ophthalmol.* **133** 820–5
[53] Mazloumi M, Entezari M, Samadikhadem S, Ramezani A, Nikkhah H and Arevalo J F 2022 Spectral domain optical coherence tomography biomarkers of retinal hyperpermeability and choroidal inflammation as predictors of short-term functional and anatomical outcomes in eyes with diabetic macular edema treated with intravitreal bevacizumab *Retina* **42** 760–6
[54] Nakano E, Ota T, Jingami Y, Nakata I, Hayashi H and Yamashiro K 2019 Correlation between metamorphopsia and disorganization of the retinal inner layers in eyes with diabetic macular edema *Graefes Arch. Clin. Exp. Ophthalmol.* **257** 1873–8
[55] Sun J K, Lin M M, Lammer J, Prager S, Sarangi R and Silva P S *et al* 2014 Disorganization of the retinal inner layers as a predictor of visual acuity in eyes with center-involved diabetic macular edema *JAMA Ophthalmol.* **132** 1309–16
[56] Chen X, Zhang L, Sohn E H, Lee K, Niemeijer M and Chen J *et al* 2012 Quantification of external limiting membrane disruption caused by diabetic macular edema from SD-OCT *Invest. Ophthalmol. Vis. Sci.* **53** 8042–8
[57] Murakami T, Nishijima K and Akagi T *et al* 2012 Optical coherence tomographic reflectivity in photoreceptors beneath cystoid spaces in diabetic macular edema *Invest. Ophthalmol. Vis. Sci.* **53** 1506–11
[58] Tao L W, Wu Z, Guymer R H and Luu C D 2016 Ellipsoid zone on optical coherence tomography: a review *Clin. Exp. Ophthalmol.* **44** 422–30
[59] Otani T, Yamaguchi Y and Kishi S 2010 Correlation between visual acuity and foveal microstructural changes in diabetic macular edema *Retina* **30** 774–80
[60] Maheshwary A S, Oster S F and Yuson R M *et al* 2010 The association between percent disruption of the photoreceptor inner segment—outer segment junction and visual acuity in diabetic macular edema *Am. J. Ophthalmol.* **150** 63–7
[61] Shin H J, Lee S H, Chung H and Kim H C 2012 Association between photoreceptor integrity and visual outcome in diabetic macular edema *Graefes Arch. Clin. Exp. Ophthalmol.* **250** 61–70

[62] Lai K, Huang C and Li L *et al* 2020 Anatomical and functional responses in eyes with diabetic macular edema treated with '1 + PRN' ranibizumab: one-year outcomes in population of mainland China *BMC Ophthalmol.* **20** 229

[63] Koc F, Güven Y Z, Egrilmez D and Aydın E 2021 Optical coherence tomography biomarkers in bilateral diabetic macular edema patients with asymmetric anti-VEGF response *Semin. Ophthalmol.* **36** 444–51

[64] Kessler L J, Auffarth G U, Bagautdinov D and Khoramnia R 2021 Ellipsoid zone integrity and visual acuity changes during diabetic macular edema therapy: a longitudinal study *J. Diabetes Res.* **2021** 8117650

[65] Saxena S, Akduman L and Meyer C H 2021 External limiting membrane: retinal structural barrier in diabetic macular edema *Int. J. Retina Vitr.* **7** 16

[66] Bolz M, Schmidt-Erfurth U, Deak G, Mylonas G, Kriechbaum K and Scholda C *et al* 2009 Optical coherence tomographic hyperreflective foci: a morphologic sign of lipid extravasation in diabetic macular edema *Ophthalmology* **116** 914–20

[67] Fragiotta S, Abdolrahimzadeh S, Dolz-Marco R, Sakurada Y, Gal-Or O and Scuderi G 2021 Significance of hyperreflective foci as an optical coherence tomography biomarker in retinal diseases: characterization and clinical implications *J. Ophthalmol.* **2021** 6096017

[68] Deák G G, Bolz M, Kriechbaum K, Prager S, Mylonas G and Scholda C *et al* 2010 Effect of retinal photocoagulation on intraretinal lipid exudates in diabetic macular edema documented by optical coherence tomography *Ophthalmology* **117** 773–9

[69] Yamada Y, Suzuma K, Fujikawa A, Kumagami T and Kitaoka T 2013 Imaging of laser-photocoagulated diabetic microaneurysm with spectral domain optical coherence tomography *Retina* **33** 726–31

[70] Horii T, Murakami T, Nishijima K, Akagi T, Uji A and Arakawa N *et al* 2012 Relationship between fluorescein pooling and optical coherence tomographic reflectivity of cystoid spaces in diabetic macular edema *Ophthalmology* **119** 1047–55

[71] Yoshitake S, Murakami T, Uji A, Ogino K, Horii T and Hata M *et al* 2014 Association between cystoid spaces on indocyanine green hyperfluorescence and optical coherence tomography after vitrectomy for diabetic macular oedema *Eye* **28** 439–48

[72] Murakami T, Suzuma K, Dodo Y, Yoshitake T, Yasukura S and Nakanishi H *et al* 2018 Decorrelation signal of diabetic hyperreflective foci on optical coherence tomography angiography *Sci. Rep.* **8** 8798

[73] Lee H, Jang H, Choi Y A, Kim H C and Chung H 2018 Association between soluble CD14 in the aqueous humor and hyperreflective foci on optical coherence tomography in patients with diabetic macular edema *Invest. Ophthalmol. Vis. Sci.* **59** 715–21

[74] Fehér J, Taurone S, Spoletini M, Biró Z, Varsányi B and Scuderi G *et al* 2018Dec Ultrastructure of neurovascular changes in human diabetic retinopathy *Int. J. Immunopathol. Pharmacol.* **31** 394632017748841

[75] Vujosevic S, Bini S, Midena G, Berton M, Pilotto E and Midena E 2013 Hyperreflective intraretinal spots in diabetics without and with nonproliferative diabetic retinopathy: an *in vivo* study using spectral domain OCT *J. Diabetes Res.* **2013** 491835

[76] Framme C, Schweizer P, Imesch M, Wolf S and Wolf-Schnurrbusch U 2012 Behavior of SD-OCT-detected hyperreflective foci in the retina of anti-VEGF-treated patients with diabetic macular edema *Invest. Ophthalmol. Vis. Sci.* **53** 5814–8

[77] Niu S, Yu C, Chen Q, Yuan S, Lin J and Fan W *et al* 2017 Multimodality analysis of hyper-reflective foci and hard exudates in patients with diabetic retinopathy *Sci. Rep.* **7** 1568

[78] Kodjikian L, Bellocq D, Bandello F, Loewenstein A, Chakravarthy U and Koh A *et al* 2019 First-line treatment algorithm and guidelines in center-involving diabetic macular edema *Eur. J. Ophthalmol.* **29** 573–84
[79] Busch C, Okada M, Zur D, Fraser-Bell S, Rodríguez-Valdés P J and Cebeci Z *et al* 2020 Baseline predictors for visual acuity loss during observation in diabetic macular oedema with good baseline visual acuity *Acta Ophthalmol.* **98** e801–6
[80] Davoudi S, Papavasileiou E, Roohipoor R, Cho H, Kudrimoti S and Hancock H *et al* 2016 Optical coherence tomography characteristics of macular edema and hard exudates and their association with lipid serum levels in type 2 diabetes *Retina* **36** 1622–9
[81] Chung Y-R, Lee S Y, Kim Y H, Byeon H-E, Kim J H and Lee K 2020 Hyperreflective foci in diabetic macular edema with serous retinal detachment: association with dyslipidemia *Acta Diabetol.* **57** 861–6
[82] De Benedetto U, Sacconi R, Pierro L, Lattanzio R and Bandello F 2015 Optical coherence tomographic hyperreflective foci in early stages of diabetic retinopathy *Retina* **35** 449–53
[83] Frizziero L, Midena G, Longhin E, Berton M, Torresin T and Parrozzani R *et al* 2020 Early retinal changes by OCT angiography and multifocal electroretinography in diabetes *J. Clin. Med.* **9** E3514
[84] Roy R, Saurabh K, Shah D, Chowdhury M and Goel S 2019 Choroidal hyperreflective foci: a novel spectral domain optical coherence tomography biomarker in eyes with diabetic macular edema *Asia Pac. J. Ophthalmol.* **8** 314–8
[85] Saurabh K, Roy R, Herekar S, Mistry S and Choudhari S 2021 Validation of choroidal hyperreflective foci in diabetic macular edema through a retrospective pilot study *Indian J. Ophthalmol.* **69** 3203–6
[86] Arrigo A, Capone L, Lattanzio R, Aragona E, Zollet P and Bandello F 2020 Optical coherence tomography biomarkers of inflammation in diabetic macular edema treated by fluocinolone acetonide intravitreal drug-delivery system implant *Ophthalmol. Ther.* **9** 971–80
[87] Chatziralli I P, Sergentanis T N and Sivaprasad S 2016 Hyperreflective foci as an independent visual outcome predictor in macular edema due to retinal vascular diseases treated with intravitreal dexamethasone or ranibizumab *Retina* **36** 2319–28
[88] Weingessel B, Miháltz K, Gleiss A, Sulzbacher F, Schütze C and Vécsei-Marlovits P V 2018 Treatment of diabetic macular edema with intravitreal antivascular endothelial growth factor and prompt versus deferred focal laser during long-term follow-up and identification of prognostic retinal markers *J. Ophthalmol.* **2018** 3082560
[89] Uji A, Murakami T, Nishijima K, Akagi T, Horii T and Arakawa N *et al* 2012 Association between hyperreflective foci in the outer retina, status of photoreceptor layer, and visual acuity in diabetic macular edema *Am. J. Ophthalmol.* **153** 710–7 717.e1
[90] Nishijima K, Murakami T, Hirashima T, Uji A, Akagi T and Horii T *et al* 2014 Hyperreflective foci in outer retina predictive of photoreceptor damage and poor vision after vitrectomy for diabetic macular edema *Retina* **34** 732–40
[91] Li B, Zhang B, Chen Y and Li D 2020 Optical coherence tomography parameters related to vision impairment in patients with diabetic macular edema: a quantitative correlation analysis *J. Ophthalmol.* **2020** 5639284
[92] Kang J W, Chung H and Kim H C 2016 Correlation of optical coherence tomographic hyperreflective foci with visual outcomes in different patterns of diabetic macular edema *Retina* **36** 1630–9

[93] Kim K T, Kim D Y and Chae J B 2019 Association between hyperreflective foci on spectral-domain optical coherence tomography and early recurrence of diabetic macular edema after intravitreal dexamethasone implantation *J. Ophthalmol.* **2019** 3459164

[94] Park Y G, Choi M Y and Kwon J-W 2019 Factors associated with the duration of action of dexamethasone intravitreal implants in diabetic macular edema patients *Sci. Rep.* **9** 19588

[95] Ozsaygili C and Duru N 2020 Comparison of intravitreal dexamethasone implant and aflibercept in patients with treatment-naive diabetic macular edema with serous retinal detachment *Retina* **40** 1044–52

[96] Murakami T, Suzuma K, Uji A, Yoshitake S, Dodo Y and Fujimoto M *et al* 2018 Association between characteristics of foveal cystoid spaces and short-term responsiveness to ranibizumab for diabetic macular edema *Jpn. J. Ophthalmol.* **62** 292–301

[97] Schreur V, Altay L, van Asten F, Groenewoud J M M, Fauser S and Klevering B J *et al* 2018 Hyperreflective foci on optical coherence tomography associate with treatment outcome for anti-VEGF in patients with diabetic macular edema *PLoS One* **13** e0206482

[98] Chatziralli I, Theodossiadis P, Parikakis E, Dimitriou E, Xirou T and Theodossiadis G *et al* 2017Dec Dexamethasone intravitreal implant in diabetic macular edema: real-life data from a prospective study and predictive factors for visual outcome *Diabetes Ther.* **8** 1393–404

[99] Meduri A, Oliverio G W, Trombetta L, Giordano M, Inferrera L and Trombetta C J 2021 Optical coherence tomography predictors of favorable functional response in naïve diabetic macular edema eyes treated with dexamethasone implants as a first-line agent *J. Ophthalmol.* **2021** 6639418

[100] Liu S, Wang D, Chen F and Zhang X 2019 Hyperreflective foci in OCT image as a biomarker of poor prognosis in diabetic macular edema patients treating with Conbercept in China *BMC Ophthalmol.* **19** 157

[101] Yoshitake T, Murakami T, Suzuma K, Dodo Y, Fujimoto M and Tsujikawa A 2020 Hyperreflective foci in the outer retinal layers as a predictor of the functional efficacy of ranibizumab for diabetic macular edema *Sci. Rep.* **10** 873

[103] Catier A, Tadayoni R, Paques M, Erginay A, Haouchine B, Gaudric A and Massin P 2005 Characterization of macular edema from various etiologies by optical coherence tomography *Am. J. Ophthalmol.* **140** 200–6

[104] Fine B S and Brucker A J 1981 Macular edema and cystoid macular edema *Am. J. Ophthalmol.* **92** 466–481.21

[105] Hee M R, Puliafito C A and Wong C *et al* 1995 Quantitative assessment of macular edema with optical coherence tomography *Arch. Ophthalmol.* **113** 1019–29

[106] Pareja-Ríos A, Serrano-García M A, Marrero-Saavedra M D, Abraldes-López V M, Reyes-Rodríguez M A and Cabrera-López F *et al* 2009 sep Guías de práctica clínica de la SERV: Manejo de las complicaciones oculares de la diabetes. Retinopatía diabética y edema macular *Arch. Soc. Esp. Oftalmol.* **84** 9

[108] Liang M C, Vora R A and Duker J S *et al* 2013 Solid-appearing retinal cysts in diabetic macular edema: a novel optical coherence tomography finding *Retin. Cases Br. Rep.* **7** 255–8

[109] Kashani A H, Green K M and Kwon J *et al* 2018 Suspended scattering particles in motion: a novel feature of OCT angiography in exudative maculopathies *Ophthalmol. Retina* **2** 694–702

[110] Al Faran A, Mousa A and Al Shamsi H *et al* 2014 Spectral domain optical coherence tomography predictors of visual outcome in diabetic cystoid macular edema after bevacizumab injection *Retina* **34** 1208–15

[111] Grauslund J and Blindbaek S L 2017 Diabetic macular oedema: what to fear? How to treat? *Acta Ophthalmol.* **95** 117–8

[112] Scalinci S Z, Scalinci S Z and Scorolli L *et al* 2011 Potential role of intravitreal human placental stem cell implants in inhibiting progression of diabetic retinopathy in type 2 diabetes: neuroprotective growth factors in the vitreous *Clin. Ophthalmol.* **5** 691–6

[113] Schmidt-Erfurth U, Garcia-Arumi J and Bandello F *et al* 2017 Guidelines for the management of diabetic macular edema by the European society of retina specialists (EURETINA) *Ophthalmologica* **237** 185–222

[114] García Layana A, Adán A and Ascaso F J *et al* 2020 Use of intravitreal dexamethasone implants in the treatment of diabetic macular edema: expert recommendations using a Delphi approach *Eur. J. Ophthalmol.* **30** 1042–52

[115] Vujosevic S, Toma C, Villani E, Muraca A, Tort E and Florimbi G *et al* 2020 Diabetic macular edema with neuroretinal detachment: OCT and OCT-angiography biomarkers of treatment response to anti-VEGF and steroids *Acta Diabetol.* **57** 287–96

[116] Otani T, Kishi S and Maruyama Y 1999 Patterns of diabetic macular edema with optical coherence tomography *Am. J. Ophthalmol.* **127** 688–93

[117] Ozdemir H, Karacorlu M and Karacorlu S 2005 Serous macular detachment in diabetic cystoid macular oedema *Acta Ophthalmol. Scand.* **83** 63–6

[118] Koleva-Georgieva D and Sivkova N 2009 Assessment of serous macular detachment in eyes with diabetic macular edema by use of spectral-domain optical coherence tomography *Graefes Arch. Clin. Exp. Ophthalmol.* **247** 1461–9

[119] Maggio E, Mete M, Sartore M, Bauci F, Guerriero M and Polito A *et al* 2022 Temporal variation of optical coherence tomography biomarkers as predictors of anti-VEGF treatment outcomes in diabetic macular edema *Graefes Arch. Clin. Exp. Ophthalmol.* **260** 807–15

[120] Catier A, Tadayoni R, Paques M, Erginay A, Haouchine B and Gaudric A *et al* 2005 Characterization of macular edema from various etiologies by optical coherence tomography *Am. J. Ophthalmol.* **140** 200–6

[121] Daruich A, Matet A, Moulin A, Kowalczuk L, Nicolas M and Sellam A *et al* 2018 Mechanisms of macular edema: beyond the surface *Prog. Retin. Eye Res.* **63** 20–68

[122] Gaucher D, Sebah C, Erginay A, Haouchine B, Tadayoni R and Gaudric A *et al* 2008 Optical coherence tomography features during the evolution of serous retinal detachment in patients with diabetic macular edema *Am. J. Ophthalmol.* **145** 289–96

[123] Weinberger D, Fink-Cohen S, Gaton D D, Priel E and Yassur Y 1995 Non-retinovascular leakage in diabetic maculopathy *Br. J. Ophthalmol.* **79** 728–31

[124] Nagaoka T, Kitaya N, Sugawara R, Yokota H, Mori F and Hikichi T *et al* 2004 Alteration of choroidal circulation in the foveal region in patients with type 2 diabetes *Br. J. Ophthalmol.* **88** 1060–3

[125] Kwon J, Kim B, Jee D and Cho Y K 2021 Aqueous humor analyses of diabetic macular edema patients with subretinal fluid *Sci. Rep.* **11** 20985

[126] Kaya M, Kaya D, Idiman E, Kocak N, Ozturk T and Ayhan Z *et al* 2019 A novel biomarker in diabetic macular edema with serous retinal detachment: serum chitinase-3-like protein 1 *Ophthalmologica* **241** 90–7

[127] Funatsu H, Noma H, Mimura T, Eguchi S and Hori S 2009 Association of vitreous inflammatory factors with diabetic macular edema *Ophthalmology* **116** 73–9

[128] Vujosevic S, Micera A, Bini S, Berton M, Esposito G and Midena E 2016 Proteome analysis of retinal glia cells-related inflammatory cytokines in the aqueous humour of diabetic patients *Acta Ophthalmol.* **94** 56–64

[129] Yenihayat F, Özkan B, Kasap M, Karabaş V L, Güzel N and Akpınar G *et al* 2019 Vitreous IL-8 and VEGF levels in diabetic macular edema with or without subretinal fluid *Int. Ophthalmol.* **39** 821–8

[130] Bandyopadhyay S, Bandyopadhyay S K, Saha M and Sinha A 2017 Study of aqueous cytokines in patients with different patterns of diabetic macular edema based on optical coherence tomography *Int. Ophthalmol.* **38** 241–9

[131] Vujosevic S, Torresin T, Bini S, Convento E, Pilotto E and Parrozzani R *et al* 2017 Imaging retinal inflammatory biomarkers after intravitreal steroid and anti-VEGF treatment in diabetic macular oedema *Acta Ophthalmol.* **95** 464–71

[132] Vujosevic S, Torresin T, Berton M, Bini S, Convento E and Midena E 2017 Diabetic macular edema with and without subfoveal neuroretinal detachment: two different morphologic and functional entities *Am. J. Ophthalmol.* **181** 149–55

[133] Arf S, Sayman Muslubas I, Hocaoglu M, Ersoz M G, Ozdemir H and Karacorlu M 2020 Spectral domain optical coherence tomography classification of diabetic macular edema: a new proposal to clinical practice *Graefes Arch. Clin. Exp.* **258** 1165–72

[134] Tsai M-J, Hsieh Y-T, Shen E P and Peng Y-J 2017 Systemic associations with residual subretinal fluid after ranibizumab in diabetic macular edema *J. Ophthalmol.* **2017** 4834201

[135] Kaya M, Karahan E, Ozturk T, Kocak N and Kaynak S 2018 Effectiveness of intravitreal ranibizumab for diabetic macular edema with serous retinal detachment *Korean J. Ophthalmol. KJO* **32** 296–302

[136] Seo K H, Yu S-Y, Kim M and Kwak H W 2016 Visual and morphologic outcomes of intravitreal ranibizumab for diabetic macular edema based on optical coherence tomography patterns *Retina* **36** 588–95

[137] Sophie R, Lu N and Campochiaro P A 2015 Predictors of functional and anatomic outcomes in patients with diabetic macular edema treated with ranibizumab *Ophthalmology* **122** 1395–401

[138] Giocanti-Aurégan A, Hrarat L, Qu L M, Sarda V, Boubaya M and Levy V *et al* 2017 Functional and anatomical outcomes in patients with serous retinal detachment in diabetic macular edema treated with ranibizumab *Invest. Ophthalmol. Vis. Sci.* **58** 797–800

[139] Ichiyama Y, Sawada O, Mori T, Fujikawa M, Kawamura H and Ohji M 2016 The effectiveness of vitrectomy for diffuse diabetic macular edema may depend on its preoperative optical coherence tomography pattern *Graefes Arch. Clin. Exp. Ophthalmol.* **254** 1545–51

[140] Korobelnik J-F, Lu C, Katz T A, Dhoot D S, Loewenstein A and Arnold J *et al* 2019 Effect of baseline subretinal fluid on treatment outcomes in VIVID-DME and VISTA-DME studies *Ophthalmol. Retina* **3** 663–9

[141] Moon B G, Lee J Y, Yu H G, Song J H, Park Y-H and Kim H W *et al* 2016 Efficacy and safety of a dexamethasone implant in patients with diabetic macular edema at tertiary centers in Korea *J. Ophthalmol.* **2016** 9810270

[142] Fickweiler W, Schauwvlieghe A-S M E, Schlingemann R O, Maria Hooymans J M, Los L I and Verbraak F D *et al* 2018 Predictive value of optical coherence tomographic features in the bevacizumab and ranibizumab in patients with diabetic macular edema (BRDME) study *Retina* **38** 812–9

[143] Brasil O F M, Smith S D, Galor A, Lowder C Y, Sears J E and Kaiser P K 2007 Predictive factors for short-term visual outcome after intravitreal triamcinolone acetonide injection for diabetic macular oedema: an optical coherence tomography study *Br. J. Ophthalmol.* **91** 761–5

[144] Ceravolo I, Oliverio G W, Alibrandi A, Bhatti A, Trombetta L and Rejdak R *et al* 2020 The application of structural retinal biomarkers to evaluate the effect of intravitreal ranibizumab and dexamethasone intravitreal implant on treatment of diabetic macular edema *Diagnostics* **10** 413

[145] Bonfiglio V, Reibaldi M, Pizzo A, Russo A, Macchi I and Faro G *et al* 2019 Dexamethasone for unresponsive diabetic macular oedema: optical coherence tomography biomarkers *Acta Ophthalmol.* **97** e540–4

[146] Meduri A, Oliverio G W, Trombetta L, Giordano M, Inferrera L and Trombetta C J 2021 Optical coherence tomography predictors of favorable functional response in naïve diabetic macular edema eyes treated with dexamethasone implants as a first-line agent *J. Ophthalmol.* **2021** 6639418

[147] Samagaio G, Estévez A, de Moura J, Novo J, Fernández M I and Ortega M 2018 Automatic macular edema identification and characterization using OCT images *Comput. Methods Programs Biomed.* **163** 47–63

[148] de Moura J, Samagaio G, Novo J, Almuina P, Fernández M I and Ortega M 2020 Joint diabetic macular edema segmentation and characterization in OCT images *J. Digit. Imaging* **33** 1335–51

IOP Publishing

Photo Acoustic and Optical Coherence Tomography Imaging, Volume 1

Diabetic retinopathy

Ayman El-Baz and Jasjit S Suri

Chapter 7

Optical coherence tomography biomarkers in diabetic macular edema: OCT biomarkers in diabetic macular edema

Nurettin Bayram

7.1 Introduction

The International Diabetes Federation (IDF) estimated the global population with diabetes mellitus (DM) to be 537 million adults in 2021 and projected it to be 643 million by 2030 and 783 million by 2045 [1]. Diabetic retinopathy (DR) is one of the main microvascular complications of DM, the leading cause of preventable visual impairment in the working-age population [2–4]. The stages of DR severity progress from mild, moderate, severe non-proliferative DR (NPDR), proliferative DR (PDR), to high-risk PDR. Diabetic macular edema (DME) is the most common form of sight-threatening retinopathy characterized by exudative fluid accumulation in the macula. DME can develop at any stage of DR, although the risk increases with the severity of the disease. DME is estimated to affect 6.7% of patients with DM [5], which means more than 30 million cases worldwide require screening and treatment for DME.

The pathogenesis of DME is complex and multifactorial, in which the hyperglycemic state caused by diabetes induces multiple pathologic processes, including angiogenic, inflammatory, hypoxic, and hemodynamic events. These events cause capillary endothelial cell proliferation, thickening of the basement membrane, loss of pericytes, and a deleterious effect on several tight junction proteins (i.e., occludin, claudins, and zonula occludens 1), leading to microaneurysm formation, increase in vessel permeability, the destruction of the blood-retinal barrier, and leakage of the intraretinal fluid in the intraretinal layers of the retina [6]. The key molecular mediators involved in the pathogenesis of DME are vascular endothelial growth factor (VEGF) and pro-inflammatory cytokines locally synthesized by the retina,

doi:10.1088/978-0-7503-2052-8ch7

leading to persistent chronic inflammation, leukocyte activation, leukostasis, and a deterioration of the function of the blood-retinal barrier [6, 7].

Although focal/grid laser photocoagulation has been considered the gold standard for many years, the management of DME has changed considerably in the last decade, especially following the development of intravitreal anti-VEGF therapy. Results of several randomized controlled studies have shown that intravitreal injections of anti-VEGF are the first-line therapy for DME [8, 9]. However, many cases with DME do not adequately respond to anti-VEGF therapy. In these cases, the sustained-release intravitreal corticosteroid options have gained importance in the management of the DME, particularly in cases with pseudophakic, at low risk for glaucoma, or with significant cardiovascular risk [10, 11]. Nevertheless, there are many unanswered questions about intravitreal agents, regarding dose, frequency and regimen of treatment, and potential long-term side effects.

DME is clinically defined by the presence of characteristic clinical findings, including hard exudates with microaneurysms and dot hemorrhages in the macular area, and slit-lamp examination helps identify retinal thickening, the hallmark feature of DME [12]. Optical coherence tomography (OCT) is a fast, non-contact, non-invasive imaging modality that provides cross-sectional images of the optical biopsy of the posterior eye segment, allowing objective qualitative and quantitative evaluation. Parallel with the development of OCT technology, various parameters have been identified to reveal the morphological features commonly seen in DME that could have prognostic relevance, creating the frame for new monitoring and treatment strategies and offering new insights into the pathogenesis of DR, such as vitreomacular traction, intraretinal cysts, subretinal fluid, subfoveal neuroretinal detachment, hyperreflective dots, and disruption of the inner and outer retinal layers [13, 14]. All of these provide important information on retinal layer structure, the morphological features of the macular edema and the morphological changes after treatment to verify the efficacy of treatment. Furthermore, improvements in the visualization of the choroid by enhanced depth imaging OCT (EDI-OCT) and swept-source OCT have allowed the investigation of choroidal features in patients with DME [13, 14]. Thus, OCT has become the most widely used modality in clinical practice for diagnosing and monitoring DME. Various OCT biomarkers have been identified for predicting clinical courses and treatment outcomes in patients with DME over the past decade, which may help predict treatment response and optimize treatment effects for individual patients. Based on Frank and Hargreaves's proposal, OCT biomarkers can be classified into three types [15]. Type 0 helps predict the longitudinal outcomes of the DR, type 1 biomarkers deal with the direct treatment results, and type 2 biomarkers are the most helpful to provide clinical endpoints of parameters such as macular thickness on DR.

This chapter provides a comprehensive summary of the recent literature that has contributed insights into the OCT biomarkers in DME that help determine DME severity, risk of progression, and treatment outcomes. It discusses a framework to assess the validity of biomarkers for treatment outcomes, which is essential in determining the prognosis before deciding on treatment in DME. OCT biomarkers are classified into three categories: the vitreoretinal interface, retina, and choroid.

7.2 OCT biomarkers of the vitreoretinal interface

It has been revealed that advanced glycosylation end-products accumulation in the hyperglycemic state due to DM causes abnormal collagen cross-linking that leads to strong adhesions between the posterior cortical vitreous and inner limiting membrane [16, 17]. OCT can provide valuable insight into vitreoretinal interface abnormalities, including epiretinal membrane (ERM) and incomplete vitreoretinal separation with either vitreomacular attachment or traction. Besides VEGF, anteroposterior and/or tangential tractional forces might contribute to the increased macular thickness in these conditions. Furthermore, intravitreal anti-VEGF injections were shown to cause an angiofibrotic switch [18], which could result in tractional forces on the retina, thereby worsening the condition. A recent study reported that 46% of eyes with DME and simple ERM responded well to anti-VEGF agents functionally and morphologically; however, none of the eyes with tractional ERM responded [19]. Another recent study investigated the influence of vitreoretinal interface abnormalities on the effectiveness of intravitreal anti-VEGF therapy in DME patients. The study showed that eyes with vitreomacular traction and ERM involving the macular center responded less to anti-VEGF therapy than eyes with a normal vitreoretinal interface or eccentric ERM [20]. In summary, tractional ERM and vitreomacular traction are the important predictors of poor functional and morphologic outcomes in center-involved-DME patients, and the surgical option seems to be the best treatment option in cases [21, 22] (figure 7.1). Vitreoretinal interface problems without traction components seem to be not associated with a difference in medical treatment responses [18, 19].

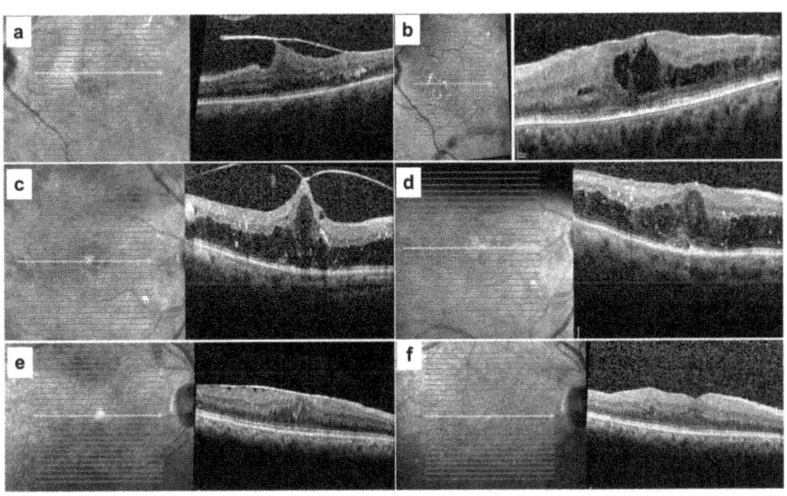

Figure 7.1. Anterior-posterior epiretinal membrane traction (a), vitreomacular traction (c) and tangential epiretinal membrane traction (e) are shown in different three eyes with diabetic macular edema on representative horizontal spectral-domain optical coherence tomography B-scan images. The appearance of these eyes after vitreoretinal surgery is observed on OCT images (b, d, f).

7.3 OCT biomarkers of the retina

DME is characterized by an abnormal intraretinal and subretinal fluid collection in the macular area caused by the alteration of the blood-retinal barrier and is presented with various OCT imaging biomarkers. Central retinal thickness (CRT) and central macular volume (CMV), the size and location of intraretinal cysts (IRC), the occurrence of disorganization of the inner retinal layers (DRIL), the presence of hyperreflective foci (HRF), serous neuroretinal detachment (SND), the state of the ellipsoid zone (EZ) and the external limiting membrane (ELM) have been used to categorize and grade DME. Although some retinal OCT biomarkers and measurements have been shown to carry clinical significance, there has been a lack of consensus and guidelines on which parameters can be used to predict the clinical course or to determine therapeutic response reliably.

7.3.1 Central macular thickness

DME is defined as a retinal thickening (⩾250 μm) within one disk diameter of the macular center or definite hard exudates in this region [23], and its ⩾300 μm value is frequently considered the cut-off to start the treatment [24, 25]. Central macular thickness (CMT) measurement is widely used as an OCT parameter to objectively diagnose and monitor DME. In a recent study, an increase in the level of CMT was shown to indicate an increase in DR severity [26]. OCT-based DME patterns can be classified as sponge-like diffuse retinal thickness, cystoid macular edema (CME), and mixed. Morphological features such as subfoveal neuroretinal detachment (SND), vitreomacular interface abnormalities, and hard exudates can also accompany these situations (figure 7.2). Castro-Navarro *et al* [27] showed that there were no statistical anatomical or functional differences among subtypes of DME refractory to anti-VEGF, treated with a dexamethasone implant at 6 months. The author observed that CME was the group with more recurrences after 6 months.

7.3.2 Central macular volume

Central Macular Volume (CMV) indicates the total cross-sectional volume of the macula calculated from all points of measurement and quantifies the degree of macular edema [26]. Although CMV is not essential to the diagnosis of DME, it provides valuable information on the whole thickness of the macula and is commonly used as an anatomical endpoint referenced for DME treatment regimens. Similar to the CMT, an increase in the level of CMV was observed to indicate an increase in DR severity [26] (figure 7.3).

7.3.3 Intraretinal cystoid spaces

Intraretinal cystoid spaces are minimally reflective, well-defined round or oval cysts within the neurosensory retina caused by microvascular damage and the blood-retinal barrier breakdown in DR. Particularly, cysts larger than 200 μm are signs of late disease and are related to visual impairment compared to smaller ones [28]. Although there is no consensus on the definition of cystoid macular degeneration,

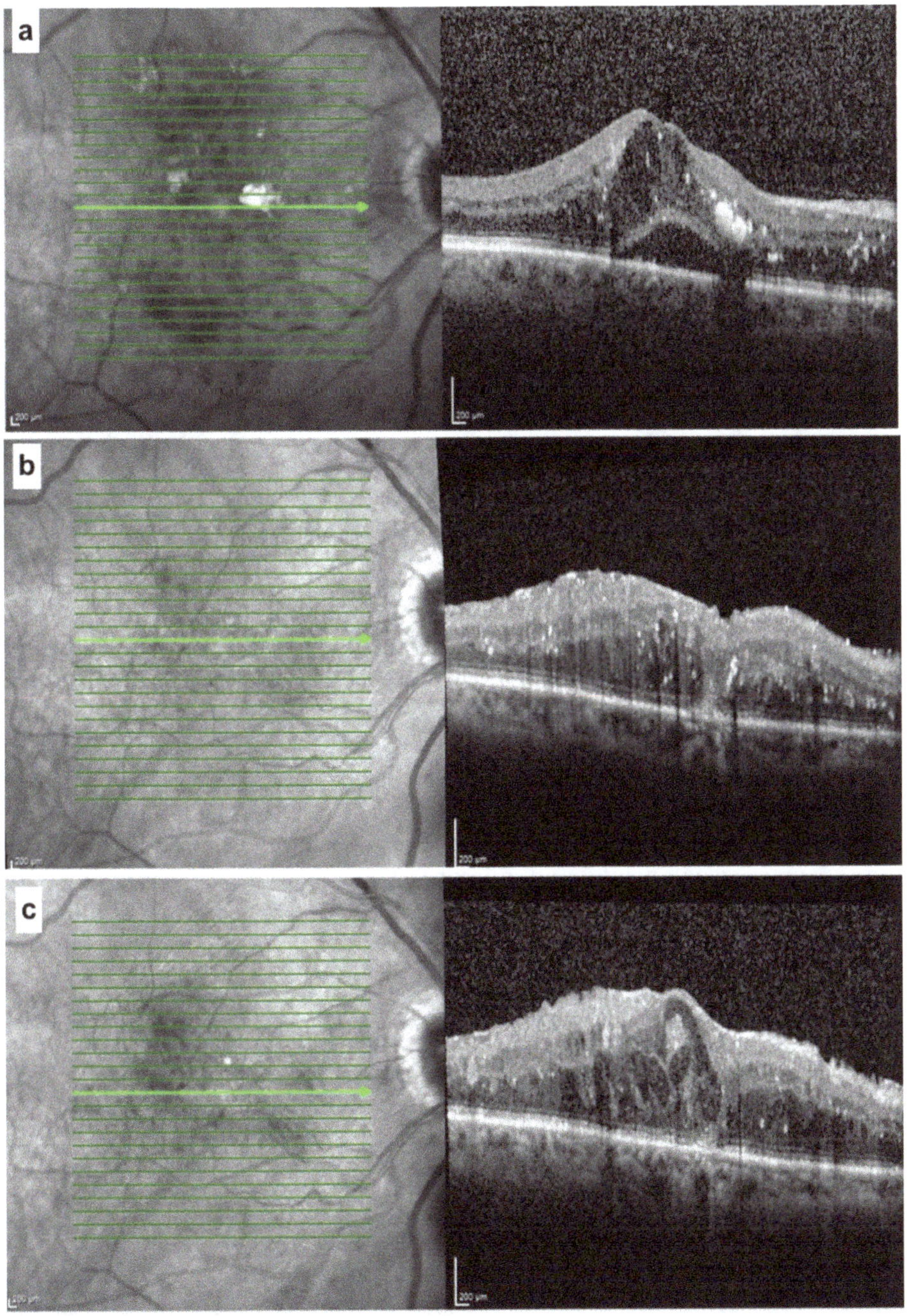

Figure 7.2. Spectral-domain optical coherence tomography demonstrates in a patient with intravitreal treatment-resistant diabetic macular edema (DME). The patient initially developed cystoids macular edema with hyperreflective foci, subfoveal neuroretinal detachment, and hard exudates (a). After eight intravitreal anti-VEGF injections, DME patterns changed to sponge-like diffuse retinal thickness with hyperreflective foci and hard exudates (b). During the follow-up period, DME did not improve and its patterns turned into a mixed type with a small number of hyperreflective foci despite the complete disappearance of subfoveal neuroretinal detachment and hard exudates (c).

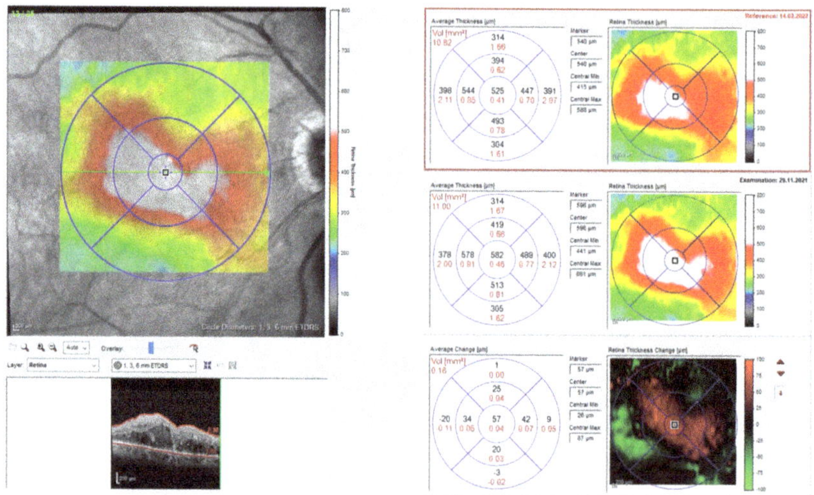

Figure 7.3. Macular thickness and macular volume compared to the values of two and a half months ago in a patient with intravitreal treatment-resistant severe diabetic macular edema are shown on spectral-Domain OCT (Heidelberg Engineering, Heidelberg, Germany) with the Early Treatment Diabetic Retinopathy Study (ETDRS) subfields. The ETDRS plot of macular topography containing variables and statistics for nine sections defines the central macula, inner macula, and outer macula. Please note that macular thickness and macular volume were slightly decreased compared to the values of two and a half months ago after intravitreal anti-VEGF treatment.

the cysts with a horizontal diameter ⩾ 600 μm are generally defined as cystoid macular degeneration (figure 7.4(a)). Outer nuclear layer cysts and their size as morphological parameters have a significant correlation with retinal function measured with visual acuity and microperimetry before and throughout anti-VEGF therapy in patients with DME [29] (figure 7.4(b)). Venkatesh *et al* reported that eyes without intracystic hyperreflective material had better visual outcomes with a lower number of intravitreal anti-VEGF injections than eyes with intracystic hyperreflective material [30] (figure 7.4(c)). A recent study showed that hyper-reflective walls in foveal cystoid spaces were often refractory to intravitreal anti-VEGF treatments and were associated with visual impairment in DME [31] (figure 7.5).

7.3.4 Disorganization of the retinal inner layers

Retinal tissue has a very organized structure and disruption of this tissue, especially in the macular region, can lead to functional impairments. Disorganization of retinal inner layers (DRIL) is the inability to distinguish between the ganglion cell, inner plexiform, and outer nuclear plexiform layers of the retina on OCT. Signals generated by photoreceptors are communicated to the brain through the axons of ganglion cells. Therefore, inner retinal layers are important for the transmission of visual signals. DRIL is significantly associated with macular capillary nonperfusion and is suggested that it causes interruptions of visual data transmission pathways

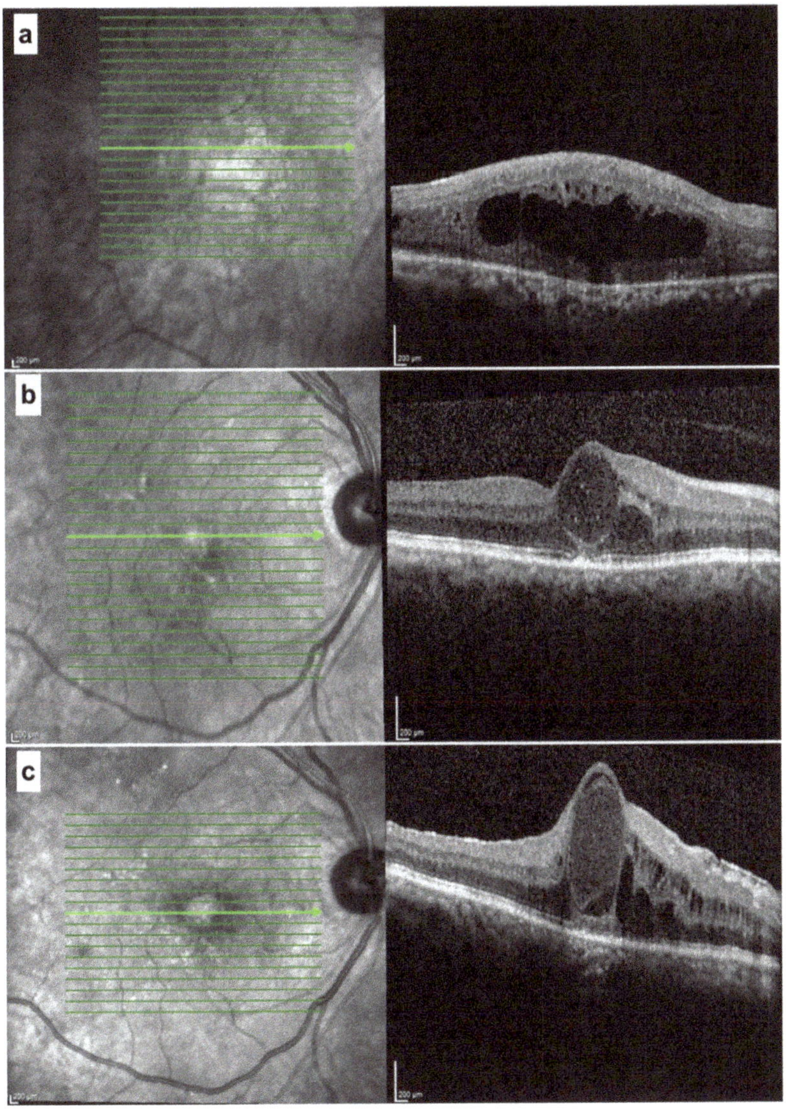

Figure 7.4. Cystoid macular degeneration (a), different sizes of two cysts affecting the outer nuclear layer (b), and intracystic hyperreflective material (c) are demonstrated in three different eyes with diabetic macular edema on representative horizontal spectral-domain optical coherence tomography B-scan images.

[32, 33]. The presence of the 1-mm foveal area DRIL was associated with worse baseline visual acuity in patients with DME, and change in DRIL had predictive value for future changes in visual acuity [34]. Das *et al* [35] observed that there was an association between DRIL and structural disruption in the outer retina, specifically the EZ and the ELM, and increasing severity of DR, specifically proliferative DR. Most recently, Koc *et al* [19] observed that eyes with mild to moderate baseline DRIL usually responded to the anti-VEGF loading dose, while

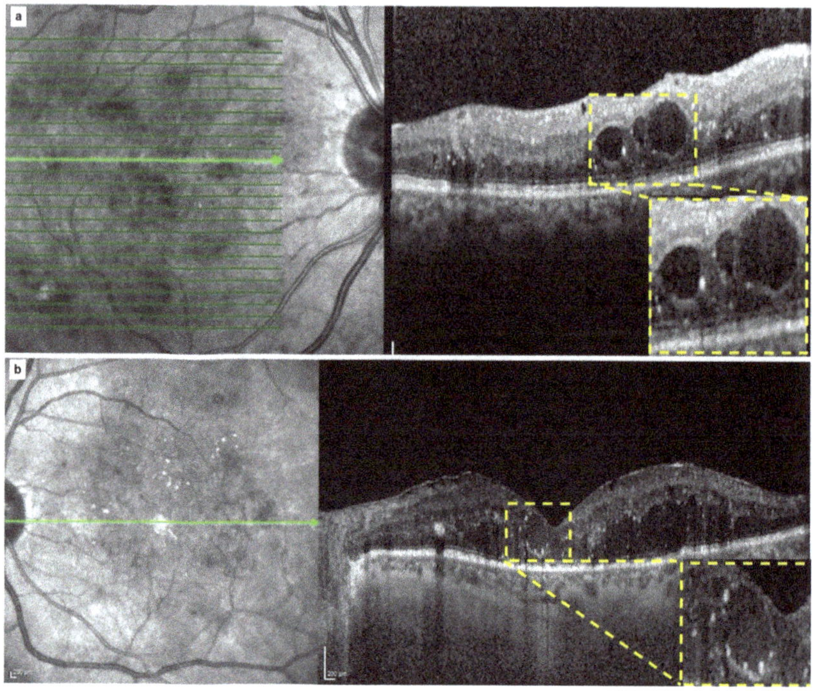

Figure 7.5. Representative horizontal optical coherence tomography B-scan images through the fovea from two eyes with intravitreal treatment-resistant diabetic macular edema are shown the hyperreflective walls in the foveal cystoid spaces (a), (b). A magnified visualization of the hyperreflective walls in the foveal cystoid spaces is illustrated in the lower right corner.

eyes with extensive DRIL tended to resist or respond slowly to the treatment. In summary, the presence of DRIL seems to be a reliable predictive tool for visual acuity and capillary perfusion and is associated with structural disruption in the outer retina, specifically the EZ and the ELM, and increasing severity of DR (figure 7.6).

7.3.5 Hyperreflective foci

HRF have specific characteristics on OCT, such as small size (less than 30 μm) with high reflectivity similar to the nerve fiber layer. They are initially present in the inner retinal layers and subsequently distributed over all retinal layers without back-shadowing. Various hypotheses of HRF have been proposed regarding the exact nature of HRF, such as activated microglia, the precursor of hard exudates or lipoprotein, and migration of retinal pigment epithelium (RPE) [19, 25, 36–38]. A most recent literature review including 36 studies [39] reported no correlation between baseline HRF and anatomical and functional outcomes in six studies, while a significant correlation between these variables was found in 12 studies. Eight studies observed that baseline HRF could predict poor visual outcomes, while HRF had a predictive value for visual improvement in six studies. Fifteen out of 17 studies reported that HRF was reduced after intravitreal anti-VEGF and steroid

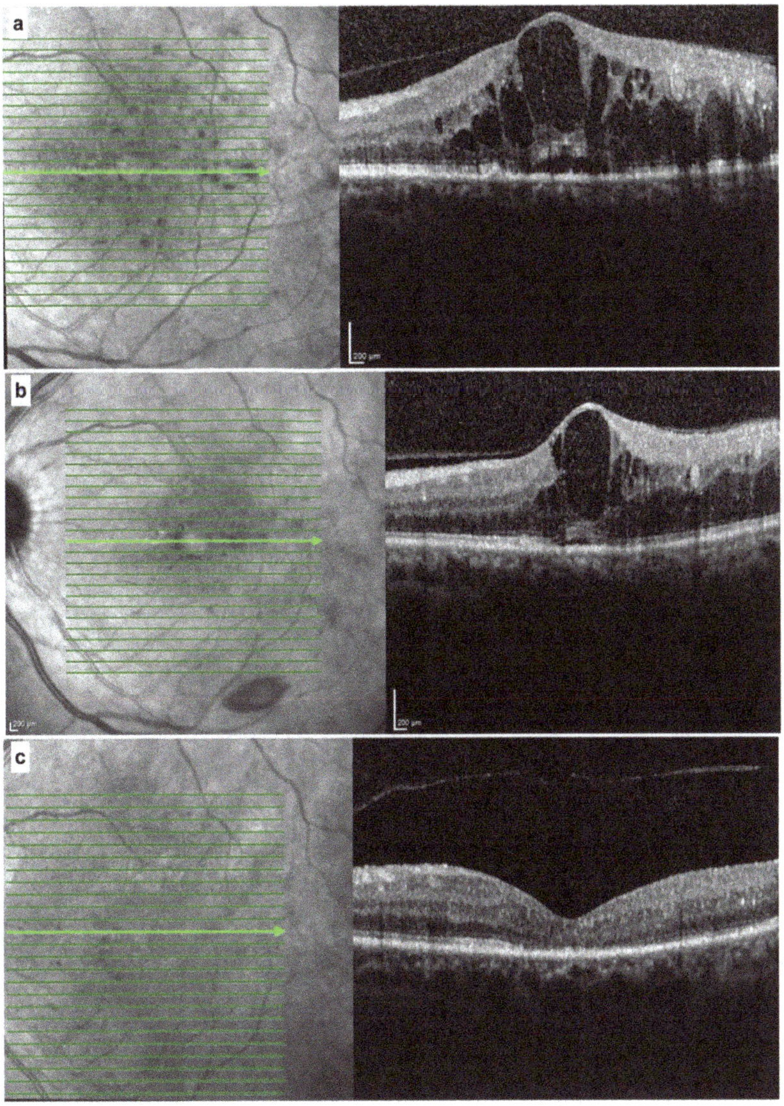

Figure 7.6. Disorganization of retinal inner layers (DRIL) affecting the fovea and temporal macula is observed on spectral-domain optical coherence tomography in a patient who is under intravitreal anti-VEGF therapy due to severe diabetic macular edema (a)–(c). Although the macular edema completely resolved with treatment lasting approximately 4 years, the DRIL persisted and no significant increase was observed in the patient's visual acuity due to foveal ischemia and structural disruption in the outer retina including the ellipsoid zone and the external limiting membrane (c).

treatments, confirming HRF as an inflammatory biomarker. In a meta-analysis [40], eyes with treatment naïve DME showed a positive correlation between HRF and visual improvement and CMT reduction, implying inflammation's significant role in the treatment naïve DME. Based on the current literature, the numbers of HRF on OCT reduce with intravitreal treatment, but whether they predict anatomical and

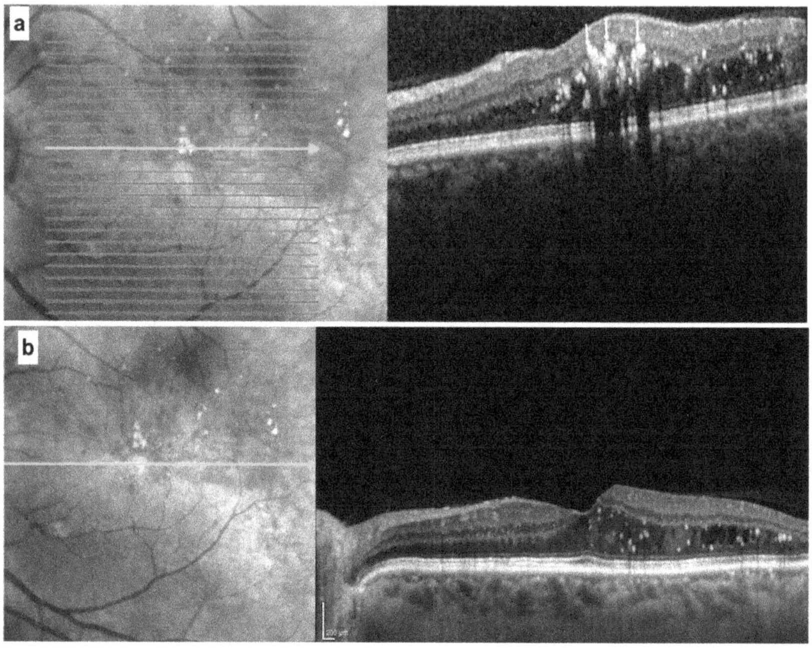

Figure 7.7. Hyperreflective foci and hard exudates (a) are seen on spectral-domain optical coherence tomography in an eye with diabetic macular edema. In the follow-up of the patient, it is observed that both hyperreflective foci and hard exudates decreased with intravitreal anti-VEGF therapy (b). Please note that yellow arrows show some hard exudates and red arrows some hyperreflective foci in image (a).

functional outcomes in DME is controversial. However, it is clear that they themselves can be a biomarker of treatment response in DME (figure 7.7).

7.3.6 Subfoveal neuroretinal detachment

SND appears on OCT as a hyporeflective area beneath the neuroretina with a 15%–30% prevalence in eyes with DME [41] (figures 7.8(a) and (b)). Various hypotheses have been proposed regarding the pathogenesis of SND. The leakage from the retinal or choroidal circulation into the subretinal space exceeding the reabsorption capacity is considered in the pathophysiology [41, 42]. In DR, reduced RPE capacity due to local hypoxia and ELM status seem to be the other important factors in the development of SND [41, 42]. A breakdown of the inner blood-retinal barrier causes extravasation of lipids and proteins in DME eyes, when ELM is compromised, proteins and fluid may move through it into the subretinal space. Sonoda *et al* [43] found a significant relationship between SND and intravitreal IL-6 levels, indicating inflammation may play an important role in the development of SND in DME. Recent studies [44–46] reported that DME with SND may represent a specific ‘more inflammatory’ pattern of DME, with a high number of HRF and a better response to intravitreal steroids rather than to anti-VEGF treatment. There was a significant association between SND and HRF. However, it is not clear whether SND predicts

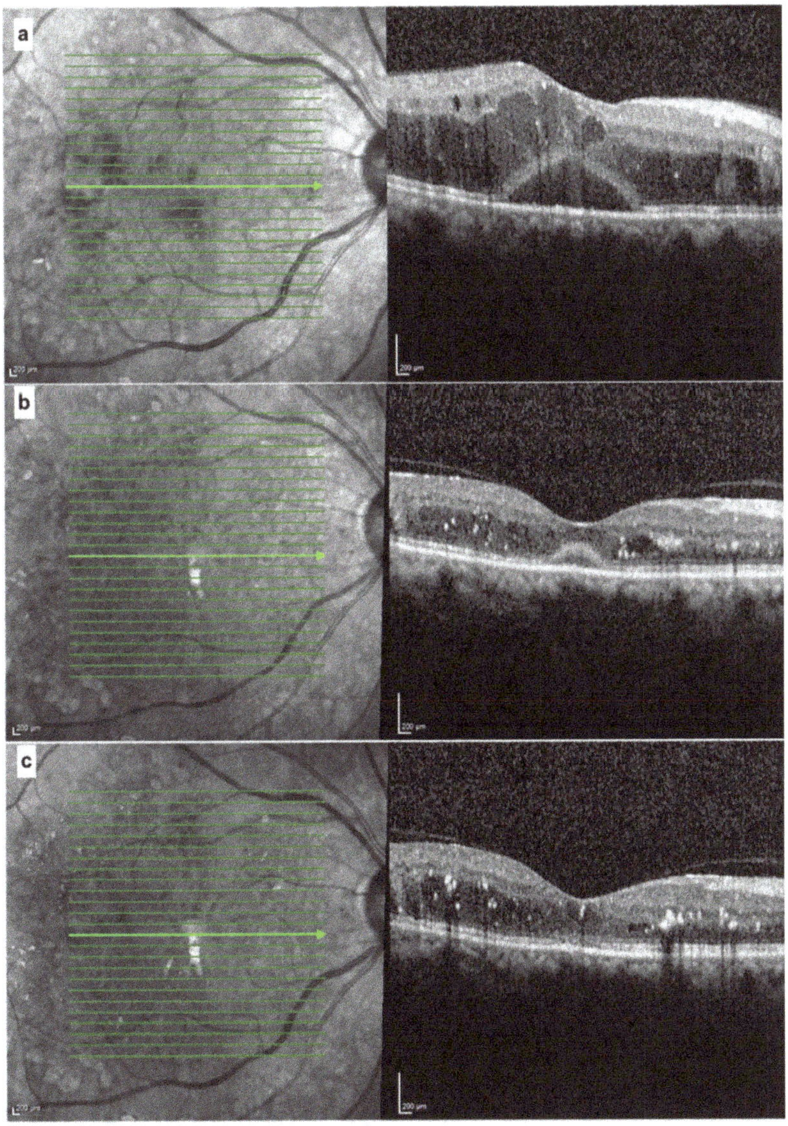

Figure 7.8. Subfoveal neuroretinal detachment with hyperreflective foci appears on spectral-domain optical coherence tomography in an eye with severe diabetic macular edema (a). In the follow-up of the patient, it is observed that subfoveal neuroretinal detachment shrinks (b) and disappears in parallel to the decrease in macular edema (c).

anatomical and functional outcomes in DME. Ichiyama *et al* [47] reported that the presence of SND was associated with better visual gains at the end of one year. In contrast, Seo *et al* [48] have observed a poorer visual outcome after treatment of DME with SDN. A most recent study found no role of SND and HRF as baseline predictors of visual and anatomic outcomes, while a worse visual outcome was significantly correlated with a higher incidence of relapsing SND and HRF [49].

Taken together, SND is an important inflammatory OCT biomarker and has a significant association with HRF (figures 7.8(a)–(c)), but its prognostic role in the anatomic and functional outcomes needs further investigation.

7.3.7 Ellipsoid zone disruption

EZ disruption is defined as having any discontinuity of the second hyperreflective layer of the fovea on OCT and indicates photoreceptor damage (figures 7.9(a) and (b)). Previous studies have shown that EZ disruptions might be reinstated to some extent with treatment; nevertheless, residual EZ interruptions visible on OCT are associated with poorer visual outcomes [48, 50, 51]. Maheshwary *et al* [52] showed a statistically significant correlation between the percentage of EZ disruption and the best-corrected visual acuity. Zur *et al* [53] described three grades of EZ disruption aspects: continuous, partly disrupted, and completely disrupted, and concluded that eyes with intact EZ have better outcomes following treatment with a dexamethasone implant. A recent study showed that eyes with extensive EZ disruption were

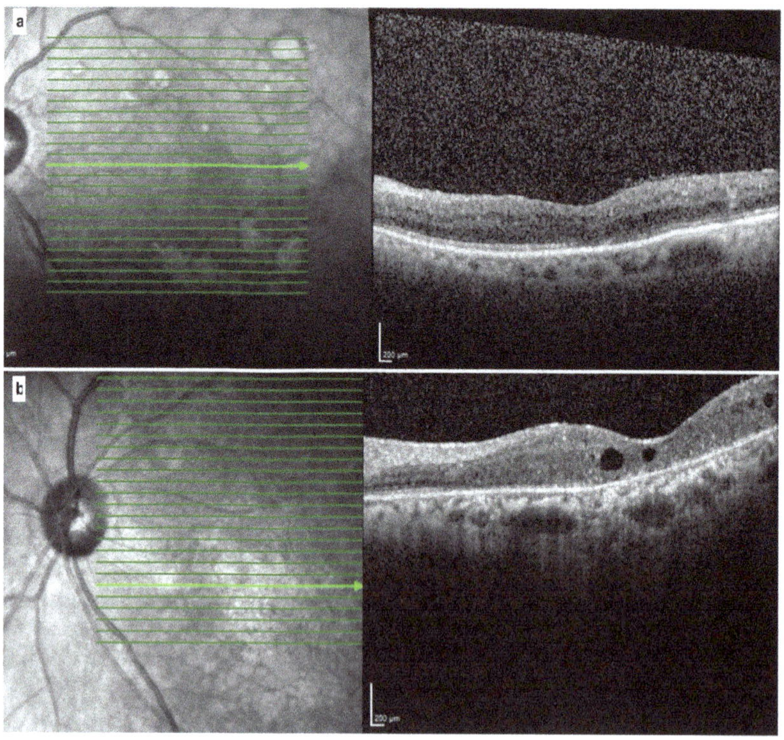

Figure 7.9. Spectral-domain optical coherence tomography images demonstrate the external limiting membrane and ellipsoid zone disruption in two eyes with resolved diabetic macular edema after approximately 2.5 years (a), and 5 years of anti-VEGF therapy (b). It is observed that disruption of the outer retinal layers is accompanied by retinal pigment epithelium atrophy and choroidal hyperreflective foci in the eye received 5 years of anti-VEGF therapy (b). Please note that the eye with retinal pigment epithelium atrophy and choroidal hyperreflective foci had noticeably poorer visual acuity (b) than the other eye (a).

predominantly resistant to anti-VEGF treatment and had a predictive role in the anti-VEGF response [19].

7.3.8 External limiting membrane disruption

ELM is not a true membrane but comprises an intercellular junction between Müller and photoreceptor cells that maintain the integrity of the inner segments of the photoreceptors by serving as a barrier to macromolecules between the subretinal space and the inner retina [54, 55]. Extensive disruption of the ELM in eyes with DME has been postulated to be secondary to mechanical compression by intracellular fluid accumulation [56]. The ELM disruption indicates Müller cells damage as well as photoreceptors. Müller cells manage the homeostatic and metabolic support of retinal neurons, control the composition of the extracellular space fluid, and regulate the tightness of the blood-retinal barrier [55–57]. Moreover, the foveal Müller cell cone structure on OCT images seems to be a predictor of the response to initial anti-VEGF injection in treatment-naive patients with DME [58]. Thus, ELM injury may allow blood components to pass into the outer retinal layers and aggravate the photoreceptor damage. Muftuoglu *et al* [56] demonstrated baseline ELM damage as the most important variable in predicting functional nonresponse to DME with several treatment options. Borrelli *et al* [59] also reported that RPE and ELM disruptions represent OCT biomarkers of long-term visual outcomes in eyes with DME treated with anti-VEGF. A recent study indicated that extensive ELM disruption was a strong predictor of anti-VEGF resistance [19] (figures 7.9(a) and (b)).

7.3.9 Retinal pigment epithelium status

RPE is composed of single-layer cells lining photoreceptor cells, forming the outermost layer of the retina. It is responsible for complex functions such as the maintenance of photoreceptor physiology, light absorption, blood-retinal barrier formation, and phagocytosis of shed outer segments of photoreceptors. Eyes with proliferative DR and DME have decreased thickness of the RPE layer, but RPE thickness could not be identified as a biomarker to correlate significantly with visual acuity [60]. However, RPE atrophy was the parameter that has a significant correlation with visual acuity in the eyes with DME treated with anti-VEGF therapy [60] (figure 7.9(b)).

7.4 OCT biomarkers of the choroid

The choroid is an important vascular structure, supplying oxygen and nutrients to the outer retinal layers, RPE, and the photoreceptors, and has a vital role in the pathogenesis of DR and DME. Technological advancements have enhanced the signal penetration depth and deep structure visualization on OCT over the last decade. The enhanced depth imaging (EDI) mode is a feature that is integrated into most SD-OCT devices, providing the visualization of the choroid by moving the choroid closer to the zero-delay line [61]. Swept-source OCT is a novel development imaging technique that uses a tunable laser source with a higher wavelength of light than conventional SD-OCT [62]. Damage to the choroidal vasculature may cause severe functional damage in the retina, resulting in visual impairment. With these recent advances, the structural changes in the choroid have been analyzed in patients with DME and correlated with ocular factors.

7.4.1 Subfoveal choroidal thickness

Subfoveal choroidal thickness was measured as the distance between Bruch's membrane and the choroid-scleral interface (figures 7.10(a) and (e)). Various studies have compared the subfoveal choroidal thickness between patients with DR and healthy controls [63–65]. However, there is no consensus on the results of these studies. A review article summarizing the recent literature concerning changes in choroidal structure in diabetic patients concluded that it is unclear if choroidal

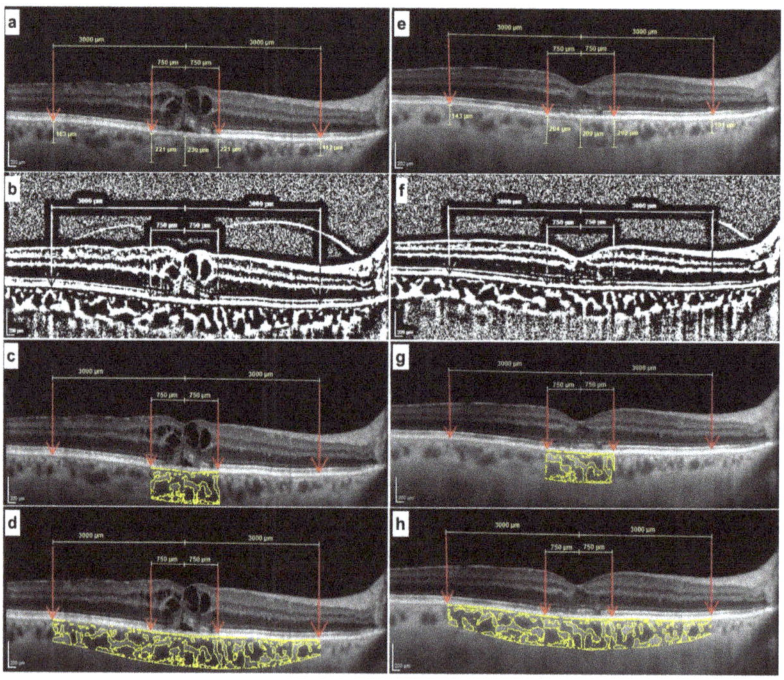

Figure 7.10. Enhanced depth imaging (EDI) optical coherence tomography (OCT) images show a comparison of choroidal thickness and choroidal vascularity index in an eye with diabetic retinopathy treated with intravitreal anti-VEGF, during macular edema (a), (b), (c), (d) and after improvement (e), (f), (g), (h). Measurement of choroidal thickness from the central subfoveal area and nasal and temporal 3000 microns using the spectral-domain OCT software's caliper tool reveals thick choroid during macular edema (a) compared to after improvement (e). The OCT sections are presented after binarization by ImageJ software using the Niblack auto local threshold technique (b), (f). Black areas observed in the choroid refer to the vascular area and white areas refer to the stromal area. A superimposed image of the binarized segment over the EDI-OCT scan demonstrates the optically segmented choroid into two parts, vascular and stromal areas (c), (d), (g), (h). These areas were calculated separately for both the fovea (c), (g) and the macula (d), (h). Please note that the choroidal vascularity index increases in both the fovea and the macula during macular edema compared to after improvement in this patient (for fovea, 66.0% vs. 64.8% and for macula, 65.9% vs. 65.0%).

changes in patients with DR are predictive factors, and there are the inconclusive results from the clinical studies [66]. These inconsistent results may be because the choroidal thickness is a gross indicator and can be influenced by many factors.

7.4.2 Choroidal vascularity index

Based on the EDI-OCT image, the choroidal vascularity index (CVI) presents valuable information on the relative change between the stromal and vascular components and is a more stable parameter than the subfoveal choroidal thickness [67]. It refers to the ratio of the vascular choroidal area to the total choroidal area (figures 7.10(b)–(d) and (f)–(h)). The CVI significantly correlates with the DR severity and decreases significantly in DME, which may be a sensitive marker of choroidal vascular change in DME [68, 69]. Dou *et al* [70] reported that the baseline CVI was significantly lower in eyes with unresponsive to anti-VEGF treatment than in responsive eyes, and a higher CVI at baseline was more likely to have visual improvement after anti-VEGF treatment in DME patients. However, prospective studies with longer follow-up periods are needed to confirm whether the CVI has a productive value in visual response to anti-VEGF therapy.

7.4.3 Choroidal hyperreflective foci

Choroidal hyperreflective foci (CHF) are novel SD-OCT biomarkers in DME. The HRF are suggested to be activated resident microglia that migrate from the inner to the outer retinal layers along with the progression of DR under the influence of inflammatory mediators [36–38]. While migrating from the inner to the outer retina, these HRF have been imaged in the inner choroid as well, where they are termed CHF [71] (figures 7.11(a)–(d)). A cross-sectional study has reported that CHF was associated with DR severity, high CMT, and poor visual acuity [71]. A recent retrospective pilot study showed that the presence of HCF meant significantly worse initial visual acuity compared to the eye that had HRF alone [72]. The final visual acuity was also worse in eyes with HCF compared to those with HRF and without HCF. However, the difference did not reach a significant level, probably due to the small sample size or pointing to the fact that HCF and HRF have the same pathophysiological mechanism. Further prospective studies with a larger sample size are needed to advance these findings.

7.5 Conclusion

OCT has become a precious imaging technique in the DR. It is helpful in the diagnosis and monitoring of DME as well as decision-making regarding the treatment of DME. It also provides information regarding the vitreoretinal interface and therefore helps with the surgical decision. Recent progress in OCT imaging techniques offers the identification of new parameters with the potential to be biomarkers in DME. This chapter summarizes the prognostic value of different OCT parameters in DME. An updated classification system with comprehensive inclusion of the validated OCT-based biomarkers will be essential for optimal therapy in DME and has to be tailor-made to patients' needs. Analysis of newly

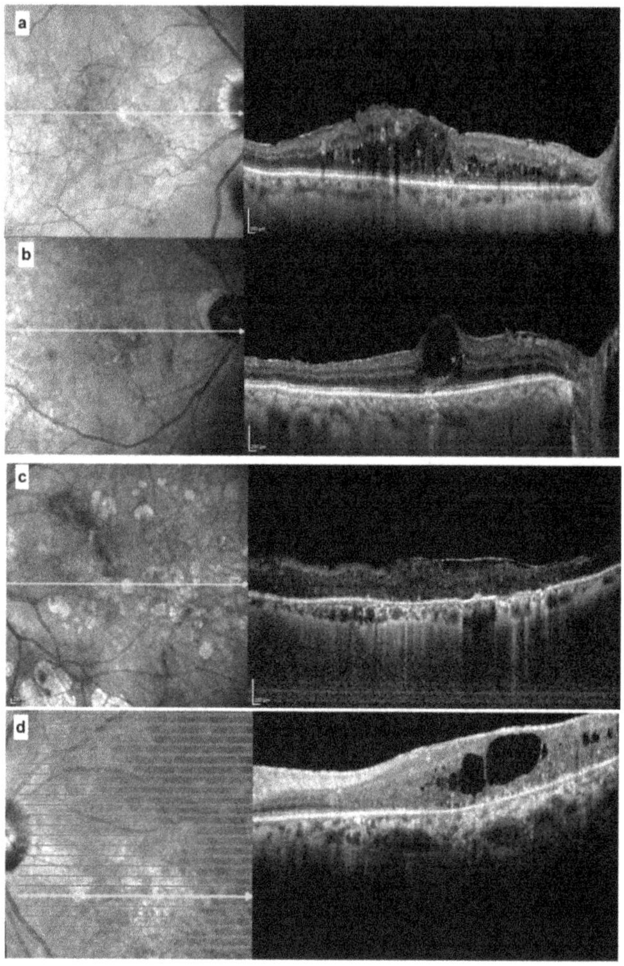

Figure 7.11. Spectral-domain optical coherence tomography images are shown the choroidal hyperreflective foci in four different eyes with intravitreal treatment-resistant severe diabetic macular edema (a)–(d). Please note that numerous choroidal hyperreflective foci are seen within choroidal vessel layers (a)–(d). Epiretinal membrane and the disruption of the external limiting membrane and the ellipsoid zone are seen in all eyes with different subtypes of macular edema.

defined biomarkers and their association with those already known provide new insights into the pathogenesis, early diagnosis, and monitoring of DR and DME and opened new avenues of investigation.

Financial disclosure

I have no financial affiliation with the license/permit holders or products mentioned in this book chapter.

References

[1] International Diabetes Federation 2021 *IDF Diabetes Atlas* 10th edn (Brussels, Belgium) (https://diabetesatlas.org)

[2] Khalil H 2017 Diabetes microvascular complications-a clinical update *Diabetes Metab. Syndr* **11** S133–9

[3] Sivaprasad S, Gupta B, Crosby-Nwaobi R and Evans J 2012 Prevalence of diabetic retinopathy in various ethnic groups: a worldwide perspective *Surv. Ophthalmol.* **57** 347–70

[4] Teo Z L *et al* 2021 Global prevalence of diabetic retinopathy and projection of burden through 2045: systematic review and meta-analysis *Ophthalmology* **128** 1580–91

[5] Yau J W *et al* 2012 Global prevalence and major risk factors of diabetic retinopathy *Diabetes Care* **35** 556–64

[6] Klaassen I, Van Noorden C J and Schlingemann R O 2013 Molecular basis of the inner blood-retinal barrier and its breakdown in diabetic macular edema and other pathological conditions *Prog. Retin. Eye Res.* **34** 19–48

[7] Tang J and Kern T S 2011 Inflammation in diabetic retinopathy *Prog. Retin. Eye Res.* **30** 343–58

[8] Wells J A *et al* 2016 Aflibercept, Bevacizumab, or Ranibizumab for diabetic macular edema: two-year results from a comparative effectiveness randomized clinical trial *Ophthalmology* **123** 1351–9

[9] Glassman A R, Wells J A, Josic K, Maguire M G, Antoszyk A N, Baker C, Beaulieu W T, Elman M J, Jampol L M and Sun J K 2020 Five-year outcomes after initial Aflibercept, Bevacizumab, or Ranibizumab treatment for diabetic macular edema (protocol t extension study) *Ophthalmology* **127** 1201–10

[10] Furino C, Boscia F, Reibaldi M and Alessio G 2021 Intravitreal therapy for diabetic macular edema: an update *J. Ophthalmol.* **2021** 6654168

[11] Lattanzio R, Cicinelli M V and Bandello F 2017 Intravitreal steroids in diabetic macular edema *Dev. Ophthalmol.* **60** 78–90

[12] Tan G S, Cheung N, Simó R, Cheung G C and Wong T Y 2017 Diabetic macular oedema *Lancet Diabetes Endocrinol.* **5** 143–55

[13] Kim B Y, Smith S D and Kaiser P K 2006 Optical coherence tomographic patterns of diabetic macular edema *Am. J. Ophthalmol.* **142** 405–12

[14] Virgili G, Menchini F, Murro V, Peluso E, Rosa F and Casazza G 2011 Optical coherence tomography (OCT) for detection of macular oedema in patients with diabetic retinopathy *Cochrane Database Syst. Rev.* CD008081

[15] Frank R and Hargreaves R 2003 Clinical biomarkers in drug discovery and development *Nat. Rev. Drug Discov.* **2** 566–80

[16] Lundquist O and Osterlin S 1994 Glucose concentration in the vitreous of nondiabetic and diabetic human eyes *Graefes Arch. Clin. Exp. Ophthalmol.* **232** 71–4

[17] Sebag J, Buckingham B, Charles M A and Reiser K 1992 Biochemical abnormalities in vitreous of humans with proliferative diabetic retinopathy *Arch. Ophthalmol.* **110** 1472–6

[18] Van Geest R J, Lesnik-Oberstein S Y, Tan H S, Mura M, Goldschmeding R, Van Noorden C J, Klaassen I and Schlingemann R O 2012 A shift in the balance of vascular endothelial growth factor and connective tissue growth factor by Bevacizumab causes the angiofibrotic switch in proliferative diabetic retinopathy *Br. J. Ophthalmol.* **96** 587–90

[19] Koc F, Güven Y Z, Egrilmez D and Aydın E 2021 Optical coherence tomography biomarkers in bilateral diabetic macular edema patients with asymmetric anti-VEGF response *Semin. Ophthalmol.* **36** 444–51

[20] Kulikov A N, Sosnovskii S V, Berezin R D, Maltsev D S, Oskanov D H and Gribanov N A 2017 Vitreoretinal interface abnormalities in diabetic macular edema and effectiveness of anti-VEGF therapy: an optical coherence tomography study *Clin. Ophthalmol.* **11** 1995–2002

[21] Diabetic Retinopathy Clinical Research Network Writing Committee *et al* 2010 Vitrectomy outcomes in eyes with diabetic macular edema and vitreomacular traction *Ophthalmology* **117** 1087–93.e3

[22] Shah S P, Patel M, Thomas D, Aldington S and Laidlaw D A 2006 Factors predicting outcome of vitrectomy for diabetic macular oedema: results of a prospective study *Br. J. Opthalmol.* **90** 33–6

[23] Early Treatment Diabetic Retinopathy Study Research Group 1987 Treatment techniques and clinical guidelines for photocoagulation of diabetic macular edema. Early Treatment Diabetic Retinopathy Study Report Number 2. *Ophthalmology* **94** 761–74

[24] Massin P *et al* 2010 Safety and efficacy of ranibizumab in diabetic macular edema (RESOLVE Study): a 12-month, randomized, controlled, double-masked, multicenter phase II study *Diabetes Care* **33** 2399–405

[25] Chou H D *et al* 2022 Optical coherence tomography and imaging biomarkers as outcome predictors in diabetic macular edema treated with dexamethasone implant *Sci. Rep.* **12** 3872

[26] Saxena S, Caprnda M, Ruia S, Prasad S, Ankita, Fedotova J, Kruzliak P and Krasnik V 2019 Spectral domain optical coherence tomography based imaging biomarkers for diabetic retinopathy *Endocrine* **66** 509–16

[27] Castro-Navarro V, Cervera-Taulet E, Navarro-Palop C, Hernández-Bel L, Monferrer-Adsuara C, Mata-Moret L and Montero-Hernández J 2020 Analysis of anatomical biomarkers in subtypes of diabetic macular edema refractory to anti-vascular endothelial growth factor treated with dexamethasone implant *Eur. J. Ophthalmol.* **30** 764–9

[28] Deák G G, Bolz M, Ritter M, Prager S, Benesch T and Schmidt-Erfurth U 2010 Diabetic Retinopathy Research Group Vienna A systematic correlation between morphology and functional alterations in diabetic macular edema *Invest. Ophthalmol. Vis. Sci.* **51** 6710–4

[29] Reznicek L, Cserhati S, Seidensticker F, Liegl R, Kampik A, Ulbig M, Neubauer A S and Kernt M 2013 Functional and morphological changes in diabetic macular edema over the course of anti-vascular endothelial growth factor treatment *Acta Ophthalmol.* **91** e529–36

[30] Venkatesh R, Sangai S, Reddy N G, Sridharan A, Pereira A, Aseem A, Gadde S G K, Yadav N K and Chhablani J 2021 Intracystic hyperreflective material in centre-involving diabetic macular oedema *Graefes Arch. Clin. Exp. Ophthalmol.* **259** 2533–44

[31] Terada N, Murakami T, Uji A, Dodo Y, Mori Y and Tsujikawa A 2020 Hyperreflective walls in foveal cystoid spaces as a biomarker of diabetic macular edema refractory to anti-VEGF treatment *Sci. Rep.* **10** 7299

[32] Dodo Y, Murakami T, Uji A, Yoshitake S and Yoshimura N 2015 Disorganized retinal lamellar structures in nonperfused areas of diabetic retinopathy *Invest. Ophthalmol. Vis. Sci.* **56** 2012–20

[33] Nicholson L, Ramu J, Triantafyllopoulou I, Patrao N V, Comyn O, Hykin P and Sivaprasad S 2015 Diagnostic accuracy of disorganization of the retinal inner layers in detecting macular capillary non-perfusion in diabetic retinopathy *Clin. Exp. Ophthalmol.* **43** 735–41

[34] Sun J K, Lin M M, Lammer J, Prager S, Sarangi R, Silva P S and Aiello L P 2014 Disorganization of the retinal inner layers as a predictor of visual acuity in eyes with center-involved diabetic macular edema *JAMA Ophthalmol.* **132** 1309–16

[35] Das R, Spence G, Hogg R E, Stevenson M and Chakravarthy U 2018 Disorganization of inner retina and outer retinal morphology in diabetic macular edema *JAMA Ophthalmol.* **136** 202–8

[36] Lee H, Jang H, Choi Y A, Kim H C and Chung H 2018 Association between soluble CD14 in the aqueous humor and hyperreflective foci on optical coherence tomography in patients with diabetic macular edema *Invest. Ophthalmol. Vis. Sci.* **59** 715–21

[37] Ashraf M, Souka A and Adelman R 2016 Predicting outcomes to anti-vascular endothelial growth factor (VEGF) therapy in diabetic macular oedema: a review of the literature *Br. J. Ophthalmol.* **100** 1596–604

[38] Vujosevic S *et al* 2017 Hyperreflective retinal spots in normal and diabetic eyes: B-scan and en face spectral domain optical coherence tomography evaluation *Retina* **37** 1092–103

[39] Huang H, Jansonius N M, Chen H and Los L I 2022 Hyperreflective dots on OCT as a predictor of treatment outcome in diabetic macular edema: a systematic review *Ophthalmol. Retina* **6** 814–27

[40] Ganne P, Krishnappa N C, Karthikeyan S K and Raman R 2021 Behavior of hyperreflective spots noted on optical coherence tomography following intravitreal therapy in diabetic macular edema: a systematic review and meta-analysis *Indian J. Ophthalmol.* **69** 3208–17

[41] Suciu C I, Suciu V I and Nicoara S D 2020 Optical coherence tomography (angiography) biomarkers in the assessment and monitoring of diabetic macular edema *J. Diabetes Res.* **2020** 6655021

[42] Vujosevic S, Torresin T, Berton M, Bini S, Convento E and Midena E 2017 Diabetic macular edema with and without subfoveal neuroretinal detachment: two different morphologic and functional entities *Am. J. Ophthalmol.* **181** 149–55

[43] Sonoda S, Sakamoto T, Yamashita T, Shirasawa M, Otsuka H and Sonoda Y 2014 Retinal morphologic changes and concentrations of cytokines in eyes with diabetic macular edema *Retina* **34** 741–8

[44] Vujosevic S, Toma C, Villani E, Muraca A, Torti E, Florimbi G, Leporati F, Brambilla M, Nucci P and De Cilla' S 2020 Diabetic macular edema with neuroretinal detachment: OCT and OCT-angiography biomarkers of treatment response to anti-VEGF and steroids *Acta Diabetol.* **57** 287–96

[45] Bonfiglio V, Reibaldi M, Pizzo A, Russo A, Macchi I, Faro G, Avitabile T and Longo A 2019 Dexamethasone for unresponsive diabetic macular oedema: optical coherence tomography biomarkers *Acta Ophthalmol.* **97** e540–e4

[46] Muftuoglu I K, Tokuc E O, Sümer F and Karabas V L 2021 Evaluation of retinal inflammatory biomarkers after intravitreal steroid implant and Ranibizumab injection in diabetic macular edema *Eur. J. Ophthalmol.* **32** 1627–35

[47] Ichiyama Y, Sawada O, Mori T, Fujikawa M, Kawamura H and Ohji M 2016 The effectiveness of vitrectomy for diffuse diabetic macular edema may depend on its preoperative optical coherence tomography pattern *Graefes Arch. Clin. Exp. Ophthalmol.* **254** 1545–51

[48] Seo K H, Yu S Y, Kim M and Kwak H W 2016 Visual and morphologic outcomes of intravitreal ranibizumab for diabetic macular edema based on optical coherence tomography patterns *Retina* **36** 588–95

[49] Maggio E, Mete M, Sartore M, Bauci F, Guerriero M, Polito A and Pertile G 2022 Temporal variation of optical coherence tomography biomarkers as predictors of anti-VEGF treatment outcomes in diabetic macular edema *Graefes Arch. Clin. Exp. Ophthalmol.* **260** 807–15

[50] Guyon B, Elphege E, Flores M, Gauthier A S, Delbosc B and Saleh M 2017 Retinal reflectivity measurement for cone impairment estimation and visual assessment after diabetic macular edema resolution (RECOVER-DME) *Invest. Ophthalmol. Vis. Sci.* **58** 6241–7

[51] Shin H J, Lee S H, Chung H and Kim H C 2012 Association between photoreceptor integrity and visual outcome in diabetic macular edema *Graefes Arch. Clin. Exp. Ophthalmol.* **250** 61–70

[52] Maheshwary A S, Oster S F, Yuson R M, Cheng L, Mojana F and Freeman W R 2010 The association between percent disruption of the photoreceptor inner segment-outer segment junction and visual acuity in diabetic macular edema *Am. J. Ophthalmol.* **150** 63–7

[53] Zur D, Iglicki M, Busch C, Invernizzi A, Mariussi M, Loewenstein A and International Retina Group 2018 OCT biomarkers as functional outcome predictors in diabetic macular edema treated with dexamethasone implant *Ophthalmology* **125** 267–75

[54] Bunt-Milam A H, Saari J C, Klock I B and Garwin G G 1985 Zonulae adherentes pore size in the external limiting membrane of the rabbit retina *Invest. Ophthalmol. Vis. Sci.* **26** 1377–80

[55] Williams D S, Arikawa K and Paallysaho T 1990 Cytoskeletal components of the adherens junctions between the photoreceptors and the supportive Müller cells *J. Comp. Neurol.* **295** 155–64

[56] Muftuoglu I K, Mendoza N, Gaber R, Alam M, You Q and Freeman W R 2017 Integrity of outer retinal layers after resolution of central involved diabetic macular edema *Retina* **37** 2015–24

[57] Coughlin B A, Feenstra D J and Mohr S 2017 Müller cells and diabetic retinopathy *Vision Res.* **139** 93–100

[58] Choi M, Yun C, Oh J H and Kim S W 2022 Foveal müller cell cone as a prognostic optical coherence tomography biomarker for initial response to antivascular endothelial growth factor treatment in cystoid diabetic macular edema *Retina* **42** 129–37

[59] Borrelli E, Grosso D, Barresi C, Lari G, Sacconi R, Senni C, Querques L, Bandello F and Querques G 2022 Long-term visual outcomes and morphologic biomarkers of vision loss in eyes with diabetic macular edema treated with anti-VEGF therapy *Am. J. Ophthalmol.* **235** 80–9

[60] Damian I and Nicoara S D 2020 Optical coherence tomography biomarkers of the outer blood-retina barrier in patients with diabetic macular oedema *J. Diabetes Res.* **2020** 8880586

[61] Spaide R F, Koizumi H and Pozzoni M C 2008 Enhanced depth imaging spectral-domain optical coherence tomography *Am. J. Ophthalmol.* **146** 496–500

[62] Zafar S, Siddiqui M R and Shahzad R 2016 Comparison of choroidal thickness measurements between spectral-domain OCT and swept-source OCT in normal and diseased eyes *Clin. Ophthalmol.* **10** 2271–6

[63] Querques G, Lattanzio R, Querques L, Del Turco C, Forte R, Pierro L, Souied E H and Bandello F 2012 Enhanced depth imaging optical coherence tomography in type 2 diabetes *Invest. Ophthalmol. Vis. Sci.* **53** 6017–24

[64] Regatieri C V, Branchini L, Carmody J, Fujimoto J G and Duker J S 2012 Choroidal thickness in patients with diabetic retinopathy analyzed by spectral-domain optical coherence tomography *Retina* **32** 563–8

[65] Kim J T, Lee D H, Joe S G, Kim J G and Yoon Y H 2013 Changes in choroidal thickness in relation to the severity of retinopathy and macular edema in type 2 diabetic patients *Invest. Ophthalmol. Vis. Sci.* **54** 3378–84

[66] Melancia D, Vicente A, Cunha J P, Abegão Pinto L and Ferreira J 2016 Diabetic choroidopathy: a review of the current literature *Graefes Arch. Clin. Exp. Ophthalmol.* **254** 1453–61

[67] Iovino C *et al* 2020 Choroidal vascularity index: an in-depth analysis of this novel optical coherence tomography parameter *J. Clin. Med.* **9** 595

[68] Gupta C, Tan R, Mishra C, Khandelwal N, Raman R, Kim R, Agrawal R and Sen P 2018 Choroidal structural analysis in eyes with diabetic retinopathy and diabetic macular edema-a novel OCT based imaging biomarker *PLoS One* **13** e0207435

[69] Kim M, Ha M J, Choi S Y and Park Y H 2018 Choroidal vascularity index in type-2 diabetes analyzed by swept-source optical coherence tomography *Sci. Rep.* **8** 70

[70] Dou N, Yu S, Tsui C K, Yang B, Lin J, Lu X, Xu Y, Wu B, Zhao J and Liang X 2021 Choroidal vascularity index as a biomarker for visual response to antivascular endothelial growth factor treatment in diabetic macular edema *J. Diabetes Res.* **2021** 3033219

[71] Roy R, Saurabh K, Shah D, Chowdhury M and Goel S 2019 Choroidal hyperreflective foci: a novel spectral domain optical coherence tomography biomarker in eyes with diabetic macular edema *Asia Pac. J. Ophthalmol.* **8** 314–8

[72] Saurabh K, Roy R, Herekar S, Mistry S and Choudhari S 2021 Validation of choroidal hyperreflective foci in diabetic macular edema through a retrospective pilot study *Indian J. Ophthalmol.* **69** 3203–6

IOP Publishing

Photo Acoustic and Optical Coherence Tomography Imaging, Volume 1
Diabetic retinopathy
Ayman El-Baz and Jasjit S Suri

Chapter 8

Optical coherence tomography and OCTA for the diagnosis of diabetic macular edema

Eugene Hsu, Nayan Sanjiv, Pawarissara Osathanugrah, Josh Agranat and Manju Subramanian

8.1 Introduction to DME

Background and epidemiology: Diabetes mellitus (DM) and its vascular complications, including retinal damage known as diabetic retinopathy (DR), have been studied for the past century. The CDC predicts up to 33% of the United States (US) population will have DM by the year 2050 and a portion of these patients will suffer sight-threatening complications such as retinopathy and macular edema [1]. Over the last decade, technological advances in retinal imaging and novel therapeutics have vastly improved the evaluation, management, and outcomes of patients with DR and diabetic macular edema (DME) [2]. For most diabetic patients, retinopathy develops within 10–15 years after initial diagnosis [2]. DR is the leading cause of vision loss in the US in patients between 25 and 74 years of age affecting approximately 4.2 million adults, with approximately 655 000 having vision-threatening DR [3]. Progression of DR can lead to vision loss, often secondary to macular edema, hemorrhage from neovascularization, retinal detachment, or neovascular glaucoma.

Diabetic macular edema, a complication of DR, is one of the leading causes of visual impairment in the US and is the most common cause of vision loss in patients with DR [1, 4]. In the US, the overall prevalence of DME is 3.8% and affects approximately 750 000 individuals [5]. The prevalence of DME in patients with type 1 DM (T1DM) and type 2 DM (T2DM) is 14% and 6%, respectively [6]. Prolonged diabetes, elevated hemoglobin A1c concentrations, and longstanding hyperglycemia are associated with a greater risk for development of DR and DME [5].

doi:10.1088/978-0-7503-2052-8ch8

Definitions: Diabetic retinopathy encompasses a spectrum of diseases. The classification of DR is categorized by assessment for abnormal retinal vascularization and its severity [7]. Although the stratification of the disease has been useful for studying the effects of treatment and early diagnosis, each patient with DR exhibits a unique presentation of symptoms, imaging, clinical findings, and rate of progression, necessitating a personalized approach to their management [8].

Nonproliferative diabetic retinopathy (NPDR) is defined by abnormalities in the retinal vasculature, ranging from mild cases that only involve microaneurysms to moderate and severe cases that involve additional vascular abnormalities described in table 8.1 [8] The pathophysiology of retinal neovascularization stems from a state of chronic hyperglycemia. Hyperglycemia damages retinal capillaries, weakening

Table 8.1. Diagnostic criteria for nonproliferative diabetic retinopathy, proliferative diabetic retinopathy, clinically significant macular edema, and diabetic macular edema.

Nonproliferative diabetic retinopathy		
	Mild	– Microaneurysms with no other retinal findings
	Moderate	– Increased number of microaneurysms and dot-blot hemorrhages – May have presence of cotton wool spots, venous beading, and hard exudates – Does not meet severe NPDR criteria
	Severe	– Microaneurysms and severe intraretinal hemorrhages **and** – microaneurysms in all 4 retinal quadrants **or** – definite venous beading in ⩾2 quadrants **or** – moderate intraretinal microvascular abnormalities in ⩾1 quadrant
	Very Severe	– If ⩾2 of severe NPDR criteria and does not meet PDR criteria
Proliferative diabetic retinopathy		
	Early	– Neovascularization **without** meeting high-risk PDR criteria
	High-risk	– Neovascularization of the disk ⩾1/3 to ½ disk area **or** – Neovascularization of the disk and vitreous hemorrhage or the presence of preretinal hemorrhage **or** – Neovascularization in other parts of the retina that is >1/2 disk area **with** the presence of vitreous or preretinal hemorrhage

	Severe	– -Preretinal or vitreous hemorrhage obscuring the posterior fundus – -Detachment of the center of the macula
Clinically significant macular edema		– Retinal thickening ⩽500 microns from the center of the macula **or** – Presence of hard exudates and retinal thickening **or** – Presence of retinal thickening that is ⩾ 1 disc area in size that is located ⩽ 1 disc diameter from the foveal center – Can occur at any stage of NPDR and PDR
Center-involved DME		– Retinal thickening in the macula **involving** the central subfield zone (1mm in diameter)
Non-center involved DME		– Presence of retinal thickening in the macula **NOT** involving the central subfield zone (1mm in diameter)

the vessel walls which can lead to the formation of microaneurysms. When the microaneurysms rupture, they can cause hemorrhages in the deeper layers of the retina (i.e., inner nuclear and outer plexiform layers) confined superficially by the internal limiting membrane.

- Mild NPDR—microaneurysms with no other retinal findings.
- Moderate NPDR—presence of microaneurysms and other vascular abnormalities like dot-blot hemorrhages, cotton wool spots, venous beading, and hard exudates, but does not meet the criteria for severe NPDR classification.
- Severe NPDR—microaneurysms and severe intraretinal hemorrhages with microaneurysms in all four retinal quadrants, definite venous beading in more than two quadrants, or moderate intraretinal microvascular abnormalities (IRMA) in at least one quadrant. Severe NPDR does not meet the additional criteria for proliferative diabetic retinopathy. Within one year, 52%–75% of patients in the severe NPDR category will progress to PDR [8].

Proliferative diabetic retinopathy (PDR) is defined by the presence of retinal neovascularization. In PDR, the pathophysiology of neovascularization is a response to retinal ischemia. Angiogenic factors, one of which being vascular endothelial growth factor (VEGF), are released to stimulate the formation of new retinal vessels to compensate for the hyperglycemia-induced vessel damage. An important characteristic of PDR is that neovascularization extends beyond the internal limiting membrane and leads to growth into the vitreous. Fibrovascular contractile membranes can form in the vitreous and cause traction on the retina which can progress into tractional retinal detachment. Severity of PDR is further categorized by degree of neovascularization of the disc, presence of vitreous or preretinal hemorrhage, and detachment of the macula.

Diabetic macular edema (DME) is defined by thickening of the retina near the macula or the presence of hard exudates with adjacent retinal thickening. DME was historically classified as focal versus diffuse based on diagnostic criteria for clinically significant macular edema (CSME), as defined by the ETDRS (table 8.1) using information gathered by slit-lamp exam. DME involving the fovea was shown in the Early Treatment of Diabetic Retinopathy Study (ETDRS) to have a 24% risk of developing moderate visual loss if left untreated.

With the development of optical coherence tomography, DME is now defined by the presence of foveal edema, categorized as 'center-involved' or 'not center-involved'. DME can occur at any stage of DR, in both nonproliferative and proliferative diabetic retinopathy, with increasing frequency as the severity of DR progresses [1]. Symptoms of vision loss are dependent on the location of the edema. DME that is not center-involved is usually asymptomatic. However, patients with center-involved DME often experience progressive vision loss on the timescale of weeks to years after the initial onset of symptoms [1].

Pathophysiology: The development and progression of diabetic retinopathy occurs as a direct consequence of diabetes mellitus and hyperglycemia causing pathology of the retinal vessels [1]. The two principal changes in the retinal vasculature from DR are abnormal permeability and occlusion leading to ischemia and subsequent neovascularization [1, 9]. Early in the disease, hyperglycemia-induced damage to the vasculature can occur prior to the onset of clinical signs. Later in the disease, overactivity of angiogenic factors like vascular endothelial growth factor (VEGF) and insulin-like growth factor (IGF-1) lead to the progression from NPDR into PDR or DME [10]. The variability in these factors and their subsequent response to hyperglycemia is thought to explain the spectrum of the severity, rate of development, and rate of progression of DR and DME in patients with DM despite adequate glucose control [9, 10].

Structural anatomic retinal changes—Anatomic changes in the retina due to diabetes include retinal pericyte loss, thickening of the capillary basement membrane, and microaneurysms of the capillary walls. The anatomical changes increase the susceptibility of the retinal capillaries and arterioles to become nonperfused, leading to retinal ischemia and disruption of the blood-retinal barrier. In response to the disruption of the vasculature, there is an increase in vascular permeability leading to retinal edema. As the retinal ischemia worsens, the risk for retinal traction and potential detachment increases [7].

Retinal microthrombosis—the formation of retinal microthrombosis in the retinal vessels can lead to occlusion of the capillaries and capillary leakage. One of the first signs of change in the retina prior to clinically detectable DR is adhesion of leukocytes to the retinal vascular endothelium, hypothesized to contribute to increased vascular permeability. The loss of endothelial integrity contributes to retinal ischemia, leading to release of endothelial growth factors such as IGF-1, VEGF, fibroblast growth factor (FGF), and platelet derived growth factor (PDGF) that induce vascular proliferation and angiogenesis [7].

Altered retinal blood flow—Physiologic retinal blood flow is kept constant by an autoregulated mechanism until the mean arterial pressure is raised 40% above baseline. In patients with chronic hyperglycemia, this autoregulation mechanism is thought to be impaired, leading to a sustained increase in retinal blood flow causing increased shear stress on the retinal blood vessels. The pathologic environment is also thought to increase the stimulation of vasoactive substances leading to vascular leakage and accumulation of fluid in the outer layers of the retina which can progress to macular edema [11].

Growth factors: Damage to the retinal vessels leads to release of growth factors and subsequent neovascularization. The process of neovascularization is thought to be from the interaction of IGF-1 and VEGF. The role of IGF-1 was shown in experiments that demonstrated an increase in neovascularization in the year following the initiation of intensive insulin therapy, as insulin stimulates IGF-1 [10]. Furthermore, studies have shown that decreases in IGF-1 concentrations via injury to the pituitary or hypophysectomy leads to decreases in neovascularization, with the potential to reverse advanced retinal disease [10].

VEGF is an angiogenic growth factor and vascular permeability factor that is thought to play a significant role in the pathogenesis of DR and DME. Studies have demonstrated a positive correlation between the immunostaining intensity for VEGF and the severity of retinopathy [11].

Elevated erythropoietin (EPO) has also been shown to cause retinal angiogenesis in DR with increased concentrations found in the vitreous fluid of patients with DR [5].

Currently known risk factors for the development and progression of DR and DME include chronicity of diabetes, hemoglobin A1c, hypertension, nephropathy, and dyslipidemia.

Modifiable Risk Factors: Clinical trials and epidemiological studies show that the two most important modifiable risk factors in the development of DR and DME are blood sugar and blood pressure control [12]. Strict blood glucose control with HbA1c <7% can lead to sustainable and potent protective effects against the development of DR and DME in both type 1 and 2 DM. Moreover, the Diabetes Control and Complications Trial showed that strict blood glucose control decreases the incidence of retinopathy by up to 76% and decreases the progression of mild DR to severe DR by up to 54% [12]. Thus, control of hyperglycemia has a great effect on preventing or delaying onset as well as preventing progression of DR. Likewise, strict blood glucose control demonstrated a 46% reduction in the incidence of DME [12]. Controlling hyperglycemia should be carefully managed to avoid causing hypoglycemia, particularly in the elderly population. The ADVANCE and the Veterans Affairs Diabetes Trial studies concluded that more intensive glycemic control, with target HbA1c < 6.5%, did not have a significant further reduction in the incidence of DR or disease progression [13, 14].

Although the link between hypertension and dyslipidemia and the development of DR and DME are weak, careful management can reduce the risk of other diabetic complications such as nephropathy and cardiovascular diseases [13]. The UK Prospective Diabetes Trial concluded that tight blood pressure control (BP < 150/85 mmHg) in patients with type 2 DM reduced the risk of developing microvascular disease by 37% and decreased the rate of DR progression by 34% [15]. However, the risk reduction in disease progression waned after stopping intensive blood pressure control and studies have not been able to show a risk reduction in DR incidence with management of hypertension [15]. Thus, the treatment of hypertension for the sole purpose of reducing DR incidence and progression is not currently recommended but should be initiated for control and prevention of microvascular complications from DM, particularly nephropathy.

Smoking is a modifiable risk factor that has been shown to increase the prevalence and rate of progression of DR in patients with type 1 DM. However, its role as a risk factor in patients with T2DM is disputed, with studies from the United Kingdom Prospective Diabetes Study and the Wisconsin Epidemiologic Study of Diabetic Retinopathy demonstrating no association of smoking with increased prevalence or progression of DR [16].

Non-modifiable risk factors: Studies have shown that patients with both type 1 and type 2 DM have an increased risk of developing DR and DME with increased duration of disease, independent of their glycemic control. Patients with type 1 DM will likely develop DR after a certain duration of disease, but the results are less clear for type 2 DM, likely due to competing risks of mortality [17].

Puberty has been shown to increase the risk for developing DR, particularly in patients with type 1 DM. Pre-pubertal exposure to either type 1 or type 2 DM increases the risk of developing DR. However, type 1 DM during puberty has a greater impact on risk for developing DR [17]. Moreover, the onset of type 1 DM during or post puberty increases the risk of developing severe DR compared to patients with disease onset during their pre-pubertal years.

Pregnancy increases the risk of developing DR by 3 times in patients with type 1 DM versus patients with type 2 DM. However, the development of DR in pregnancy is often transient and patients typically demonstrate resolution and regression of DR in the postpartum period. There is no significant difference in the prevalence of DR when comparing patients who had been pregnant compared to those who had not been pregnant [17].

Symptomatic presentation and physical examination—Patients with DM should receive annual ophthalmic evaluations if they either do not have retinopathy or have mild NPDR to monitor progression, as DR is often asymptomatic regardless of disease severity. It is important to note that mild to severe DME can be present even without symptoms noticeable to the patient, and the possibility of progression to DME in any stage of DR further necessitates ophthalmic examinations. A thorough documentation of the patient's history of diabetes, medications, A1c trends, diet, and the presence of other complications of diabetes such as hypertension, hypercholesterolemia, renal disease, and thyroid disease is critical in the evaluation. As most early stages of DR are asymptomatic, the presence of symptoms may be a sign of more advanced disease, thus screening examinations are critical to achieve early diagnosis. Symptomatic patients may experience blurred vision, narrowed field of vision, flashes, floaters, elevated intraocular pressure in the case of neovascular glaucoma, and difficulty with night vision. The index for suspicion of DR and DME increases in patients with recent vision impairment, particularly in the setting of chronic diabetes and hyperglycemia, nephropathy, hypertension, or dyslipidemia.

On physical examination, patients receive a biomicroscopic examination using a slit lamp microscope and indirect ophthalmoscope. Historically, DME was categorized by guidelines established by the ETDRS to define clinically significant macular edema (CSME). The criteria for the diagnosis of CSME are in table 8.1. The diagnosis of CSME using the ETDRS criteria was thus a clinical diagnosis made via slit lamp exam findings. The importance of detecting CSME via conducting a careful physical exam was highlighted from ETDRS data showing that patients with type 2 DM with severe NPDR had a 50% chance of developing high-risk characteristics for vision loss if they did not receive laser treatment. Consequently, a diagnosis of CSME using ETDRS criteria was used to prevent a high percentage of visual loss by initiating laser, pharmacological, or surgical treatment.

With the advent of anti-VEGF therapy in recent years and advances in imaging with optical coherence tomography, the ETDRS criteria has become relatively obsolete. DME is classified more practically now as fovea-involving vs. non fovea-involving, as fovea-involving DME poses a greater risk for development and progression of visual loss and are candidates for anti-VEGF therapy.

Overview of diagnostic procedures, laboratory testing, differential diagnosis: Diagnosis of diabetic retinopathy is based on the findings from the retinal exam by direct/indirect fundoscopy, fundus photography, slit lamp biomicroscopy, and undilated gonioscopy (to assess neovascularization of the iris neovascularization or in the presence of elevated intraocular pressure). A full ocular exam should be performed to assess visual acuity, pupillary response, and intraocular pressure. Dilated visual exams are performed to assess for degree of retinopathy, including microaneurysms, intraretinal hemorrhages, venous dilation, cotton wool spots, retinal thickening, lipid deposits, venous dilation/beading/loops, and intraretinal microvascular abnormalities. Further assessment of the retina if deemed necessary can include fluorescein angiography (FA), ultrasound, optical coherence tomography (OCT), and optical coherence tomography angiography (OCTA) [9, 18, 19].

FA is a valuable imaging tool to aid in the diagnosis and management of DR and DME as it can detect microaneurysms, dot and blot hemorrhages, areas of nonperfusion, and presence of neovascularization. FA images distinguish microaneurysms, which appear hyperfluorescent, from dot and blot hemorrhages, which appear hypofluorescent. Nonperfused areas are detected on FA as homogenous hypofluorescent patches. Neovascular tufts often leak dye because of their weaker vessel walls and higher vascular permeability, which are identified on FA as hyperfluorescent areas that increase in size and intensity as the test phase progresses. FA also helps to identify the location of microaneurysm leakage for laser therapy.

Screening: The American Diabetes Association (ADA) and American Academy of Ophthalmology (AAO) have the following DR and DME screening recommendations for patients with DM:

- For patients with type 1 DM, an initial dilated and comprehensive eye exam is recommended within 5 years of onset of diabetes.
- For patients with type 2 DM, a comprehensive dilated eye exam is recommended at the onset of diabetes.

- For type 1 and type 2 DM, annual follow-up eye exams are recommended.
- Screening should continue regularly during pregnancy, at each trimester of pregnancy, and 1 year postpartum depending on the severity of retinopathy. Gestational diabetes does not appear to increase the risk of developing DR [18, 19].

Laboratory testing: Routine measurements of A1c and lipid panel (LDL, HDL, TG) levels (table 8.2).

Differential diagnosis: The differential diagnosis for DR and DME are presented in table 8.2.

Follow-Up Care: Frequent follow-up visits and dilated ophthalmic examinations, especially in at-risk patients with DR, can prevent vision loss and reduce the cost of treating more advanced disease [20]. The AAO guidelines for follow-up exams are listed in table 8.3.

Complications and Prognosis: Diabetic retinopathy is the most common cause of new onset blindness in adults. The vision loss due to DR is associated with a significant economic burden for patients and their families [18]. Diabetic retinopathy can progress in severity if left untreated. Progression includes transition from NPDR to PDR, worsening vision, and increased risk of blindness. The risk of progression to proliferative diabetic retinopathy within 1 year is 5% with mild NPDR, 20% with moderate NPDR, and 50% with severe NPDR [7].

Table 8.2. Differential diagnosis for DR and DME[19].

Leading Diagnosis	Differential Diagnosis
Macular Edema	– Hypertensive retinopathy – Retinal vein occlusion – Microaneurysm rupture – Radiation – Irvine-Gass syndrome – Subfoveal choroidal neovascularization
Nonproliferative DR	– Central/branch retinal vein occlusion – Ocular ischemic syndrome – Hypertensive retinopathy – Radiation retinopathy
Proliferative DR	– Neovascular complications of central retinal artery or vein occlusion – Branch artery or vein occlusion – Sickle cell retinopathy – Embolization from IV drug use – Sarcoidosis – Inflammatory conditions (i.e. SLE) – Complications from ocular ischemic syndrome – Hypercoagulable states (i.e. antiphospholipid syndrome) – Radiation retinopathy

Table 8.3. AAO guidelines for follow-up care based on severity of DR and DME.

Stage of DR	No DME or CSME	DME present	CSME present
No Diabetic Retinopathy/ Mild NPDR	Annual dilated exams	Follow-up every 4–6 months	Follow-up every month
Moderate NPDR	Follow-up every 6–12 months	Follow-up every 3–6 months	Follow-up every month
Severe or Very Severe NPDR	Follow-up every 4 months	Follow-up every 2–4 months	Follow-up every month
PDR	Follow up every 4 months	Follow-up every 4 months	Follow-up every month
High-Risk PDR	Follow-up every 4 months	Follow-up every 4 months	Follow-up every month

Vitreous hemorrhage as a complication of proliferative diabetic retinopathy is a cause of severe vision loss and occurs due to rupture of blood vessels in the vitreous. Vitreous hemorrhage occurs from upregulation of angiogenic factors like IGF-1, VEGF, and FGF in the vitreous, fibrovascular membranes, and from the retina due to hypoxic conditions of DR leading to formation of neovascular buds from retinal blood vessels. The neovascularized tissue can proliferate and invade the space between the retina and the posterior hyaloid face producing firm adhesions. As the vessels continue to proliferate, their fibrous component subsequently increases. Contraction of the fibrous component generates traction on the posterior hyaloid face and pulls on the weak and friable neovascular tissue on the retina leading to a vitreous hemorrhage [21].

Vitreous hemorrhage can lead to further fibrosis and contraction of the vitreous, which can lead to another vision-threatening complication of DR: tractional retinal detachment (TRD). TRD is the separation of the retina from the retinal pigment epithelium due to traction from the vitreous or retinal surface. Diabetes and DR are the most common causes of RTD [22].

8.2 Development/history of OCT use in DME

The role of imaging for the management of diabetic macular edema began with fluorescein angiography (FA), first described in 1961 and later adapted into standard practice within the field of ophthalmology [23]. In patients with DME, FA demonstrates patterns of hyper-fluorescence, representing the breakdown of the blood-retinal barrier. In eyes with CSME that may be candidates for laser photocoagulation therapy, FA was utilized to identify the leaking microaneurysms and locate the areas of diffuse leakage to be targeted for therapy. However, major limitations of FA are the inability to quantify the amount of leakage and discriminate depth-dependent pathology between the superficial versus the deep capillary plexuses, rendering it less clinically useful when monitoring response to therapy [24].

Optical coherence tomography (OCT) is a noninvasive imaging modality that produces cross-sectional images of optical reflectivity of the retina [25]. The advent of OCT into the clinical setting revolutionized the management of DME. OCT has been increasingly used for the diagnosis of DR and for monitoring disease progression because of its ability to measure retinal thickness directly and reliably from the tomogram by either measuring the inner and outer boundaries of the retina manually, or by using computer-aided image processing algorithms [25]. OCT also provides the capability for morphologic assessment of the macula, with several anatomic patterns that have been reported in DME. Moreover, the objective output of OCT allows for quantification of cell layer thickness and macular volume, enabling longitudinal assessments to demonstrate and monitor response to therapy, a quality that makes it superior to FA in its clinical utility.

The development of OCT began with time-domain OCT, which uses interferometry-based technology to generate images with an axial resolution of 10 um, and was first used in a clinical setting in 1993 for retinal imaging [26, 27]. TD-OCT was utilized in clinical trials to monitor and quantify the benefits of anti-VEGF therapy for DME by evaluating central subfield thickness of the retina to aid in making retreatment decisions.

In the 2000s, spectral-domain OCT (SD-OCT) was developed as the next-generation alternative to TD-OCT. SD-OCT utilizes a spectrophotometer and the Fourier transform to generate images with a higher axial resolution of 5 um along with faster acquisition speeds, making the cell layers of the macula more distinctive [28]. With the advances brought forth by SD-OCT, ophthalmologists have been provided with more information to help guide clinical management than what was previously available with TD-OCT [28]. SD-OCT is currently the most commonly used and commercially available imaging device for the diagnosis and management of DME.

The breakthrough with SD-OCT along with the improvements in available laser technology in recent years paved the way for the development of swept-source-OCT (SS-OCT), a variation of SD-OCT. SS-OCT utilizes a photodetector instead of a spectrometer, and uses swept-source lasers that have increased scanning speeds capable of up to a million A-scans per second, which provides a high-density scan that produces higher resolution en face OCT images. SS-OCT has the benefit of improved visualization of the choroidal-scleral surfaces and identifying structures below the retinal pigment epithelium. The high-resolution properties of SS-OCT are promising but its use is currently limited to research settings and is not yet commercially available, limited by high cost, availability, and lack of clear clinical advantages over SD-OCT. SS-OCT devices may see increased prevalence in clinical applications beyond ophthalmology, including cardiology, dermatology, and gastroenterology.

8.3 Pathology and mechanism of OCT findings in DME

Optical coherence tomography (OCT) is a diagnostic imaging technique that is analogous to B-scan ultrasound by using optical interferometry to resolve the distances of reflective structures within tissues by using light instead of sound [29]. OCT images are two-dimensional data sets that represent the optical backscattering within a cross-sectional plane through the target tissue. The first step in reconstructing a tomographic image is the measurement of the axial distance, or the range of information within the target tissue. OCT essentially images by measuring the echo time delay and intensity of backscattered or backreflected light from the unique internal microstructure in materials or tissues of interest [28, 29]. The measured optical backscattering or backreflection in a cross-sectional plane or volume allows OCT images to reflect a two-or-three-dimensional data set, respectively. Two important imaging characteristics are image resolution and imaging depth. The resolution of OCT depends directly on the frequency or wavelength of the light source used; thus, the inherently high resolution of OCT allows identification of tissue architectural morphology features [27, 29]. Unlike ultrasound, where the velocity of sound in water is approximately 1500 m s^{-1}, the velocity of light is approximately 3×10^8 m s^{-1}, which means the measurement of the echo time delay with light cannot be measured with direct electronic detection. Therefore, OCT utilizes low-coherence interferometry to measure the echo time delay. In the older generation time-domain OCT (TD-OCT), low-coherence interferometry measures the echo time delay and intensity of the backscattered light from tissue by comparing it to light that has traveled a known reference path length and time delay [30]. This is accomplished with a Michelson-type interferometer, where a light source is directed into a beam splitter that sends one of the beams as the incident light source into the tissue sample to be imaged, while the second beam travels a reference path with a variable path length and time delay [30]. The backscattered light from the sample is then interfered with the reflected signal from the reference arm and detected with a photodetector at the output of the interferometer. Based on the coherence pattern, the echo time delay and intensity of backscattered light from the sample arm can be measured by comparing it with the reference path length [30].

In spectral-domain OCT, the schematic is similar to TD-OCT, but the reference arm is immobilized, and the detector is replaced by a spectrometer, allowing for the capability to collect signals from all depths of the sample throughout the entire acquisition time. A Fourier transform of the spectral measurement converts the detected backscattered signals from the imaged tissue into the frequency domain, which represents the tissue depth reflectivity profile. SD-OCT includes a host of advantages over TD-OCT including that of faster scanning times, higher resolution, greater field-of-view, and increased signal-to-noise ratio [31].

8.4 Findings identified on OCT in the diagnosis of DME

See figures 8.1–8.9 for findings identified on OCT in the diagnosis of DME.

ILM = internal limiting membrane
NFL = nerve fiber layer
GCL = ganglion cell layer
IPL = inner plexiform layer
INL = inner nuclear layer

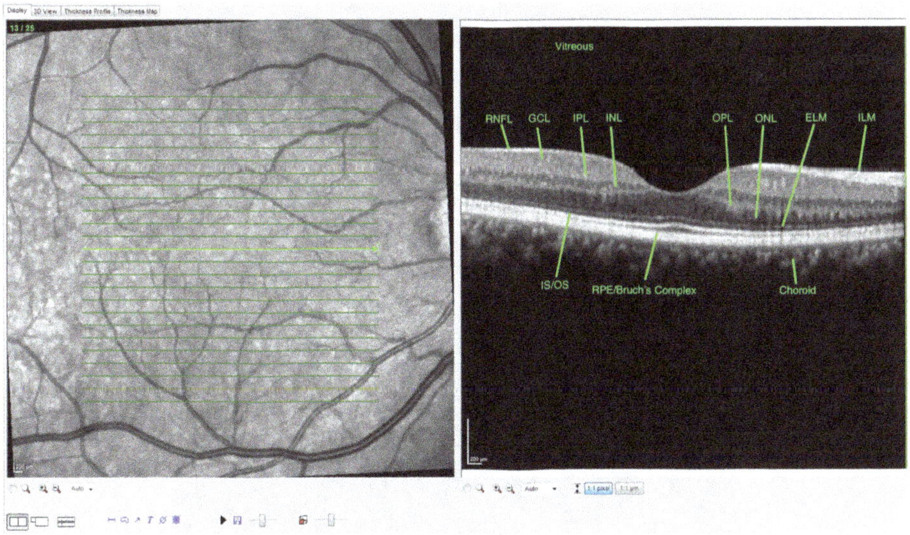

Figure 8.1. OCT showing a normal foveal contour. The layers of the retina from anterior to posterior are listed below.

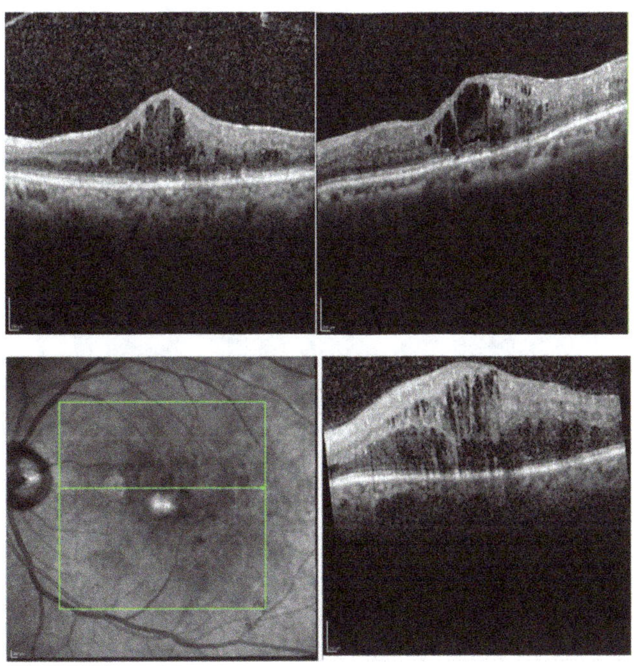

Figure 8.2. Intraretinal fluid (IRF) is a result of damaged, leaky vessels as a result of diabetic microvascular changes. IRF appears as dark cystic spaces within the retinal layers which can often distort the foveal contour. Center-involving edema warrants treatment.

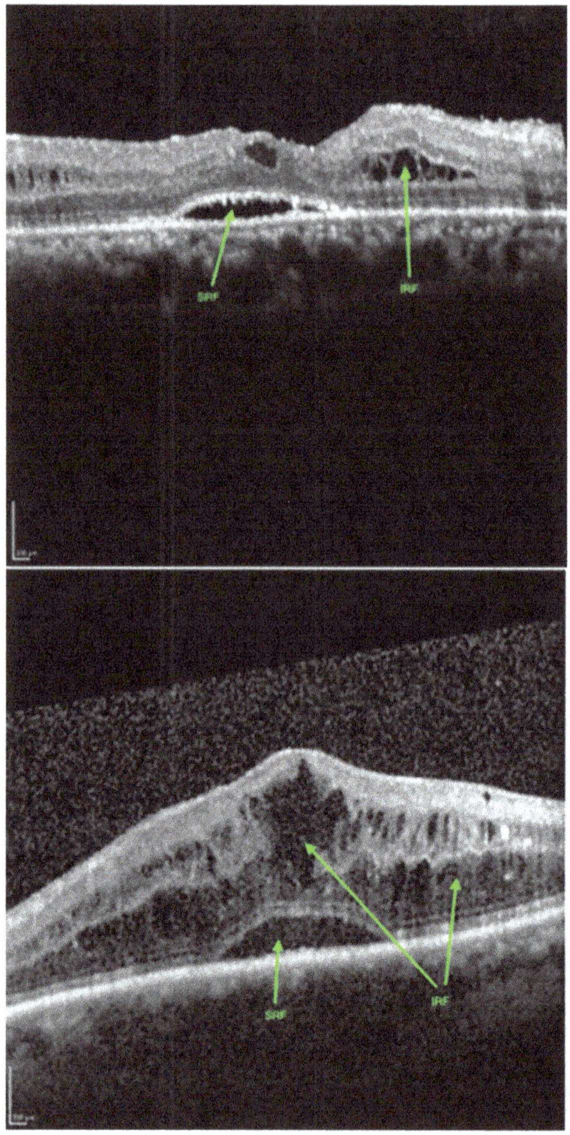

Figure 8.3. Fluid can also build up underneath the retina above the RPE. Below are examples of **subretinal fluid** (SRF) with adjacent and overlying intraretinal fluid with foveal contour distortion.

OPL = outer plexiform layer
ONL = outer nuclear layer
ELM = external limiting membrane
IS/OS junction = photoreceptor layer
RPE/Bruch's complex
Choroid

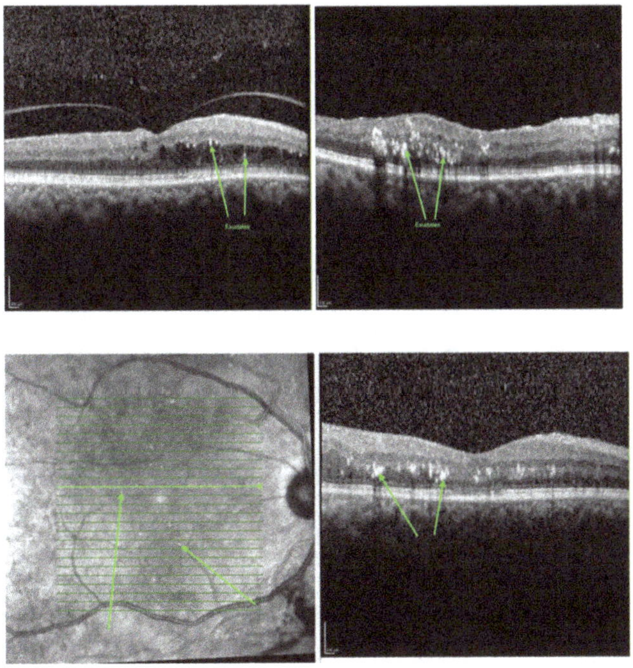

Figure 8.4. Exudates can be seen on OCT as hyperreflective foci. They are composed of lipid and protein materials that leak from impaired blood vessels or aneurysms and are commonly found in the outer plexiform layer.

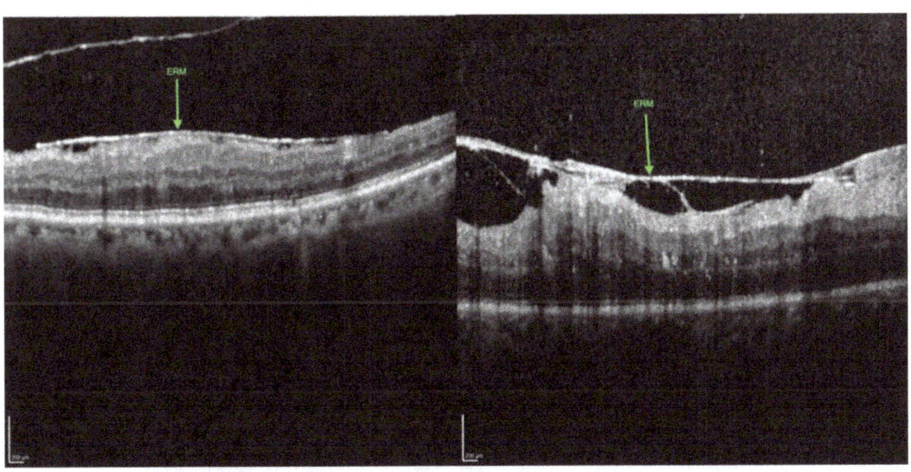

Figure 8.5. Epiretinal membranes are often seen as a complication of diabetic retinopathy from inflammation. They represent the proliferation of glial cells that overtime can cause traction of the retina leading to various degrees of visual distortion and decreased vision.

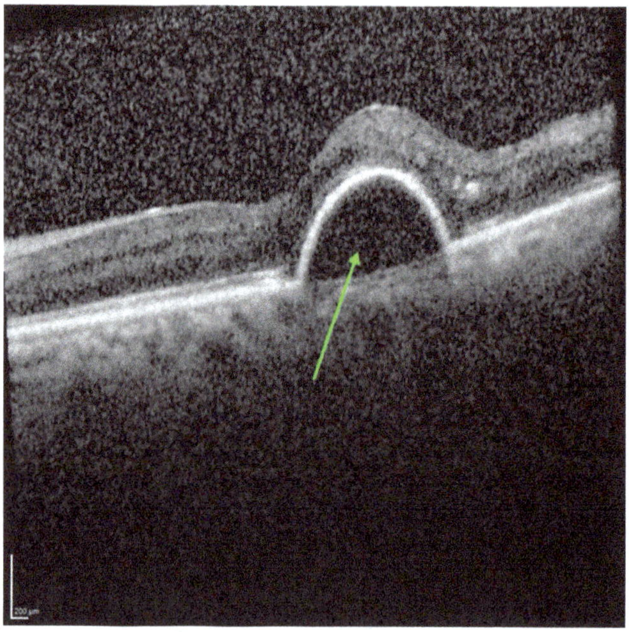

Figure 8.6. Fluid can also develop underneath the pigment epithelium and is known as a **pigment epithelium detachment** (PED)

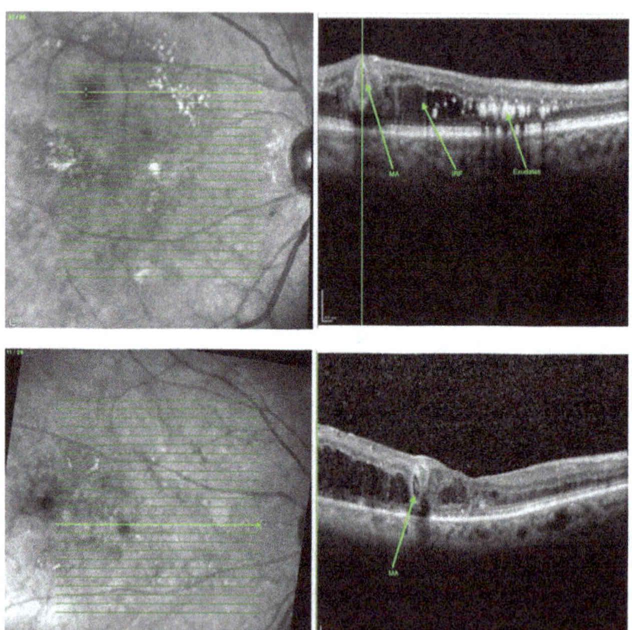

Figure 8.7. **Microaneurysms** (MA) can be seen clinically as well as on OCT. Leaking aneurysms will develop surrounding intraretinal fluid often combined with exudates in a circinate pattern surrounding the MA.

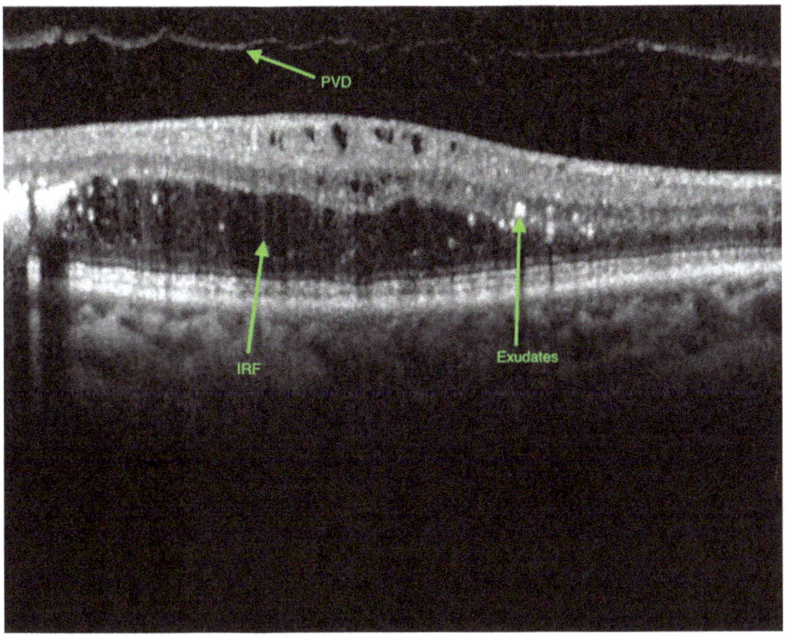

Figure 8.8. Separation of the posterior hyaloid membrane from the retina is known as a **posterior vitreous detachment (PVD)** which can often be seen on OCT within the vitreous. This is an example of a complete PVD.

8.5 Classification and grading of OCT findings in DME

Prior to OCT technology, edema was graded clinically on fundoscopic examination for clinically significant macular edema (CSME), strict criteria used in the ETDRS trials. With the widespread availability of OCT technology, DME involving the center of the macula (defined as retinal thickening in the macula involving the central subfield zone 1 mm in diameter) is the main criteria for treatment of DME, versus non-center involved (retinal thickening that does not involve the central subfield zone).

Additionally, several OCT biomarkers have demonstrated prognostic value including disorganization of the retinal layers (DRIL), ellipsoid zone disruption, and vitreomacular interface abnormalities. These features have been shown to be associated with poor outcomes despite anti-VEGF treatment.

8.6 Introduction and development of OCTA in the diagnosis of DME

OCT has revolutionized retinal imaging for the diagnosis and management of diabetic retinopathy (DR) and diabetic macular edema (DME). However, OCT is unable to provide direct information about structural and functional changes in the retinal and choroidal vasculature such as blood flow velocity, differentiating afferent from efferent vessels, and detecting changes in vasculature permeability. Consequently, fluorescein angiography (FA) and indocyanine green angiography remain the gold standard for structural visualization and monitoring of dynamic changes of the retinal vasculature in DR and DME [32].

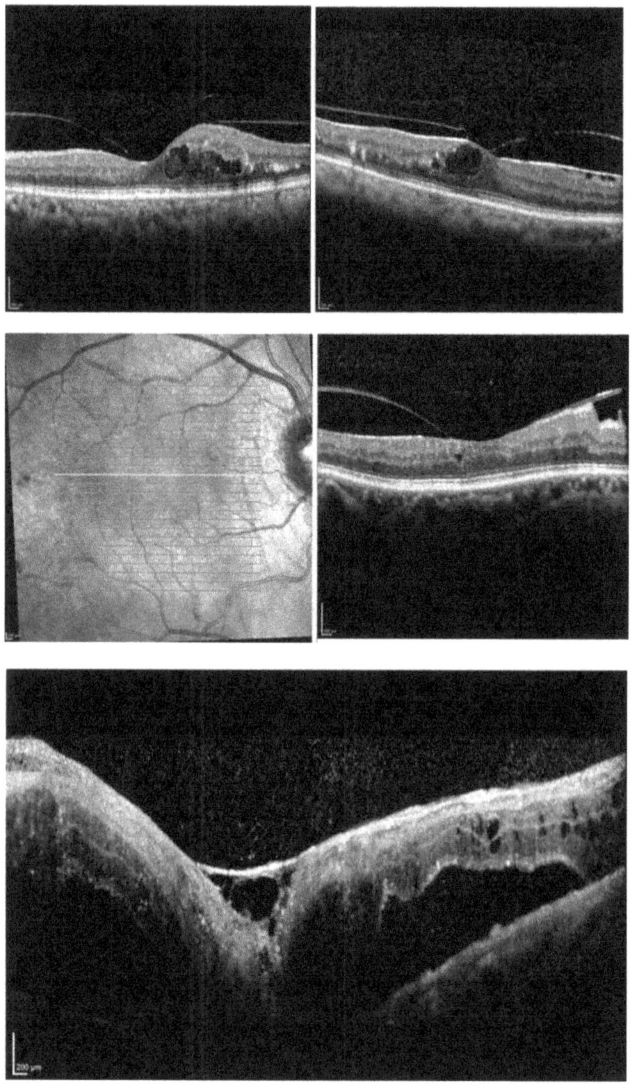

Figure 8.9. Vitreomacular adhesion (VMA) and **vitreomacular traction** (VMT) is when the posterior hyaloid has not fully detached and is still adherent to the retina. Traction in VMT can distort the foveal contour.

However, FA has several limitations including the inability to quantify leakage, discriminate between the superficial and deep capillary plexuses, administration of intravenous dye, and long acquisition times. To address the limitations of FA, OCT Angiography (OCTA) has been developed in recent years as a novel use of OCT technology to visualize the microvasculature of the retina and the choroid non-invasively. It is distinct from FA in that it does not necessitate the injection of fluorescein dye into a peripheral vein. With repeated scans at the same anatomic location, OCTA can detect the changes in OCT reflectance signal from the flow through blood vessels, allowing for depth-resolved, motion-contrast imaging of the retinal vasculature. OCTA operates under the assumption that the only moving objects in the retina are the blood cells circulating through the vasculature. The identification of the moving blood cells translates to blood vessels in the final output images.

OCTA generates angiograms from different segments of the retina, typically including the superficial capillary plexus, deep capillary plexus, and the choriocapillaris. In healthy eyes, the deeper vasculature plexuses have a higher density of blood vessels compared to the superficial layers. In patients with DR, vascular abnormalities can be present in the different retinal layers. The ability to generate images of the retinal layers from superficial to deep vascular plexuses is the primary advantage of OCTA over fluorescein angiography, the latter of which focuses on superficial and larger retinal blood vessels [33]. In patients with DME and DR, OCTA can delineate the foveal avascular zone and identify areas of impaired blood flow when compared to FA. OCTA has been reported to identify changes in the microvasculature before clinically detectable DR or DME.

8.7 Pathology and mechanism of OCTA findings in DME

OCTA is designed based on OCT technology that allows for visualization of the functional vasculature in the eye. The mechanism of OCTA utilizes the variation in OCT signal from moving particles, such as red blood cells, as a contrast mechanism for imaging blood flow. By repeating multiple scans at the same anatomic location, temporal changes of the OCT signal caused by the moving particles generate the angiographic contrast and outputs the microvasculature visualization [33, 34].

OCTA has been shown to be comparable with FA and the clinical examination to identify vascular changes such as microaneurysms, impaired perfusion, retinal edema, vascular loops, intraretinal microvasculature abnormalities, and neovascularization [35]. Microaneurysms are detected on OCTA as hyperreflective spots, with further differentiation of the structural characteristics not seen by FA including fusiform, saccular, curved, or coiled shapes.

OCTA can assist in differentiating retinal neovascularization from intraretinal microvasculature abnormalities. Retinal and disc neovascularization is identified as interwoven vessels above the surface of the retina and the optic nerve [35]. Retinal neovascularization has been found to be adjacent to retinal capillary areas of nonperfusion, demonstrating that OCTA may have a role in distinguishing severe NPDR from PDR, and can be used for monitoring progression of severe NPDR (figures 8.10 and 8.11).

OCTA is also capable of identifying and distinguishing the foveal avascular zone (FAZ). The FAZ is the avascular area surrounded by foveal capillary circles at the center of the macula. Therefore, structural or perfusion abnormalities in the FAZ can greatly hinder visual acuity. In patients with DR, the FAZ is increased in size and is non-symmetrical due to loss of blood vessel integrity and damage to the capillary plexuses. Studies have shown that the grading of FAZ abnormalities is correlated with severity of DR. The FAZ was first identified by FA, but dye leakage

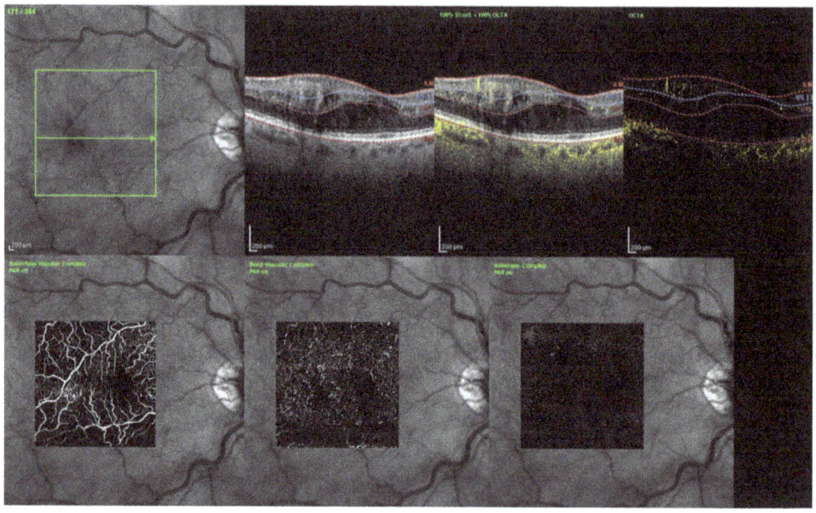

Figure 8.10. OCTA images from a patient with diabetic retinopathy with microaneurysms, impaired perfusion, and retinal edema.

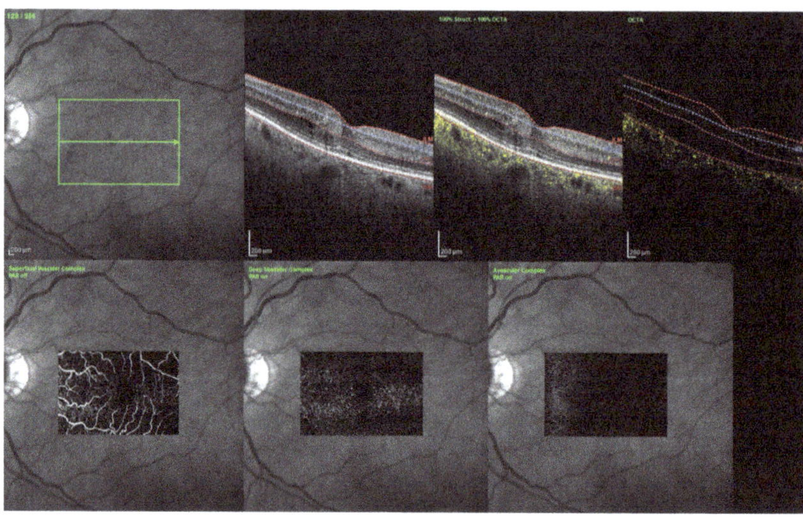

Figure 8.11. Example of OCTA in a healthy control eye.

often covered the FAZ. OCTA is now considered a superior option to FA to distinguish the FAZ and to characterize the central and parafoveal macular microvasculature [35].

OCTA's ability to distinguish areas of nonperfusion, alterations to the FAZ, and characteristics of microaneurysms may better help clinicians understand the pathophysiology of DME, improve grading of DME, and guide therapy decisions and management.

References

[1] Holekamp N M 2016 Overview of diabetic macular edema *Am. J. Manag. Care* **22** s284–91

[2] Jampol L M, Glassman A R and Sun J 2020 Evaluation and care of patients with diabetic retinopathy *New Engl. J. Med.* **382** 1629–37

[3] Willis J R *et al* 2017 Vision-related functional burden of diabetic retinopathy across severity levels in the United States *JAMA Ophthalmol.* **135** 926–32

[4] Duphare C *et al* 2021 Diabetic Macular Edema *StatPearls [Internet]* (StatPearls Publishing)

[5] Varma R *et al* 2014 Prevalence of and risk factors for diabetic macular edema in the United States *JAMA Ophthalmol.* **132** 1334–40

[6] Coyne K S *et al* 2004 The impact of diabetic retinopathy: perspectives from patient focus groups *Fam. Pract.* **21** 447–53

[7] Wong T Y, Cheung C M, Larsen M, Sharma S and Simó R 2016 Diabetic retinopathy *Nat. Rev. Dis. Primers* **2** 16012

[8] Aiello L M 2003 Perspectives on diabetic retinopathy *Am. J. Ophthalmol.* **136** 122

[9] Frank R N 2004 Medical progress: diabetic retinopathy *New Engl. J. Med.* **350** 48–58

[10] Ruberte J *et al* 2004 Increased ocular levels of IGF-1 in transgenic mice lead to diabetes-like eye disease *J. Clin. Invest.* **113** 1149–57

[11] Boulton M *et al* 1998 VEGF localisation in diabetic retinopathy *Br. J. Ophthalmol.* **82** 561–8

[12] Nathan D M and DCCT/Edic Research Group 2014 The diabetes control and complications trial/epidemiology of diabetes interventions and complications study at 30 years: overview *Diabetes Care* **37** 9–16

[13] Beulens J W J *et al* 2009 Effects of blood pressure lowering and intensive glucose control on the incidence and progression of retinopathy in patients with type 2 diabetes mellitus: a randomised controlled trial *Diabetologia* **52** 2027–36

[14] Duckworth W *et al* 2009 Glucose control and vascular complications in veterans with type 2 diabetes *New Engl. J. Med.* **360** 129–39

[15] UK Prospective Diabetes Study Group 1998 Tight blood pressure control and risk of macrovascular and microvascular complications in type 2 diabetes: UKPDS 38 *Br. Med. J* **317** 703–13

[16] Campagna D *et al* 2019 Smoking and diabetes: dangerous liaisons and confusing relationships *Diabetol. Metab. Syndr.* **11** 1–12

[17] Lee R, Wong T Y and Sabanayagam C 2015 Epidemiology of diabetic retinopathy, diabetic macular edema and related vision loss *Eye Vis.* **2** 1–25.

[18] American Academy of Ophthalmology Diabetic Retinopathy Preferred Practice Pattern—Updated 2019. AAO 2019 Oct

[19] Yanoff M and Duker J 2019 *Ophthalmology* 5th edn (Elsevier) ed N Bagheri, B Wajda, C Calvo and A Durrani 2017 *The Wills Eye Manual: Office and Emergency Room Diagnosis and Treatment of Eye Disease* 7th edn (Philadelphia, PA: Wolters Kluwer)

[20] Harris Nwanyanwu K, Talwar N and Gardner T W *et al* 2013 Predicting development of proliferative diabetic retinopathy *Diabetes Care* **36** 1562–68

[21] Jaafar E A and Carvounis P E 2014 Current management of vitreous hemorrhage due to proliferative diabetic retinopathy *Int. Ophthalmol. Clin.* **54** 141

[22] Mishra C and Tripathy K 2021 Retinal traction detachment *Statpearls [Internet]* (StatPearls Publishing)

[23] Gass J D M, Sever R J, Sparks D and Goren J 1967 A combined technique of fluorescein funduscopy and angiography of the eye *Arch Ophthalmol.* **78** 455–61

[24] Acón D and Wu L 2018 Multimodal imaging in diabetic macular edema *Asia-Pac. J. Ophthalmol.* **7** 22–7

[25] Virgili G *et al* 2015 Optical coherence tomography (OCT) for detection of macular oedema in patients with diabetic retinopathy *Cochrane Database Syst. Rev.* **1**

[26] Liu M M *et al* 2014 Comparison of time-and spectral-domain optical coherence tomography in management of diabetic macular edema *Invest. Ophthalmol. Vis. Sci.* **55** 1370–7

[27] Drexler W *et al* 2001 Ultrahigh-resolution ophthalmic optical coherence tomography *Nat. Med.* **7** 502–7

[28] Hee M R *et al* 1998 Topography of diabetic macular edema with optical coherence tomography *Ophthalmology* **105** 360–70

[29] Fujimoto J G *et al* 2000 Optical coherence tomography: an emerging technology for biomedical imaging and optical biopsy *Neoplasia* **2** 9–25

[30] Yaqoob Z, Wu J and Yang C 2005 Spectral domain optical coherence tomography: a better OCT imaging strategy *Biotechniques* **39** S6–S13

[31] Khadamy J, Abri Aghdam K and Falavarjani K G 2018 An update on optical coherence tomography angiography in diabetic retinopathy *J. Ophthalm. Vis. Res.* **13** 487–97

[32] Tey K Y *et al* 2019 Optical coherence tomography angiography in diabetic retinopathy: a review of current applications *Eye Vis.* **6** 1–10.

[33] Gabriele M L *et al* 2011 Optical coherence tomography: history, current status, and laboratory work *Invest. Ophthalmol. Vis. Sci.* **52** 2425–36

[34] Matsunaga D R, Yi J J, De Koo L O, Ameri H, Puliafito C A and Kashani A H 2015 Optical coherence tomography angiography of diabetic retinopathy in human subjects *Ophthalm. Surg. Lasers Imaging Retina.* **46** 796–805

[35] Yu S, Lu J and Cao D *et al* 2016 The role of optical coherence tomography angiography in fundus vascular abnormalities *BMC Ophthalmol.* **16** 107

IOP Publishing

Photo Acoustic and Optical Coherence Tomography Imaging, Volume 1

Diabetic retinopathy

Ayman El-Baz and Jasjit S Suri

Chapter 9

Optical coherence tomography and OCTA for the treatment of diabetic macular edema

Eugene Hsu, Nayan Sanjiv, Pawarissara Osathanugrah, Josh Agranat and Manju Subramanian

9.1 Introduction

Vascular complications from diabetic retinopathy (DR) are the leading cause of vision loss in the US in patients between 25 and 74 years of age. For most diabetic patients, retinopathy develops within 10–15 years after initial diagnosis [1]. Diabetic macular edema (DME), a complication of DR, is one of the leading causes of visual impairment in the US and is the most common cause of vision loss in patients with DR. Patients with DR and DME can often present asymptomatically, highlighting the importance of screening for early detection of the disease and for tools to diagnose and monitor treatment response to prevent vision loss.

Over the last decade, technological advances in retinal imaging and novel therapeutics have vastly improved the evaluation, management, and outcomes of patients with DR and DME [1]. Optical coherence tomography (OCT) is a noninvasive imaging modality that produces cross-sectional images of optical reflectivity of the retina [2]. The advent of OCT into the clinical setting in 1993 revolutionized the management of DME. OCT has been increasingly used to monitor disease progression and treatment response because of its ability to measure retinal thickness directly and reliably [2]. OCT also provides morphologic assessment of the macula, with several anatomic patterns that have been reported in DME. The objective output of OCT allows for quantification of cell layer thickness and macular volume, enabling longitudinal assessments to demonstrate and monitor response to therapy to help preserve sight in patients with DM.

doi:10.1088/978-0-7503-2052-8ch9

9.2 Preventative measures for DR and DME

Strict control of diabetes, hypertension, and hypercholesterolemia: For patients with diabetes, the American Academy of Ophthalmology (AAO) and the American Diabetes Association (ADA) recommends counseling on compliance with lifestyle modifications and medications to control hyperglycemia, hypertension, and hypercholesterolemia, although diabetic complications may not be prevented in every case. As DR and DME can present asymptomatically at any stage of disease, annual dilated and comprehensive eye examinations are recommended for patients without retinopathy or with mild NPDR to monitor disease progression. Frequency of follow-up dilated ophthalmic exams from the AAO Guidelines are shown in table 9.1.

Maintaining blood glucose control is considered the primary preventative measure for the development of diabetic retinopathy, as the disease develops as a direct consequence of chronic hyperglycemia. A decrease in hemoglobin A1c levels can improve outcomes in DR and DME. The recommended acceptable A1c target is <7% [3]. A decrease in the A1c by 1% has been shown to reduce the incidence of DR by 35% and progression of DR by 15%–25% [3].

The ADA recommends blood pressure goals of <140 mmHg systolic and <90 mm Hg diastolic [4].

Many patients with diabetes mellitus, especially type 2, will require statins for blood cholesterol control. Although the benefit of controlling hypercholesterolemia does not significantly decrease the incidence of DR, studies have shown that statins reduce the incidence of DME [5].

Routine exercise and increased physical activity are lifestyle modifications that can reduce the incidence of retinopathy. Treatment of secondary conditions, such as obstructive sleep apnea, has also been shown to reduce the risk of DR development and progression [4].

Role of Anti-VEGF Agents in Disease Prevention: The DCRC Protocol W studied patients with moderate to severe baseline NPDR to determine if Aflibercept therapy

Table 9.1. AAO guidelines for follow-up care based on severity of DR and DME.

Stage of DR	No DME or CSME	DME present	CSME present
No diabetic retinopathy/ mild NPDR	Annual dilated exams	Follow-up every 4–6 months	Follow-up every month
Moderate NPDR	Follow-up every 6–12 months	Follow-up every 3–6 months	Follow-up every month
Severe or very severe NPDR	Follow-up every 4 months	Follow-up every 2–4 months	Follow-up every month
PDR	Follow up every 4 months	Follow-up every 4 months	Follow-up every month
High-risk PDR	Follow-up every 4 months	Follow-up every 4 months	Follow-up every month

would reduce their risk of developing DME after 2 years. Although the probability of developing center-involved DME and vision loss was 4.1% in the Aflibercept group and 14.8% in the sham group, there were no significant differences in visual acuity change. Thus, administration of anti-VEGF therapy for prevention of DR and DME is not currently recommended [6, 7].

9.3 Treatment options for DR and DME

Indications and current criteria to initiate treatment: Treatment of DME includes control of blood glucose, hypertension, and hypercholesterolemia and intervention with intravitreal anti-vascular endothelial growth factor (anti-VEGF) injection, laser photocoagulation, and combination therapy. Current indications for the initiation of treatment are summarized below: [1]

Intravitreous anti-VEGF injections:

- Recommended for center-involved DME, unless there is no DR or only minimal NPDR
- Can be considered for non-center-involved DME in the presence of moderate or severe NPDR
- Can be considered if there is no DME, but in the presence of severe NPDR or PDR
- Prophylactic treatment of DME in the absence of impaired visual acuity is not recommended.

Focal Laser Photocoagulation:

- Can be considered in the presence of DME and mild or worsening NPDR or PDR
- Can be considered in the presence of high-risk PDR
- Not recommended in the absence of DME.

Intravitreal corticosteroids:

- Can be considered as second-line agents for treating DME that may not be responsive to anti-VEGF injections
- Considered in DME that may include an inflammatory component, such as uveitis and surgery
- Addition of corticosteroids to anti-VEGF therapy may not improve visual outcomes.

9.3.1 Treatment options

Intravitreal anti-VEGF injection: Given the central role of VEGF in the pathogenesis of DR and DME, VEGF represents a crucial target for therapeutic inhibition. Anti-VEGF agents are therapeutic antibodies developed against VEGF that are injected into the vitreous for treatment of many retinal diseases including age macular degeneration, macular edema, retinal vein occlusions, DR, and DME.

The anti-VEGF agents for DME approved by the FDA are aflibercept, bevacizumab, brolucizumab, faricimab, ranibizumab, and ranibizumab-nuna.

Aflibercept is an intravitreal injection dosed at 2 mg monthly for the first five injections, followed by 2 mg once every two months. Studies have not shown increased overall efficacy with injection every month after the first five months. Despite this, clinicians may assess patients on a case-by-case basis as some patients may benefit from continued monthly injections.

Bevacizumab is an FDA-approved agent used for many types of cancer, but is commonly used 'off-label' for use in DR or DME. Suggested dosing is intravitreal injections of 1.25 mg at 0, 6, and 12 weeks. Additional injections can be considered every 6 weeks, with a maximum of 9 injections over the span of the first 12 months [8]. Since bevacizumab was initially approved and packaged by the manufacturer for treatment of cancer, its off-label use for DR and DME requires that it be dosed in aliquots that contain 1/500 of the systemic therapeutic dose by pharmacies with expertise in compounding, allowing it to be the most cost-effective anti-VEGF agent on the market in the United States [9]. However, its use as an off-label medication may prevent it from being universally accessible and distributed in many countries around the world, as well as in health systems within the US.

Brolucizumab is an anti-VEGF agent with a low molecular weight, allowing more of the drug to be delivered intraocularly per injection to increase tissue penetration and durability of effect. Phase III non-inferiority studies for Brolucizumab demonstrated that more than 50% of patients could maintain stability with dosing every 12 weeks. There have been reports of adverse effects of retinal vascular occlusion and retinal vasculitis with intraocular inflammation with this drug [10].

Faricimab is the most recent FDA-approved anti-VEGF agent. In its phase II study, 6.0mg doses showed a significant improvement in visual acuity over ranibizumab. Faricimab also demonstrated improvements in DR severity, reduced central subfield thickness, and longer intervals between retreatment compared to ranibizumab. Non-inferiority trials of Faricimab compared to ranibizumab showed that more than 50% of patients could receive dosing every 16 weeks with non-inferior results, thereby demonstrating a potential to decrease the treatment burden for patients [11].

Ranibizumab is FDA approved for the treatment of DME and DR. Intravitreal injection of 0.3 mg per month is currently recommended. Ranibizumab has also been recently approved to be administered via surgical ocular implant [12].

A clinician's choice of agent must consider the cost of care and treatment efficacy for each patient's specific ophthalmic history and retinal examination. Moreover, patients may see improvement with a second or third choice drug at any visual acuity level even after poor responses to initial treatment with any of the anti-VEGF agents [1, 9]. The CATT trial was an NIH-sponsored comparative effectiveness study of bevacizumab versus ranibizumab, which found that both drugs had equivalent effects for the treatment of age-related macular degeneration on visual acuity after 1 year when administered on the same schedule. It also reported that as-needed injections of ranibizumab had equivalent results after 1 year compared to monthly ranibizumab injections. The risk of serious adverse events was similar in patients receiving either Ranibizumab or Bevacizumab. Later, the protocol T study by the DRCR aimed to compare the efficacy of bevacizumab, ranibizumab, and aflibercept for the treatment of DME. The study demonstrated similar efficacy of all three drugs when initial visual acuity loss was mild (20/32 to 20/40). However, in patients with moderate or worse initial visual acuity loss (20/50 to 20/320), aflibercept was superior to bevacizumab and ranibizumab in improving visual acuity. Aflibercept ($1,850 for 2.0mg) and ranibizumab ($1,170 for 0.3mg) are significantly more expensive compared to bevacizumab ($60 for 1.25mg), and studies have shown that aflibercept and ranibizumab are not cost-effective alternatives to bevacizumab [13].

The limitations of anti-VEGF agents include the cost and frequency of injections on the patient, as studies demonstrated that the largest gains in visual acuity are associated with more frequent (i.e., monthly) injections of anti-VEGF [1, 9]. Prior to injection, adequate anesthesia and broad-spectrum microbicide are given. As an invasive procedure, consideration should be given to risks of anti-VEGF therapy for the treatment of any eye disease. Risks fall into two categories: injection (procedure)-related and medication-related risk. Most risks are related to the procedure, the most common of which are mild, such as irritation, foreign body sensation, burning, lacrimation, intraocular inflammation, iritis, visual disturbances, conjunctival bleeding, and corneal abrasion, all of which may result from anesthetic sterilization of the eye prior to the procedure with betadine eyewash, or the injection itself. Moderate to severe adverse events include bleeding in the vitreous cavity, cataract formation from direct trauma to the lens, infection leading to endophthalmitis, and retinal detachment. The more severe risks, while rare in incidence, may lead to vision loss and necessitate further management with surgical intervention. Receiving multiple injections long term can also lead to glaucoma. Patients should be counseled on all these risks prior to recommending treatment [14].

Current research and development for treatment of DME focuses on reduction of treatment burden for patients. For many, monthly visits to the retina specialist poses an undue burden on patients who have limited options for time, transportation, and caregivers. Studies have shown that failure to adhere to follow up results in greater need for injections and treatments over a longer period [1].

Laser photocoagulation: Prior to the development of anti-VEGF agents for the treatment of DME, laser photocoagulation was the standard of care for CSME starting in 1985 after the ETDRS study was published identifying a decrease in the risk of developing moderate visual loss within 3 years from 24% to 12% [15]. While anti-VEGF therapy is currently considered first-line, laser/focal photocoagulation is an alternate treatment option particularly for patients who have difficulty with follow-up. Photocoagulation may also be used as an adjunctive therapy to anti-VEGF agents in patients with incomplete response to anti-VEGF therapy [9]. Laser photocoagulation can be done directly on microaneurysms or in areas of retinal edema without targeting specific lesions in the retinal vasculature. Although the mechanism of photocoagulation on reduction of edema is unknown, many studies suggest some dependence on the creation of a grid pattern of light burns on the retinal pigment epithelium [9]. Limitations on the use of photocoagulation include center-involving DME. Laser to the central fovea is contraindicated because it burns the retina and destroys the retinal architecture at the site responsible for central vision. Thus, its primary benefit is for non-center involving DME. Adverse effects of laser photocoagulation for DME include scars from the laser that accumulate over time, paracentral scotomas, accidental foveal photocoagulation resulting in potential decreases in visual acuity, choroidal neovascularization, and subfoveal fibrosis.

Combined therapy: In patients that have incomplete or no response to anti-VEGF therapy, it may be beneficial to add focal laser photocoagulation, panretinal photocoagulation, or steroids in combination with anti-VEGF agents. The use of combination therapy appears to improve immediate short-term effect that can be maintained in the long term and is associated with decreased frequency of intravitreal injections. The data for these studies observed results at one year, so additional long-term studies are needed to further elucidate the optimal treatment regimen [16].

Focal laser + anti-VEGF: In a study comparing combination therapy to laser alone, the combination of ranibizumab and laser photocoagulation almost doubled the chances of gaining three or more lines of vision and decreased the risk of losing three or more lines of vision [1]. Timing of focal/grid laser treatment can either be prompt (within 1 week) or deferred (within 24 weeks), and injections occur monthly for the first 4 months, followed by continued monthly injections if OCT central subfield thickness is greater than 250 um and visual acuity is worse than 20/20.

Steroids + Laser: Intravitreal steroids (triamcinolone) with the addition of focal laser within one week is more effective than laser monotherapy short-term at 4 months, however the long-term benefit was not significant in combination therapy vs. monotherapy.

Anti-VEGF + Steroids: Although intravitreal steroid injections have not been shown to be as efficacious as anti-VEGF therapy in the treatment of DME, it has been demonstrated to have a positive effect for DME in some eyes. Multiple studies have shown that combining steroids with anti-VEGF agents have an anatomic benefit with a significant reduction in retinal thickening measured on OCT. However, combination therapy has not been shown to provide superior visual outcomes, measured by visual acuity changes, compared to anti-VEGF agents as monotherapy [16, 17]. Although combined steroid and anti-VEGF therapy has demonstrated improvement in anatomic outcomes, long-term studies are required to investigate the benefit of combination therapy in the absence of improvement in visual acuity and is thus not currently recommended [16, 17]. Some studies have pointed out that combination therapy was shown to increase the risk of developing cataracts along with increased intraocular pressure compared to eyes treated with anti-VEGF as monotherapy [18].

Anti-VEGF + Peripheral Targeted Retinal Photocoagulation: Peripheral targeted retinal photocoagulation (TRP) is retinal photocoagulation on regions of retinal capillary nonperfusion outside of the macula, an area that is hypothesized to synthesize and release VEGF and EPO. Although some studies suggested that TRP could decrease the treatment burden of anti-VEGF agents, there have not been significant improvements in combination therapy of anti-VEGF with peripheral TRP [19].

9.4 Monitoring trcatmcnt rcsponsc with OCT

Prior to the widespread implementation of OCT in the clinical setting, fluorescein angiography (FA) was commonly used in both patient care and clinical research to diagnose, plan treatment, and monitor disease progression or treatment response in patients with DR and DME. FA was used to evaluate the retinal vasculature based on its ability to assess perfusion, leakage, and staining. However, limitations of FA include invasive intravenous needle injection of fluorescein dye, adverse reactions to the dye, lengthy acquisition time of 20–30 min, and inability to quantify thickness measurements of the retina. OCT and OCT angiography address these limitations by acquiring images non-invasively, with faster acquisition speed, and with the capability of quantifying retinal thickness to monitor disease progression and treatment response.

Parameters Monitored: With advances in OCT technology and the development of SD-OCT and SS-OCT, the qualitative evaluation and diagnosis of DR and DME quickly improved. OCT can identify subretinal fluid, intraretinal cystoid fluid, disorganization of retinal inner layers (DRIL) and assess the vitreomacular interface. The speed and accuracy of OCT scans of the macula has led to its rapid integration into the clinical workflow for the evaluation of DR and DME disease severity and to monitor disease/treatment progression. OCT's capabilities in qualitative assessment of the retinal morphology has made way for detailed diagnostic criteria of DR and DME, leading to the ICO's change in DME classification in 2018 to Center-Involved DME and Non-Center-Involved DME.

OCT has also revolutionized how clinicians assess and monitor the treatment of DR and DME by detecting treatment responses in the retina after steroid, laser, or anti-VEGF therapy. OCT provides quantitative outcome measures, primarily through assessment of central retinal thickness (CRT) [1]. CRT is measured from the inner to outer borders of the retina, typically automated with computer-aided software. CRT from OCT in combination with visual acuity scores have been used in large-scale DR and DME treatment studies to determine the necessity of retreatment with anti-VEGF agents, along with assessment of treatment success, stability, or failure. As a result, OCT is now the most-used imaging modality for management of anti-VEGF treatment, providing a quantitative method for assessing individual patient responses to therapy, particularly in its ability to capture fluid leakage from active DME disease. Further advances in OCT technology may help distinguish the contributions to increased CRT from subretinal fluid, intraretinal cystoid fluid, and/or diffuse thickening (figures 9.1 and 9.2).

Beyond providing morphological and quantitative assessment of edema, OCT can also detect biomarkers representative of persistent morphological changes after treatment interventions such as DRIL, ellipsoid zone disruption, and vitreomacular interface abnormalities. These findings are almost exclusively seen on OCT and have

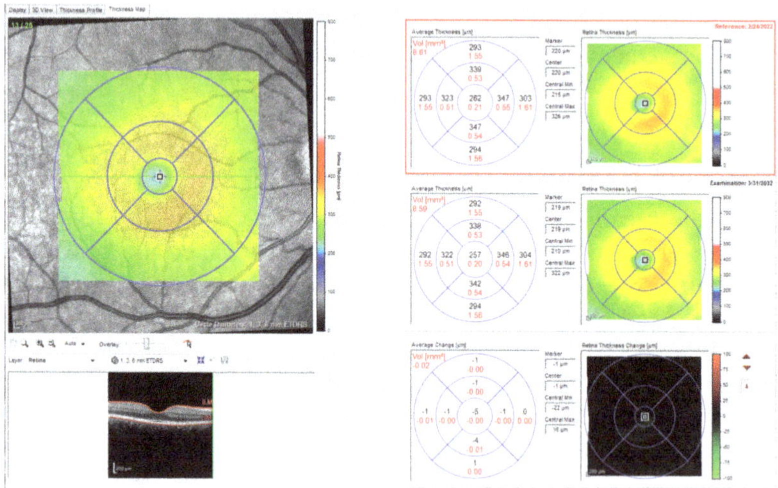

Figure 9.1. Thickness map function on OCT allows for comparison of thickening in certain areas between exams, and comparison of central macular/retinal thickness and total volume. The inferior-most map shows the average thickness change in each region, which can help to track improvement or worsening over time.

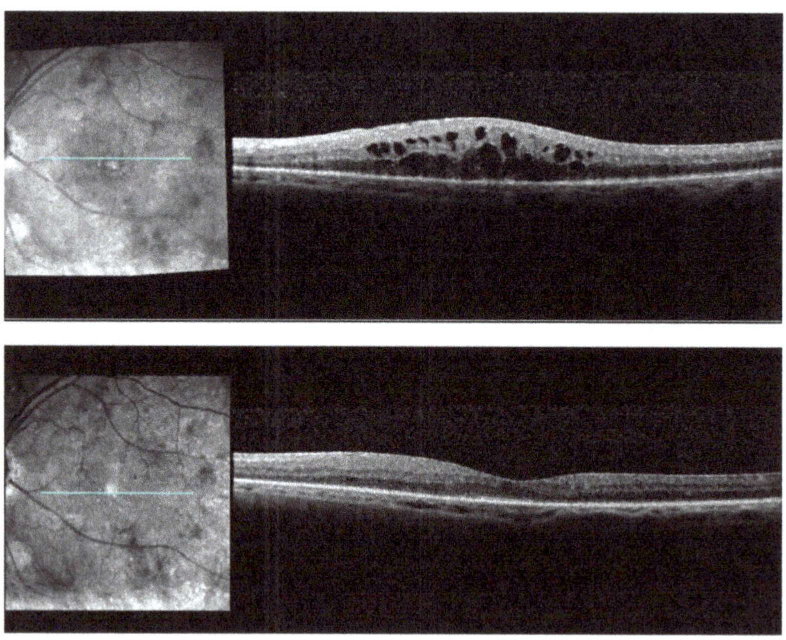

Figure 9.2. OCT of a patient with diabetic macular edema and improvement one month after intravitreal avastin injection.

the potential to predict persistence or resolution of DME following treatment and can contribute to the decision to re-treat, though large-scale trials are needed to verify the validity of these biomarkers. Other OCT modalities like OCTA and Doppler-OCT can be utilized for the visualization and differentiation between areas

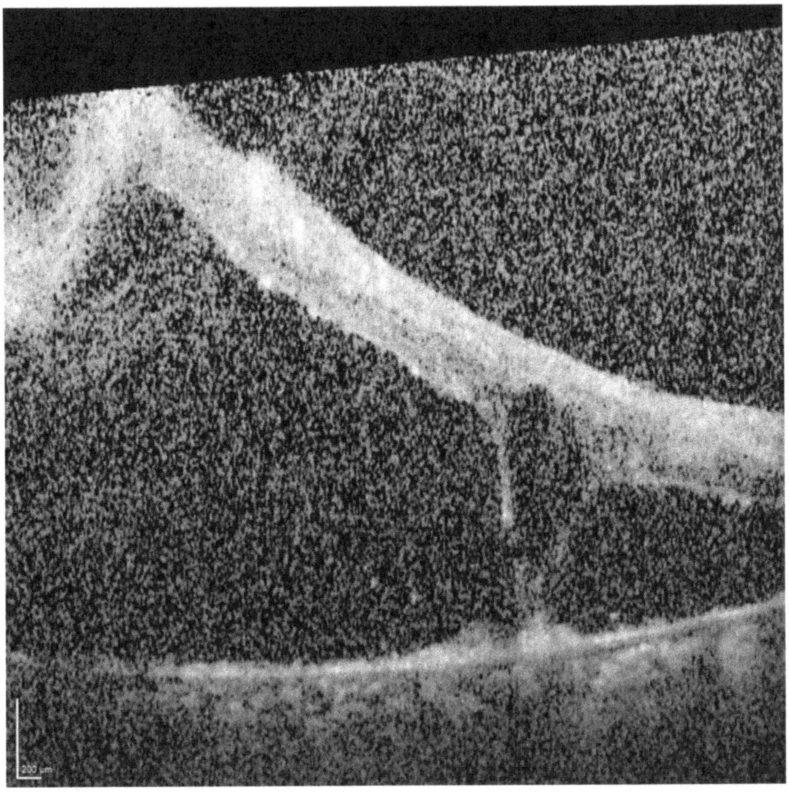

Figure 9.3. Fibrosis and preretinal membrane formation results in traction, schisis-like changes, and retinal detachments (**tractional retinal detachment**), which often require surgical intervention with vitrectomy and membrane peeling.

of perfusion and nonperfusion as adjunctive tools in the evaluation of treatment response (figures 9.3–9.7).

In the current clinical workflow, OCT and visual acuity results are recorded at baseline, and monitored real-time at each follow-up visit to determine if additional anti-VEGF injections or laser therapy would be beneficial [1]. The current recommendations to monitor DR and DME include OCT imaging, even in the absence of necessary treatment, for the early detection of morphological changes in each patient.

Definition of Treatment Response and Stability: In the DRCR Retina Network's initial study assessing treatment response of anti-VEGF agents and subsequent larger scale trials, definitions of the spectrum of treatment response were as follows:[1]

Success criteria:

- Visual acuity letter score >84 (20/20) OR
- OCT central subfield thickness <250 microns after the last anti-VEGF injection or laser photocoagulation therapy.

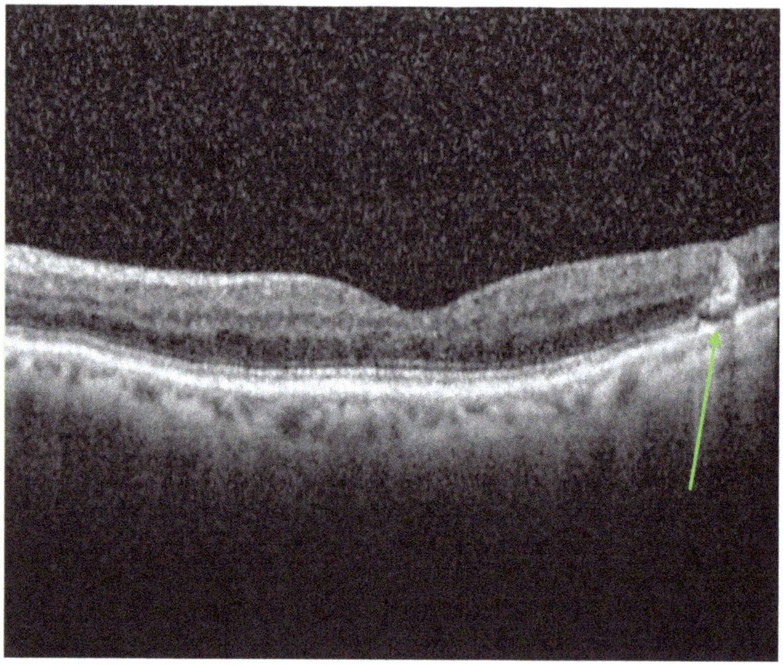

Figure 9.4. Leaking aneurysms contributing to persistent macular edema can be treated with focal laser. **Focal laser scars** can be seen as breaks in the IS/OS junction on OCT.

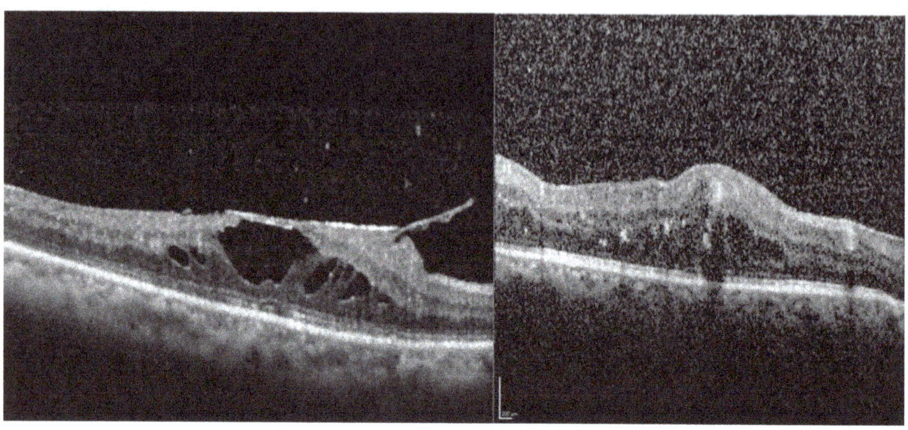

Figure 9.5. Vitreous hemorrhage can be caused by proliferative diabetic retinopathy. On OCT, vitreous hemorrhage is shown as small hyperreflective spots within the vitreous, which are representative of individual red blood cells. Other findings can include coagulated cells and the effects of shadowing.

Improvement criteria:

- Improvement of visual acuity by >5 letters OR
- OCT central subfield thickness improvement by >10% after the last anti-VEGF injection or laser photocoagulation therapy.

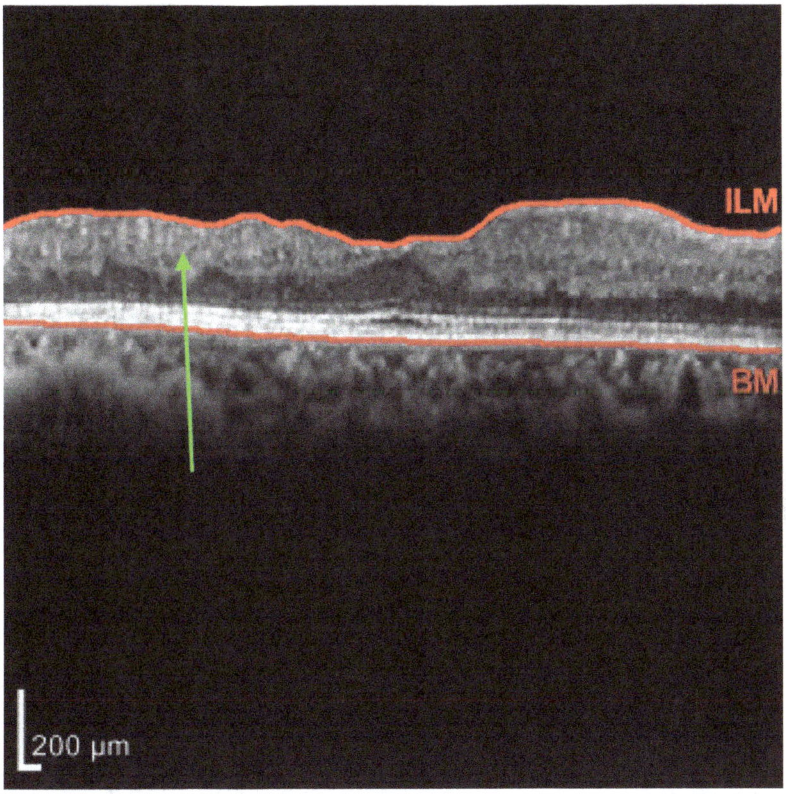

Figure 9.6. Disorganization of retinal layers is characterized by loss of the usual orderly, well demarcated layers of the inner retinal layers.

No Improvement criteria:

- Does not meet success or failure criteria
- Worsening of or improvement of visual acuity letter score by <5 letters AND
- OCT central subfield thickness decrease by <10% or increased after the last anti-VEGF injection or laser photocoagulation therapy.

Stability criteria:

- Lack of Improvement: no increase in visual acuity of ⩾5 letters
- No decrease in OCT CST ⩾10%
- No worsening
- BCVA decreases >−5 letters in the setting of persistent DME
- OCT CST increases >10%.

Failure criteria:

- Worsening of visual acuity by >10 letters AND
- OCT central subfield thickness >250 um AND
- DME as the cause of visual acuity loss.

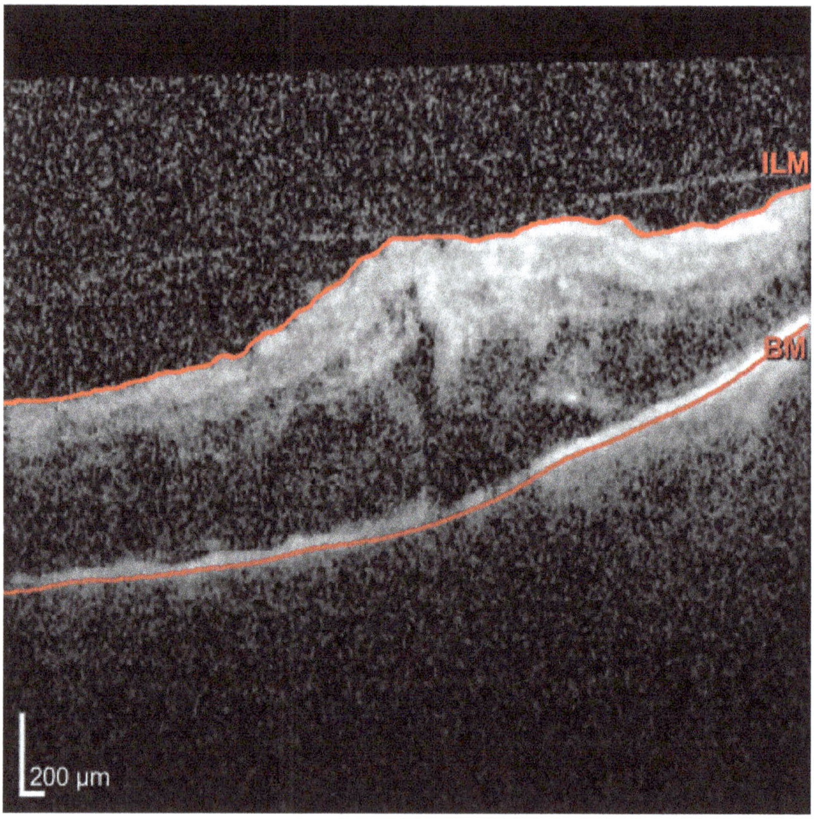

Figure 9.7. Ellipsoid zone disruption.

Risk factors for treatment failure or poor visual outcomes [20]:

- Older age
- Lower baseline best corrected visual acuity
- Increased HbA1c
- Presence of proliferative diabetic retinopathy
- Increased baseline CRT
- Increased disorganization of the retinal inner layers.

Current protocols in anti-VEGF treatment algorithm: The treatment of diabetic macular edema has seen a rapid shift with the introduction of anti-VEGF agents, leading to many studies about treatment duration and frequency. Anti-VEGF injections are cleared from the eye in about one month, but the duration of medication benefit varies between individuals. Although a standard frequency and duration of therapy are currently unknown, the DRCR Retina Network has established treatment protocols in the DRCR Protocol I for anti-VEGF agents for diabetic macular edema to reduce the overall number of injections over time in patients achieving therapeutic success. The protocol employs monthly injections for 6 months, after which treatment can be deferred if visual acuity and macular thickness measurements with OCT were stable after two consecutive injections. The results of this treatment protocol were a median use of 8, 2, 1, and 0 total injections in years 1, 2, 3, and 4, respectively, while maintaining improvement in vision over 5 years. This protocol achieved similar results compared to studies that used more frequent, including monthly, injections that can help alleviate the treatment burden (cost, injection frequency, adverse effects) on the patient and their families [1, 9, 21].

The impact of race/ethnicity on anti-VEGF treatment response in DME: While advances in diagnostic tools like OCT and treatment interventions with anti-VEGF agents continue to improve outcomes and quality of life for patients with DR and DME, the impact of race and ethnicity on treatment responses must continue to be explored and considered in disease management. Previous studies have reported that black patients have a higher susceptibility for developing DME when compared to white patients after controlling for confounding variables like HbA1c levels and length of disease. However, studies evaluating the efficacy of anti-VEGF agents in DR and DME have a predominance of white subjects, thus they are not representative of the minority groups that carry the larger burden of disease and indicate a disparity in the prevalence of disease versus the populations studied for treatment outcomes. In a retrospective review of patients with DME from an urban academic hospital, black patients had lower odds of visual acuity improvement compared to white and Hispanic patients when treated with bevacizumab. A part of these results may be due to socioeconomic status impact on health, which cannot be adequately controlled, but it demonstrates the need for more research and investigation on the differential treatment responses [22].

A follow-up study that examined clinical trials in DR and DME found that black patients are underrepresented in clinical trials compared to the burden of DME and DR in black patients. Black patients are underrepresented by a 3-fold disparity in NIH trials and a 4.5-fold disparity in industry trials for DME, while white patients are overrepresented [23]. These results indicate the need for clinical trials for DR and DME to continue recruiting patients in a more racially and ethnically inclusive way to adequately represent the populations that need treatment and continue to promote and work towards health equity.

9.5 Limitations of OCT in DME

OCT has several limitations, in both the spectral domain and time domain. It is important that the image quality, which is based on signal strength, is sufficient to produce reliable images to differentiate anatomical findings. Image artifacts can additionally limit final image data despite high resolution of images. It is not uncommon for patients with DR to develop recurrent or non-clearing vitreous hemorrhage, which obscures the OCT signal often resulting in poor quality or unusable images. Generally, artifacts observed included misidentification of the outer and inner retina, degraded scan image, cut edge artifact, incomplete segmentation error, and superior or inferior shifts of retinal images without corresponding shifts of segmentation lines. One of the most common artifact errors are segmentation errors that result when the retinal boundaries are misidentified by the machine's algorithm, which can result in errors the estimation of retinal thickness [24].

Another limitation of OCTs in DME is the reproducibility of anatomical measurements, namely retinal thickness. Reproducibility in DME poses additional challenges as edema can distort retinal layers and segmentation calculations by OCT. This is a reported finding often found by subretinal fluid in other similar diseases such as neovascular age-related macular degeneration. Different OCT devices utilize specific segmentation algorithms, which makes reproducibility impacted by both the imaging hardware as well as the algorithm, both of which can be affected by macular edema. Specifically, two of the most common OCT devices are the Cirrus HD-OCT by Carl Zeiss Meditec and the Spectralis SD-OCT by Heidelberg Engineering. When the retinal nerve fiber layers were compared in normal eyes, it was found that different devices had varying calculations on the retinal nerve fiber layer thicknesses. It was also noted that there is a variable amount of signal strength as well as different normative databases. This makes it more challenging to use OCT data from different vendors' devices interchangeably and is potentially a large limitation now that electronic health records are being merged between healthcare systems [25].

With respect to clinical significance, it is relevant to define objective markers to monitor disease progression and to gauge response to treatment. The Diabetic Retinopathy Clinical Research Network has identified a modest correlation between patients' best corrected visual acuity and the OCT center point thickness prior to focal laser photocoagulation, and additionally has identified a modest correlation between the change in visual acuity and change in OCT center point thickening through the first year after laser treatment. However, for most other measurements there was a considerable variation between the visual acuity and the retinal thickness. It is important to note that while OCT measurements can be a useful clinical tool, retinal thickness is not a surrogate for visual acuity as a primary outcome in diabetic macular edema studies [24].

In ischemic maculopathy, OCT plays a limited role in diagnosis particularly in the setting of co-existing macular edema. The presence of macular ischemia will limit response to therapy and is unable to distinguish patients who are refractory to either anti-VEGF or macular photocoagulation treatment. In these situations, fluorescein angiography remains the gold standard for diagnosis and evaluation, confirming the presence of nonperfusion in the macula by an enlarged foveal avascular zone. OCTA may demonstrate utility over fluorescein angiography as an alternative in diagnostics and monitoring for ischemic maculopathy [26].

9.6 Machine learning prospects of OCT in DME

For retinal diseases, computer-aided diagnostics has become a rapidly expanding area of innovation. As OCTs have been robustly utilized for driving diagnostic and therapeutic decisions, it is one of the key imaging modalities used to define imaging findings and run data-driven models to create machine learning algorithms. Computer-aided diagnostics and machine learning have several applications in OCT for diabetic macular edema. Given that DME can cause vision impairment and there are several treatment modalities to improve outcomes, the use of OCT to guide treatment monitoring and decisions is essential. Machine learning can be used to distinguish the presence of and the severity of disease, and can ultimately expedite and improve diagnostic and therapeutic monitoring. In DME, some of the specific findings that guide computer-aided diagnostics are diffuse retinal thickening, cystoid macular edema, and serous retinal detachments. While there are subcategories of more precise findings, such as epiretinal traction, establishing clinically significant findings are important to determine what retinal findings to focus on in computer-aided diagnostics. There exist several automated strategies to identify pathologic fluid accumulations in the retina. Certain strategies include segmenting the fluid accumulations as they can be distinguished with well-defined borders. Many of these methods are limited by filtering techniques such as size, border intensity, texture, and clustering of findings. Processing of the image involves algorithms that attempt to identify diabetic macular edema pathologies, and the processed images are passed through diagnostic evaluations by clinicians. A second tier of processing occurs to identify severity of the disease, which is again held in comparison to clinician evaluation. By repeating this process, the accuracy of detection will gradually improve, and certain techniques that combine multiple classifiers while running large sets of data will ultimately make the computer-aided diagnostic systems accurate in detecting diabetic macular edema [27].

9.7 Future directions of OCT in DME

Imaging biomarkers are a tool that may shape the future directions of OCT in diabetic macular edema to serve as predictors for treatment response. Parameters such as central foveal thickness, intraretinal cysts, subretinal fluid, disorganization of the inner retinal layers, hyperreflective dots, and vitreoretinal relationship have been studied for predicting treatment outcomes. As clinical trials study the relationship between imaging biomarkers and outcomes, the use of OCT will continue to greatly impact clinical practice [28].

Notably, swept-source OCT has been increasingly used in clinical practice. In comparison to traditional spectral-domain OCT, swept-source utilizes lights with longer wavelength along with rapid scanning that can assess deeper tissue structures that allow clinicians to examine choroidal structures. Swept-source has the added advantage of using a denser scan pattern as well as a safer higher energy to detect the deeper structures of the posterior pole [29].

One of the benefits of using swept-source OCT over spectral-domain OCT is the significantly faster scanning speed which decreases the amount of time required to capture three-dimensional images of the retina. This may overcome the hindrance of retinal fluid in visualizing posterior anatomical pathologies [30].

9.8 Limitations of OCTA in DME

OCTA is limited in current clinical practice due to its lack of broad applicability. Most physician offices have OCT devices but do not have separate OCTA devices. OCTA is more expensive, and although some OCTA capabilities can be installed as a software upgrade to a current OCT device in clinic, it is not compatible with all OCT devices. Moreover, reimbursements for OCT and OCTA are currently the same, thus physicians and offices do not receive increased compensation for obtaining an OCTA, making it more difficult and less justifiable to purchase a more expensive OCTA device. As it currently stands, clinicians are still unclear in which areas OCTA can fully replace FA as OCTA still has several limitations, lacks standardization between different OCTA devices, and still needs large population studies to provide reference ranges [15].

OCTA poses several limitations in image capture, time to acquire images, and in the segmentation analyses. OCTA utilizes information from multiple scans to exact the direction of red blood cell flow using several algorithms. When there are pathologies that result in turbulent blood flow or poor flow–for instance, in microaneurysms or polypoidal lesions–the accuracy of OCTA may fall compared to when using fluorescein angiography. Additionally, OCTA has a limited field of view compared to fluorescein angiography and indocyanine green angiography, and this inhibits the ability to detect peripheral changes such as ischemia and neovascularization on OCTA. In comparison to fluorescein angiography, OCTA typically is more prone to show artifacts because it is more reliant on careful analysis of red blood cell flow [31]. OCTA is also limited in its ability to evaluate the breakdown of the blood-retinal barrier compared to FA, as FA detects leakage of fluorescein molecules from the weaker and more permeable pathologic vessels. Patients who are unable to remain still during image acquisition can be especially challenging in the presence of diabetic macular edema due to the sensitivity to minor eye movements. This notably impacts the reproducibility of the OCTA devices as it may be subject to discrepancy from patients' ability to fixate during the imaging [31].

OCT and OCTA often use segmentation analyses to automate the measurements to make anatomic designations within the retina. For values such as retinal thickness and vascular indices, there may be a significant impact from concurrent retinal diseases in obtaining an accurate measurement. Manually segmenting the retina is particularly tedious so improved accuracy in segmentation analyses may be warranted for OCTA [31].

9.9 Future directions of OCTA in DME

Certain new algorithms have developed to correct the projection artifacts which can greatly skew visualization of the structures of the deep and superficial capillary. An algorithm known as projection-resolved OCTA utilizes the reflectance-based projection-resolved algorithm to increase flow signal to avoid projection artifacts. Similarly, adaptive optics is another variation to overcome projection artifacts with improved axial and transverse resolution to augment segmentation capabilities. Automated registration and selective merging algorithms have been suggested in conjunction with eye tracking systems to decrease motion artifacts; however, they exist with their limitations including extended acquisition time. To address the limitation of small view fields in OCTA, a concept known as montage OCTA could be utilized to create wide-field imaging and expand on the utility of OCTA over fluorescein angiography [31].

References

[1] Jampol L M, Glassman A R and Sun J 2020 Evaluation and care of patients with diabetic retinopathy *New Engl. J. Med.* **382** 1629–37

[2] Virgili G *et al* 2015 Optical coherence tomography (OCT) for detection of macular oedema in patients with diabetic retinopathy *Cochrane Database Syst. Rev.* **1** 1465–858

[3] Bain S C *et al* 2019 Worsening of diabetic retinopathy with rapid improvement in systemic glucose control: a review *Diabetes, Obes. Metab.* **21** 454–66

[4] Mohamed Q, Gillies M C and Wong T Y 2007 Management of diabetic retinopathy: a systematic review *JAMA* **298** 902–16

[5] Chung Y-R *et al* 2017 Association of statin use and hypertriglyceridemia with diabetic macular edema in patients with type 2 diabetes and diabetic retinopathy *Cardiovasc. Diabetol.* **16** 1–7

[6] Maturi R K *et al* 2018 Effect of adding dexamethasone to continued ranibizumab treatment in patients with persistent diabetic macular edema: a DRCR network phase 2 randomized clinical trial *JAMA Ophthalmol.* **136** 29–38

[7] Osaadon P *et al* 2014 A review of anti-VEGF agents for proliferative diabetic retinopathy *Eye* **28** 510–20

[8] Michaelides M *et al* 2010 A prospective randomized trial of intravitreal bevacizumab or laser therapy in the management of diabetic macular edema (BOLT study): 12-month data: report 2 *Ophthalmology* **117** 1078–86

[9] Wells J A *et al* 2015 Aflibercept, bevacizumab, or ranibizumab for diabetic macular edema *New Engl. J. Med.* **372** 1193–203

[10] Brown D M, Ou W C and Wong T P *et al* 2018 Targeted retinal photocoagulation for diabetic macular edema with peripheral retinal nonperfusion: three-year randomized DAVE trial *Ophthalmology* **125** 683–90

[11] Wykoff C C, Abreu F and Adamis A P *et al* 2022 Efficacy, durability, and safety of itnravitreal faricimab with extended dosing up to every 16 weeks in patients with diabetic macular oedema (YOSEMITE and RHINE): two randomised, double-masked, phase 3 trials *Lancet* **399** 741–55

[12] Srinivas S *et al* 2020 Effect of intravitreal ranibizumab on intraretinal hard exudates in eyes with diabetic macular edema *Am. J. Ophthalmol.* **211** 183–90

[13] Ross E L *et al* 2016 Cost-effectiveness of aflibercept, bevacizumab, and ranibizumab for diabetic macular edema treatment: analysis from the diabetic retinopathy clinical research network comparative effectiveness trial *JAMA Ophthalmol.* **134** 888–96

[14] Ghasemi Falavarjani K and Nguyen Q D 2013 Adverse events and complications associated with intravitreal injection of anti-VEGF agents: a review of literature *Eye* **27** 787–94

[15] Chua J *et al* 2020 Optical coherence tomography angiography in diabetes and diabetic retinopathy *J. Clin. Med.* **9** 1723

[16] Chawan-Saad J *et al* 2019 Corticosteroids for diabetic macular edema *Taiwan J. Ophthalmol.* **9** 233

[17] Maturi R K, Glassman A R and Josic K *et al* 2021 Effect of intravitreous anti–vascular endothelial growth factor vs sham treatment for prevention of vision-threatening complications of diabetic retinopathy: the protocol W randomized clinical trial *JAMA Ophthalmol.* **139** 701–12

[18] Mehta H *et al* 2018 Anti-vascular endothelial growth factor combined with intravitreal steroids for diabetic macular oedema *Cochrane Database Syst. Rev.* **4** 1–59

[19] Brown D M, Emanuelli A and Bandello F *et al* 2022 KESTREL and KITE: 52-week results from two Phase III pivotal trials of brolucizumab for diabetic macular edema *Am. J. Ophthalmol* **238** 157–72

[20] Chen Y-P *et al* 2019 Factors influencing clinical outcomes in patients with diabetic macular edema treated with intravitreal ranibizumab: comparison between responder and non-responder cases *Sci. Rep.* **9** 1–8

[21] Gross J G *et al* 2015 Panretinal photocoagulation vs intravitreous ranibizumab for proliferative diabetic retinopathy: a randomized clinical trial *JAMA* **314** 2137–46

[22] Osathanugrah P, Sanjiv N, Siegel N H, Ness S, Chen X and Subramanian M L 2021 The impact of race on short-term treatment response to bevacizumab in diabetic macular edema *Am. J. Ophthalmol.* **222** 310–7

[23] Sanjiv N, Osathanugrah P and Harrell M *et al* 2022 Race and ethnic representation among clinical trials for diabetic retinopathy and diabetic macular edema within the United States: a review *J. Natl. Med. Assoc.* **114** 123–40

[24] Schimel A M, Fisher Y L and Flynn H W 2011 Optical coherence tomography in the diagnosis and management of diabetic macular edema: time-domain versus spectral-domain *Ophthalmic Surg. Lasers Imaging* **42** S41–55

[25] Sohn E, John J C, Lee K, Niemeijer M, Sonka M and Abràmoff M D 2013 Reproducibility of diabetic macular edema estimates from sd-oct is affected by the choice of image analysis algorithm *Invest. Ophthalmol. Vis. Sci.* **54** 4184–8

[26] Vidal P L, de Moura J, Díaz M, Novo J and Ortega M 2020 Diabetic macular edema characterization and visualization using optical coherence tomography images *Appl. Sci.* **10** 7718

[27] Chou H D, Wu C H and Chiang W Y *et al* 2022 Optical coherence tomography and imaging biomarkers as outcome predictors in diabetic macular edema treated with dexamethasone implant *Sci. Rep.* **12** 3872

[28] Xiong K, Gong X, Li W, Yuting L, Meng J, Wang L, Wang W and Wenyong H 2021 Comparison of macular thickness measurements using swept-source and spectral-domain optical coherence tomography in healthy and diabetic subjects *Curr. Eye Res.* **46** 1567–73

[29] Fujiwara A, Kanzaki Y and Kimura S *et al* 2021 En face image-based classification of diabetic macular edema using swept source optical coherence tomography *Sci. Rep.* **11** 7665

[30] Khadamy J, Abri Aghdam K and Falavarjani K G 2018 An update on optical coherence tomography angiography in diabetic retinopathy *J. Ophthalmic Vis. Res.* **13** 487–97

[31] Chua J *et al* 2020 Optical coherence tomography angiography in diabetes and diabetic retinopathy *J. Clin. Med.* **9** 1723

IOP Publishing

Photo Acoustic and Optical Coherence Tomography Imaging, Volume 1
Diabetic retinopathy
Ayman El-Baz and Jasjit S Suri

Chapter 10

On the eye diseases diagnosis using OCT and fundus imaging techniques

Prakash Kumar Karn, Renoh Johnson Chalakkal and Waleed H Abdulla

10.1 Introduction

The retina is the eye's innermost layer, which converts the projected light into electrical signals directed to the brain. Various eye diseases affect different structures of the retina. Diabetic retinopathy (DR) causes the growth of extra blood vessels, macular oedema causes swelling of the fovea region, age-related macular degeneration (ARMD) causes the formation of drusen and cysts and many more. These initial symptoms progress to cause severe retinal illnesses such as DR, hypertensive retinopathy (HR), retinopathy of prematurity (ROP), etc. The primary microvascular consequence of diabetes is the development of DR. It also has a considerable negative influence on the global health systems. According to one study [1], the population of DR-affected individuals would rise by 50% from those in 2010 by 2030. Another study conducted in 2021 [2] predicted that the number of people with vision-threatening DR would rise from 28.54 million to 44.82 million by 2045 AD. Similarly, the World Health Organization (WHO) reports that hypertension retinopathy affects 1.13 billion individuals globally [3]. These alarming healthcare statistics have inspired numerous researchers to use all available mathematical and scientific tools, including artificial intelligence (AI), to analyse retinal images and diagnose various retinal illnesses.

The diseases mentioned above cannot be identified without a reliable retinal imaging technique. Fundus imaging can give a clear picture of the macula, optic disc, central and tertiary retinal blood vessels, and other visual anomalies. Due to a shallow learning curve to train, affordability and portable nature, fundus imaging is still widely used by specialists for diagnosing various eye diseases. However, the usability of the fundus image is limited to the retinal surface condition, as it cannot visualise the sub-retinal layers. Furthermore, the use of visible light in fundus

doi:10.1088/978-0-7503-2052-8ch10

photography cannot penetrate the cataract eye or patients having central retinal vein occlusion. It also cannot estimate the size of fluid or cyst formed beneath the retina. Considering these limitations, eye disease diagnosis using Optical Coherence Tomography (OCT) came into practice. OCT is a non-invasive technique that can capture the cross-section of the retina. Various sub-retinal layers can be visualised in an OCT image. The wide range of information offered by OCT can help ophthalmologists estimate the exact location of fluid accumulation, cyst formation, hole in the fovea, drug and laser treatment effectiveness, and many more. Generally, histopathology and visible symptoms are only seen when the disease has progressed to an advanced stage. Hence, a fundus image backed with OCT imaging which can visualise cross-sections to a micrometre level resolution allows us to analyse various sub-retinal layers and detect diseases early.

This chapter will discuss fundus and OCT image-based screening in detail by comparing their usability applied to different pathological settings. We will focus on biomarkers associated with fundus and OCT images for various retinal disease classifications, including diabetic retinopathy. Also, this chapter aims for a targeted discussion evaluating both retinal imaging methods. A comprehensive comparison of imaging techniques is carried out. The results are reported to help clinicians, researchers, and imaging technicians quickly decide which retinal method to adopt for different applications. Finally, this chapter will discuss recent advancements in commercially available fundus and OCT imaging devices.

10.2 Fundus camera and various modes of fundus examination

A fundus camera is a specialised low-power microscope with a digital camera attached to it used to take pictures of the retina, retinal vasculature, optic disc, macula, and posterior pole of the eye (i.e., the fundus). A fundus camera's optical setup is based on the idea of monocular indirect ophthalmoscopy [4]. An upright, enlarged view of the fundus is provided via a fundus camera. Figure 10.1(a) shows a DRI TRITON TOPCON camera capable of capturing the fundus, OCT, and fluorescence angiography of the retina. Figures 10.1(b) and (c) show a diseased and normal fundus image captured with the same device in figure 10.1(a). Practically, there are various modes of fundus examination, which are described as follows:

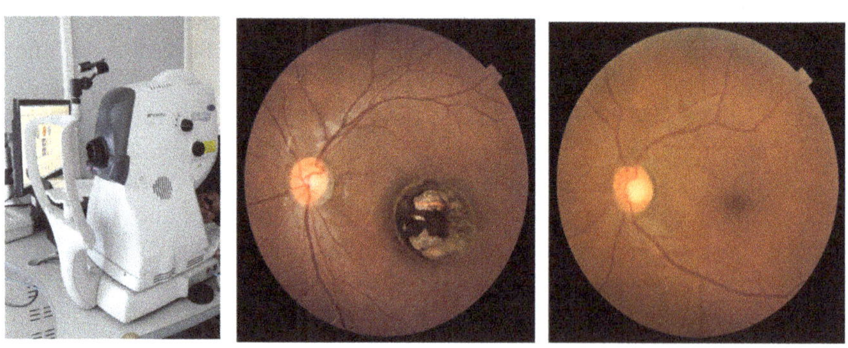

Figure 10.1. (a) Fundus camera (b) diseased fundus image (c) normal fundus image.

10.2.1 Colour fundus photography

Fundus retinal imaging is a non-invasive way of imaging the retina that is popular among the ophthalmic community and the targeted patient population. Fundus retinal images reveal vital information about the health status of a person. Over the past 15 years, extensive research and clinical studies have been carried out using fundus images for automatizing the pathology screening and diagnosing process. Various pathologies could be screened and diagnosed using fundus retinal images: macular oedema, diabetic retinopathy, and glaucoma, which are the most important causes of blindness in the developed world.

Fundus imaging is the most important method of examining the eye's surface. It gives a colour picture of the retina and is captured with white light. Sometimes the patient's pupil needs to be dilated with a 1% cyclopentolate or tropicamide drop to focus on the retina. In recent years, some handheld fundus cameras can capture without the need for dilation of the pupil. These fundus cameras are called non-mydriatic fundus cameras and use infrared light instead of white light to capture the image [5].

Fundus retinal imaging is a widely used imaging technique to record the structural information of the retina. The fundamental design of any fundus camera is based on the principles of the indirect ophthalmoscope. The retina is photographed using the pupil as the entry and exit path for the light rays traveling from the fundus camera. A fundus camera is defined in terms of its field of view (FOV). The FOV can vary from 30° to 120°. Narrower FOV offers higher magnification but a smaller photographed area of the retina.

10.2.2 Angiography fundus image

A fluorescent dye is injected into the bloodstream during an angiography procedure to photograph and record the vascular flow within the retina and the surrounding tissue. When the light of a specific wavelength strikes this dye, it shines in a different colour. A blue excitation light of 490 nm and a yellow fluorescence light of 530 nm are employed in sodium fluorescein angiography (SFA), a technique for visualising retinal vascular disease. Among other things, it is frequently used to scan cystoid macular oedema and diabetic retinopathy. Indocyanine green angiography (IGC) uses a barrier filter to capture light at 500 and 810 nm and a near-infrared diode laser at 805 nm to view more profound choroidal disorders.

10.2.3 Ultra-wide field fundus image

UWF imaging is essential to visualise the periphery in diabetic retinopathy (DR). It can capture the progress of neovascularisation elsewhere in proliferative diabetic retinopathy screening. Figure 10.2 shows an example of ultra-widefield fundus photography captured at Drishti Eye Care, Nepal using DRI OCT Triton plus with 45-degree FOV.

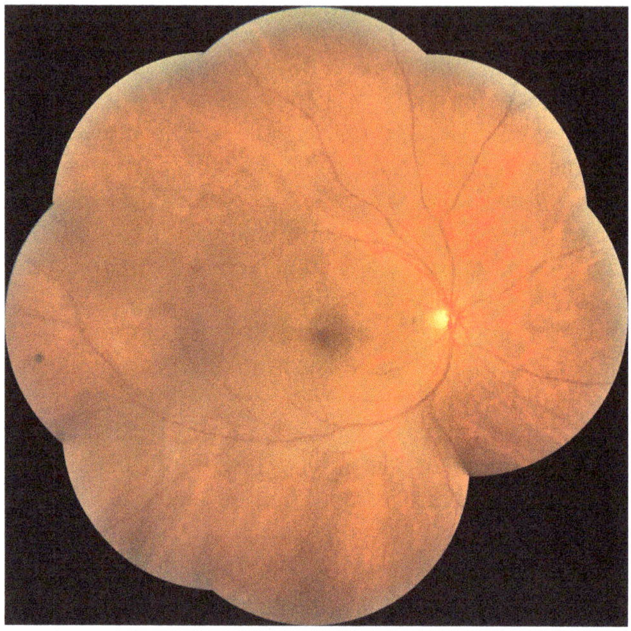

Figure 10.2. An example of ultra-wide field fundus photo.

10.2.4 Smartphone-based fundus imaging

The future devices used in retinal imaging must be mobile, portable and affordable, unlike the currently used bulky and expensive fundus cameras. A brief discussion about the mobile retinal screening facilities is given in [4]. Portable retinal screening systems can play a crucial role in eradicating preventive blindness across the globe.

The fundamental prerequisite for such a portable and mobile screening system will be a robust pre-processing unit that can standardize the camera-captured input retinal images to match the input requirements of the automatic pathology screening system, irrespective of the FOV, intensity, contrast and other nonlinearities associated with the mobile retinal image capturing process.

For examining a retina, the simplest tool that a clinician can use is an ophthalmoscope or a 20D lens. But these tools have a very steep learning curve and need a long time of practice to master the use of them. With the internet revolution, every person has access to at least one smart device in the form of a smartphone or a tablet. The prolonged use of such devices has resulted in a significant increase in the global population affected by vision-related problems. To address the increased need for specialised eye care, measures such as outreach mobile clinics and providing necessary training in using ophthalmology screening devices for optometrists, general practitioners and even nurses are being experimented with. However, due to the steep learning curve of these devices, the success of such initiatives is limited.

Fundus cameras, due to their ease of use, easier storage options, better image quality and faster options for secured electronic transfer, are widely used. However, retinal imaging using a fundus camera is a time-consuming and tedious process. Also, due to the bulk size and cost, fundus cameras are not an option for outreach mobile clinics and to be used by general practitioners or nurses at remote clinics. OCT devices also face the same problems as fundus cameras. For OCT devices, the affordability factor is far worse for small clinics and practices. Therefore, there is a need for smart, portable, and inexpensive retinal imaging systems to perform widespread screenings to detect vision-related pathologies.

Considering the complex imaging set up required for OCT imaging, since it is more feasible and practical to develop portable fundus cameras, recently there has been significant progress in developing portable retinal fundus cameras. Such devices are not only cost effective but also highly portable. Recent developments in the field of telecommunications, such as 5G, have helped us with smartphones that are small, low-power, and affordable. Making a smartphone the user interface and image processing unit of retinal fundus camera can significantly help in reducing the developing cost and make the device small and portable. First, the ophthalmoscopes were redesigned to be smartphone-based [6, 7]. Recently, new smartphone-based retinal imaging devices are released including oDocs nun IR [5, 8], D-Eye, Peek Retina, iNview, Optomed and Remedio. The requirement for the eyes to be dilated using drops for most of the smart phone-based imaging systems have limited their intended remote use by non-experts. In most countries, the dilation drops can be only administered by registered ophthalmologists and optometrists with a current drug administering licence. Considering this a few smartphone-based devices such as nun IR [5], have designed an infra-red based technique to find the view of the retina, thereby enabling non-experts to capture the retinal image without the need for dilation.

Such smartphone-based retinal imaging devices would help in conducting remote teleophthalmology consultations, making specialised eye care accessible to everyone globally. Recently, there have been a few significant developments in the field of telemedicine applied to ophthalmology [9–11]. However, more research must be done to establish the cost-effectiveness and usefulness of such telemedicine tools.

One major challenge associated with smartphone-based retinal imaging is the limited FOV of the devices. The maximum FOV reported for a smartphone-based device is around 45–50 degrees. Hence, additional image processing techniques are required to stitch the limited FOV images to create a wider montage with increased FOV. A few such retinal montaging techniques are reported in [12–14]. With such FOV-improving techniques, the usability of smartphone-based fundus imaging devices is expected to further increase.

10.3 Optical coherence tomography

A non-invasive imaging method called optical coherence tomography (OCT) can quickly record images of the retina's cross-section and microstructure. Medical diagnosis is made possible by the analysis of retinal OCT pictures. Manual analysis of each image captured takes a lot of time, and the outcomes depend on individual experience. The retinal OCT study still has certain misconceptions, despite the fact that many ophthalmologists are familiar with fundus pictures. To see the interior structure of tissues, nerves, the retina, brain neurons, the chemical structure of a compound, the flow of blood, etc., OCT is one of the most commonly used imaging techniques. OCT measures the size of backscattered light and the echo time delay to get cross-sectional images of internal microstructures in both materials and biological systems. OCT being a high-resolution imaging technique has the advantage of achieving an image resolution of 1–15 m, similar to ultrasound imaging. OCT is sometimes called sectional imaging, employing a light source with a continuous phase difference. Below is a description of various OCT imaging techniques based on their operating principles.

10.3.1 Time domain OCT (TD-OCT)

The main components of this imaging system are a coherent light source, a beam splitter, a reference arm, and a photodetector, which are based on the Michelson Interferometer [15]. With a manual adjustment of the reference arm, time domain OCT uses a light source that is totally out of phase but has a constant phase difference.

10.3.2 Fourier domain OCT (FD-OCT)

While this imaging technique uses broadband light sources that are out of phase and have variable phase differences, its operating principle is similar to that of TD-OCT. Different wavelengths of light are mixed in the light that the photodiode detects [16]. These lights run through a spectrometer, and the signal is broken down into component parts using a Fourier analysis. The reference arm is fixed in this method, while the retina's various layers are scanned using light sources with changing wavelengths. This imaging method is often referred to as SD-OCT or spectrometer-based OCT imaging.

10.3.3 Swept-source OCT

With a sweeping source as the light source, this is also a Fourier domain-based OCT imaging approach [16]. The optical setup is similar to FD-OCT, but instead of a broadband light source, it uses an optical source that quickly sweeps a short linewidth over a wide range of wavelengths. A faster A-scan with deeper penetration is made possible by swept-source OCT, which employs a wavelength-sweeping laser and a dual-balanced photodetector (EDI). Figure 10.3 classifies several OCT picture types depending on image acquisition technology and region.

10.4 Comparison between fundus and OCT imaging technique

Slit-lamp imaging or fundus imaging is still a common practice among ophthalmologists today. Despite being a standard treatment, fundus imaging has some drawbacks. It cannot accurately assess the state of the macula, which is responsible for central vision, or ensure an appropriate image during severe retinal occlusion. But at the same time, OCT imaging provides layer-by-layer detail of the retina, allowing the ophthalmologist to pinpoint the precise reason for vision loss, the likelihood that

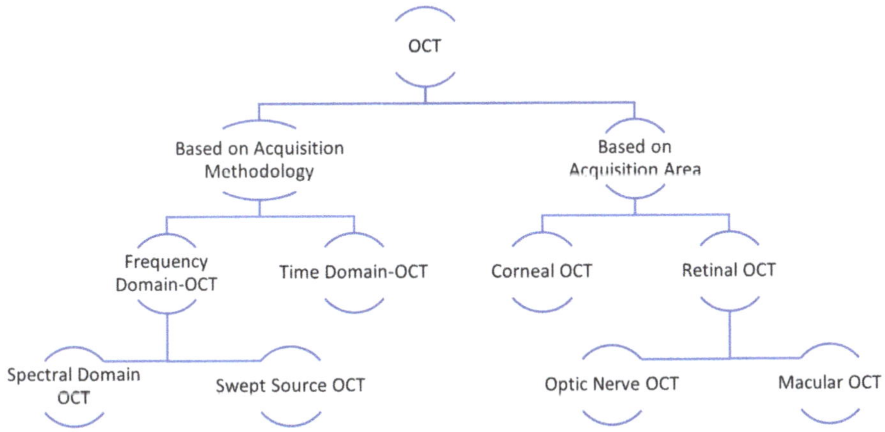

Figure 10.3. Classification of different types of OCT.

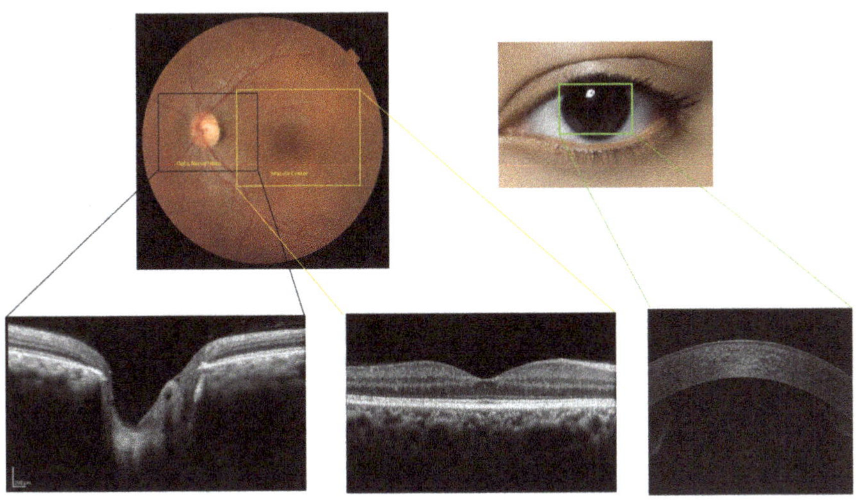

Figure 10.4. OCT image captured at various sections of the eye. (A) OCT of optic disc-centred retina (B) OCT of macula-centered retina (C) OCT of cornea.

vision will return, the status of the patient's treatment, and the efficacy of various medications. Figure 10.4 shows various OCT images captured at various sections of the eye, i.e. the optic disc-centred retina, macula-centered retina, and cornea. The retina comprises 11 layers, each playing a crucial role in maintaining clear vision. When issues such as intra-retinal cysts, sub-retinal fluid, exudates, neovascularization, and erosive layers are present, they pose a higher risk of causing damage to the retina. The specific location of these pathologies within the layers revealed by OCT determines the nature and severity of the eye disease. As a result, segmenting the 11 OCT layers is necessary to find such pathologies and categorise the abnormality using machine learning techniques.

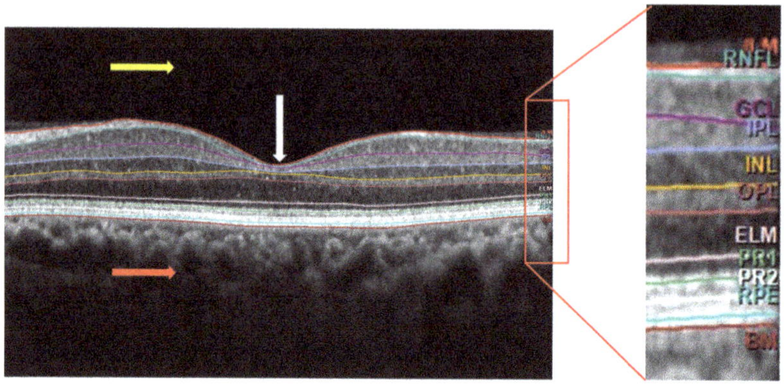

Figure 10.5. Layers of retinal OCT with various region marked by arrow.

These layers are: the internal limiting membrane (ILM), retinal nerve fibre layer (RNFL), ganglion cell layer (GCL), inner plexiform layer (IPL), outer plexiform layer (OPL), outer nuclear layer (ONL), external limiting membrane (ELM) photoreceptor layer, retinal pigment epithelium (RPE), Bruch's membrane (BM) and choroid layer. Figure 10.5 shows all the 11 sub-retinal layers of OCT. The yellow arrow shows the vitreous area, which is also called a hypo-reflective area. The deep valley-like structure shown by the white arrow is the fovea of macula. The red arrow represents the choriocapillaries or choroid region.

Each layer has its own properties, such as reflectivity and anatomical description. The detailed explanation of the different layers of OCT is given in table 10.1.

10.5 Fundus and OCT image-based diagnosis of various eye diseases

The basic screening approach for identifying the retinal structures is fundus imaging. According to published research, fundus images can diagnose several vision disorders in their early and advanced phases, such as: DR, HR, retinal holes, optic disc cupping, ARMD, etc. One critical component for identifying eye illness in all these diseases is a retinal blood vessel and sub-retinal layer segmentation.

10.5.1 Diabetic retinopathy screening

Any damage to the retina or retinal structures in the eyes is referred to as retinopathy, and it can impair vision and even result in blindness. Exudates and small aneurysms on the retina's surface are signs of the early stages of retinopathy. Therefore, an early diagnosis of exudates lessens the disease's severity and stops it from advancing. However, since exudates differ in terms of shape and size, it can be difficult to recognise exudates in its early stage. Non-proliferative DR (NPDR) and proliferative DR (PDR) are the two main stages of DR, from which it is possible to determine the severity of the condition and course of treatment. The symptoms of NPDR include micro-aneurysms, haemorrhages, and hard exudates. If this

Table 10.1. Description of various layers of OCT.

Layer no	Nomenclature	OCT reflectivity	Features
1	Inner limiting membrane (ILM)	Hyporreflective	The top layer of OCT is often misdiagnosed with epiretinal membrance (ERM)
2	Retinal nerve fiber layer (RNFL/ NFL)	Hyperreflective	Thick near-optic disc; change in thickness indicates the presence of disease
3	Ganglion cell layer (GCL)	Hyporeflective	Made of ganglion cell bodies
4	Inner plexiform layer (IPL)	Hyperreflective	Synapses between bipolar and ganglion cells
5	Inner nuclear layer (INL)	Hyporeflective	Consists of the cell bodies of horizontal cells and may displace ganglion cells
6	Outer plexiform layer (OPL	Hyperreflective	Synapses between photoreceptor, bipolar, and horizontal cells
7	External limiting membrane (ELM)	Hyperreflective	The zone between muller cells and photoreceptors
8	Photo receptor layer (PRL)	Hyporeflective	It is the brightest layer, responsible for converting light to electrical energy
9	RPE and bruch membrane	Hyperreflective	Important layer to calculate retinal thickness and Bruch membrane opening
10	Choroid	Hyporeflective	The innermost layer of OCT. Its thickness indicates central serous chorioretinopathy (CSCR)

condition is left untreated, it will eventually reach a proliferative stage in which aberrant new blood vessel growths (neovascularisation) occur. Neovascularisation is further divided into two types: neovascularisation on the disc (NVD) and neovascularisation elsewhere (NVE). With increased patient volume, it might be challenging for clinicians to manually grade fundus images because it takes a lot of time and is more likely to get errors. These difficulties with manual grading paved the way for automatic segmentation processes. Compared to the manual segmentation procedure, issues like time complexity and error probability can be resolved using automatic blood vessel segmentation. Figure 10.6 compares a healthy individual's vision to a person with DR illness.

10.5.1.1 Blood vessel segmentation method for the detection of DR

Blood vessel segmentation is the important step to monitor the proliferative DR and its progression. Blood vessels are separated from its background and are compared with the previous record to find the new blood vessels growth. Karn *et al* [17] employed a hybrid active contour model to detect retinal blood vessels using a

Figure 10.6. Normal vision versus vision with DR (PDR).

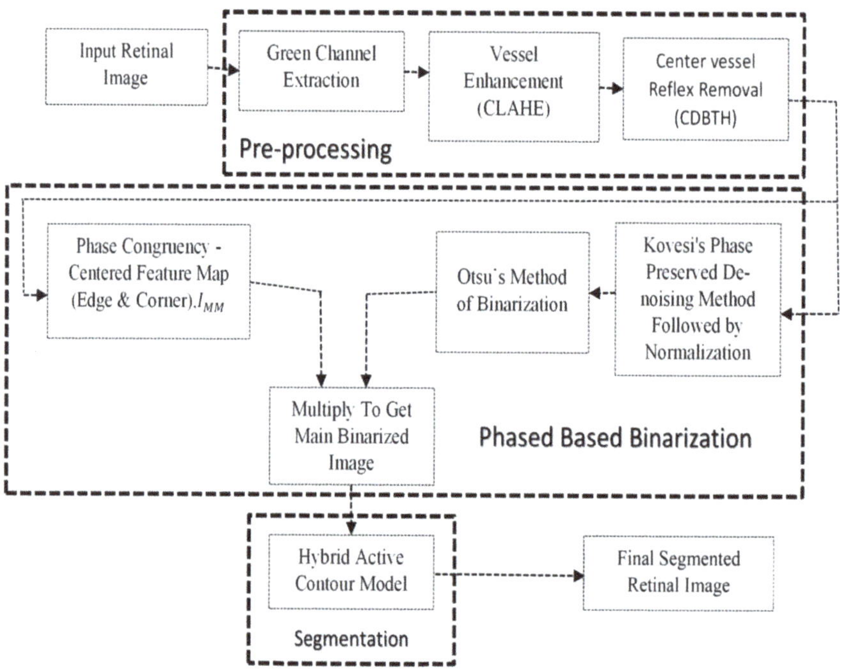

Figure 10.7. Block diagram for the segementation of blood vessels using hybrid active contour model.

fundus image. In this method, the author has reduced the snake's energy relative to the energy of the blood vessel, which combines the snake model and the balloon model. Furthermore, a balloon model is used to approach the closed loop without stepping into a local minimum that traps snakes. The proposed method given by Karn shows the combination of various image processing techniques such as image preprocessing, denoising, histogram equalisation, image binarisation, and segmentation. These all are the classical approaches to extracting the useful information from the image. Figure 10.7 is the block diagram of the algorithm proposed by Karn

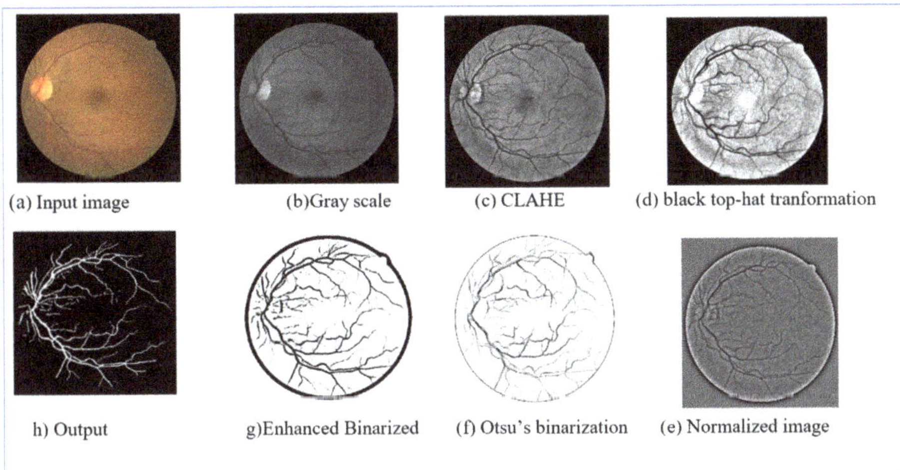

Figure 10.8. Various processes involved in retinal blood vessel segmentation for the detection of DR.

et al [17]. This approach can be used to see the progression in blood vessel growth. Figure 10.8 shows the output images after applying various preprocessing algorithms to obtain the segmented image proposed by [17].

Similar to this, Chalakkal *et al* [18] have proposed a novel method based on curvelet transform and line operators that can more accurately and sensitively segment small peripheral vessels than existing state-of-the-art approaches. In the suggested method, many image processing procedures involving colour space translation, adaptive histogram equalisation, and anisotropic diffusion filtering are used to increase the contrast between the retinal vessels and the background pixels. The difference at the retinal vascular edge is further improved by employing the updated curvelet transform coefficients. Applying the line operator response followed by appropriate thresholding produces the segmented vessels, which is the final step in segmenting the vessels. A post-processing technique is used to get rid of the scattered undesired backdrop pixels.

10.5.1.2 Exudate detection-based methods for the detection of DR

Exudates are an early biomarker for the detection of DR. Small and lesser in number are the features of first-stage NPDR, whereas the opposite properties determine the later stage. Deep M-CapsNet with Expectation-Maximization (EM) Routing is a new method proposed by [19] to segment exudates using an encoder-decoder type network. According to this paper, semantic object segmentation has memory allocation issues that can be reduced. Every child capsule in M-CapsNet communicates with every parent capsule at every location. As a result, a matrix capsule transmits the predictions to parent capsules utilising a shared kernel. The M-CapsNet removes the exudates and the optic disc from the retinal surface because the exudates and the optic disc have similar intensities. The segmented output's optic disc is removed using morphological and regional features. The literature describes a variety of computational techniques for exudate detection. Several retinal illnesses, including DR, HR, and others, are stopped from spreading by the early diagnosis of exudates. Walter *et al* [20] used a morphological closing technique for vascular removal to determine the standard deviation threshold limit for segmenting exudate regions. Hirangi *et al* [21] created a contour model for extracting exudate zones; it has a high percentage of false-positive results. As a result, machine learning algorithms are used to get around these issues in image processing methods. To detect exudates, Narang *et al* [22] used a support vector machine (SVM) as a classifier where features were detected using the Lifting Wavelet Transform (LWT) method. Xian Chen *et al* [23] identified candidate regions using SVM, which was trained on the features derived from the candidate regions of hard exudates. Similarly, other supervised algorithms such as K-Nearest Neighbour (KNN) classifier, radial basis function, multilayer perceptron (MLP), and fuzzy C-Means (FCM) clustering were also developed. Despite lowering false-positive rates, these techniques rely on manually created features taken from regions of interest (ROI) and their selection criteria.

This gap is filled by combining deep learning methodologies with an image processing algorithm to achieve greater accuracy and sensitivity. The fully connected convolutional neural network (CNN), created by Jonathan Long *et al* [24], takes images of various sizes as inputs and up-samples feature maps to provide output in the original input image size. With the development of deep learning methodologies, computer vision has made remarkable strides in medical analysis. A deep learning system was created by Sarfaraz Hussein *et al* [25] for the characterisation and localisation of lung and pancreatic tumours. Dong Yu-nan *et al* [26] suggested a deep learning network for picture recognition and classification. CNN networks were used by Zhang *et al* [27] for the segmentation of skin lesions. In their comparison of deep learning and reinforcement learning, Mufti Mahmud *et al* [28] highlighted the difficulties of mining biological data. U-net was suggested by Ronneberger *et al* [29] for segmenting physical images.

A comprehensive U-Net was used in 2016 by Cicek *et al* [30] to develop a 3D U-Net model for segmenting 3D pictures. Many hierarchical, dense-skip pathways connect the extensively trained encoder-decoder network U-Net++ created in [31]. For the segmentation of exudates, Guo *et al* [32] suggested an enhanced multi-feature fusion network. Using a dynamic routeing technique, Rodney Lalonde *et al* [33] created a network called segcaps for biomedical image segmentation. The motivations and guiding concepts for learning algorithms for deep architectures were described by Benuwa *et al* [34].

Researchers are also fascinated by classifying diabetic retinopathy using OCT images. Devi *et al* [35] have suggested using a feature-based categorisation of OCT images to find diabetic retinopathy. With the use of the Graph-Cut algorithm, they were able to segment seven levels of sub-retinal layers based on the gradient. Later, elements from the OCT image, such as retinal thickness and neovascularisation, were retrieved to categorise the patients as DR or non-DR. Ghazal *et al* [36] have suggested a CAD method to identify people with diabetes using an OCT image. The scientists employed wavelet decomposition-based edge tracking and TPS-based registration to align the images after using a grey OCT image and a probabilistic colour map as input images.

Additionally, images were divided into segments using a second-order joint Markov Gibbs random field (MGRF). A random forest classifier was used to classify the images after various features like thickness, reflectivity, and curvature had been retrieved. The evaluation measures for the provided article, such as the Dice coefficient (DC), agreement coefficient, and average deviation distance, were promising and the state of the art.

Final impression: It is found in the literature that a substantial amount of research has been carried out to detect diabetic retinopathy using fundus images, whereas OCT lags. It is because the retinal blood vessels are captured as shadows in OCT images and considered artefacts. In contrast, visualisation of the overall retinal blood vessel network is more prominent in the fundus image. But early symptoms of DR, such as cotton wools, and exudates, are visible in both imaging techniques. Hence fundus imaging is recommended for the screening of DR, and OCT is recommended for patients with DR with other pathologies.

10.5.2 Glaucoma screening

The retinal illness glaucoma impacts the ganglion cells of the optic nerves. Increased intraocular pressure (IOP) damages the optic nerve and distorts the optic nerve head (ONH). This causes the optic cup area to grow. Due to the slight or even absence of symptoms associated with this disease, many glaucoma sufferers are unaware of their condition. This is why glaucoma is often known as the 'silent thief of vision' [37]. If glaucoma is not treated in its early stages, it progresses to the near-blindness stage, or the patient remains with point vision.

In most cases, but not always, optic nerve injury is related to abnormally high IOP. If the eye's fluid drainage system isn't working correctly, the pressure inside the eye rises. The vision of a person at various stages of glaucoma is given in figure 10.9. Glaucoma can develop slowly (chronic glaucoma) or abruptly (acute glaucoma):

- Chronic glaucoma—The drainage pathways in the eye eventually become blocked when glaucoma of this type develops. It is the most typical type and causes no pain.
- Acute glaucoma—The eye's drainage passages suddenly get blocked. If left untreated, it is uncomfortable and may cause permanent eyesight loss.

There are different types of glaucoma, with a range of characteristics and causes. Some of these include:

- Primary open-angle glaucoma: A partial obstruction in the eye's drainage system allows fluid to drain out of the eye too slowly, which causes a steady rise in IOP.
- Angle-closure glaucoma: It develops when the iris enlarges and obstructs the drainage system, preventing fluid from draining properly and raising IOP.

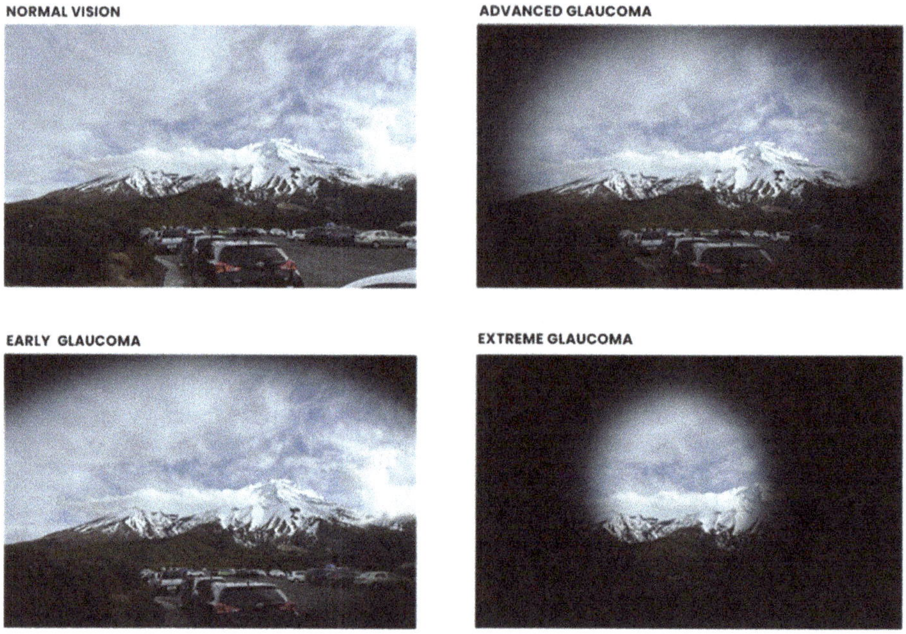

Figure 10.9. Vision of a person at different severity of glaucoma.

- Normal-tension glaucoma—Regardless of whether the pressure inside the eye is normal, this glaucoma damages the optic nerve. The cause is unknown; however, it might be because the optic nerve isn't getting enough blood flow.
- Developmental glaucoma—It typically has no symptoms and affects newborns and children.
- Pigmentary glaucoma—Granules of pigment accumulate in the eye and obstruct drainage pathways.

Evaluation of IOP, visual field, ONH damage, and central corneal thickness (CCT) are the four tests necessary for glaucoma diagnosis [7]. IOP is measured using the Goldmann Applanation Tonometer and the Schiotz Indentation Tonometer. However, it is not accurate enough to identify the condition in the case of normal-tension glaucoma. The visual field test can be used to measure the degree of vision loss in glaucoma patients. Damage to the ONH causes an expansion of the cup area, often known as cupping in medicine. The cup-to-disc ratio (CDR) can track the evolution of cupping.

The risk of glaucoma increases with increasing CDR and is the commonly used indicator by clinicians to diagnose the disease. Automated detection of a vertical CDR helps in determining glaucoma at an early stage. Eadara *et al* [38] have proposed optic disc and optic cup segmentation from fundus images using stationary wavelet transform and maximum vessel pixel sum. This approach can help calculate the cup-to-disc ratio and classify glaucoma and no-glaucoma eyes.

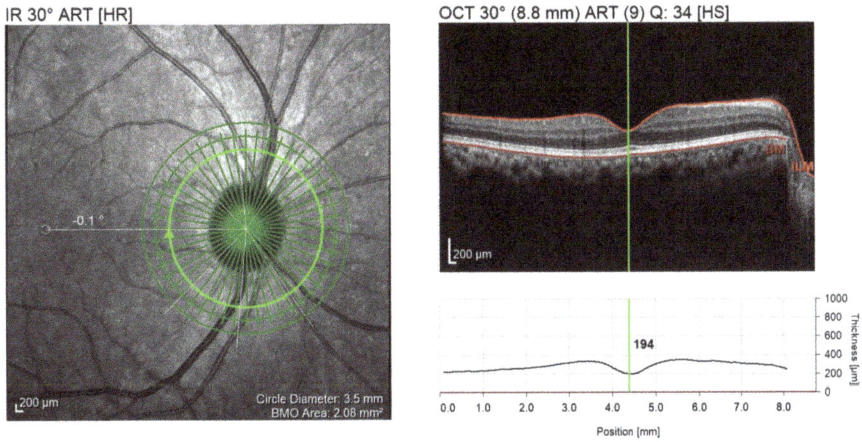

Figure 10.10. (A) Anatomical positioning of OCT on fundus photo. (B) Thickness measurement between ILM and BM.

The cup-to-disc ratio may not be helpful every time for the detection of glaucoma. A recent study suggests that choroid thickness and the distance between Bruch's membrane (BM) can be important biomarkers for glaucoma detection. Recently, OCT image analysis has been used to detect glaucoma, which mainly relies on the classification of the RNFL thickness and minimum rim width (MRW) in different zones of the retina. A report from Nepal Eye Hospital for glaucoma detection using OCT image analysis is shown in figure 10.10. It is often difficult to distinguish between optic disc and optic cup due to varying intensity, neovascularisation on disc, etc. Thus, relying on Bruch's membrane area opening (BMO) is more advantageous than calculating the cup-to-disc ratio. In figure 10.10(B), one can see the ending point of BM, which is considered as BM opening. Also, OCT image analysis can be used for estimating the thinning of the retinal nerve fibre layer (RNFL) and minimum rim width (MRW).

It has been found that thinning of RNFL and a decrease in MRW are seen in patients suffering from glaucoma. Technically vision of an eye is divided into six zones, generally called ISNT rule [39]. These zones of the eye are inferior (I), Superior (S), Nasal (N), and Temporal (T) and the zones in between them is called Superior Temporal (TS), Superior Nasal (NS), and Inferior Temporal (TI). The distribution of RNFL thickness over this zone can help the doctor to point out the severity of vision loss due to glaucoma. An example of the classification of RNFL and MRW is shown in figure 10.11. From the pie-chart of the report, RNFL thickness may be within standard limits, whereas MRW thickness indicates borderline glaucoma. To measure progressive change, OCT can offer precise data on the eye's anatomical composition. Along with the clinical examination, considering the patient's history and symptoms, and analysing the visual field results, OCT is a very helpful addition to the glaucoma diagnostic process. Figure 10.11(A) shows a patient's RNFL thickness and MRW distribution over various vision zones.

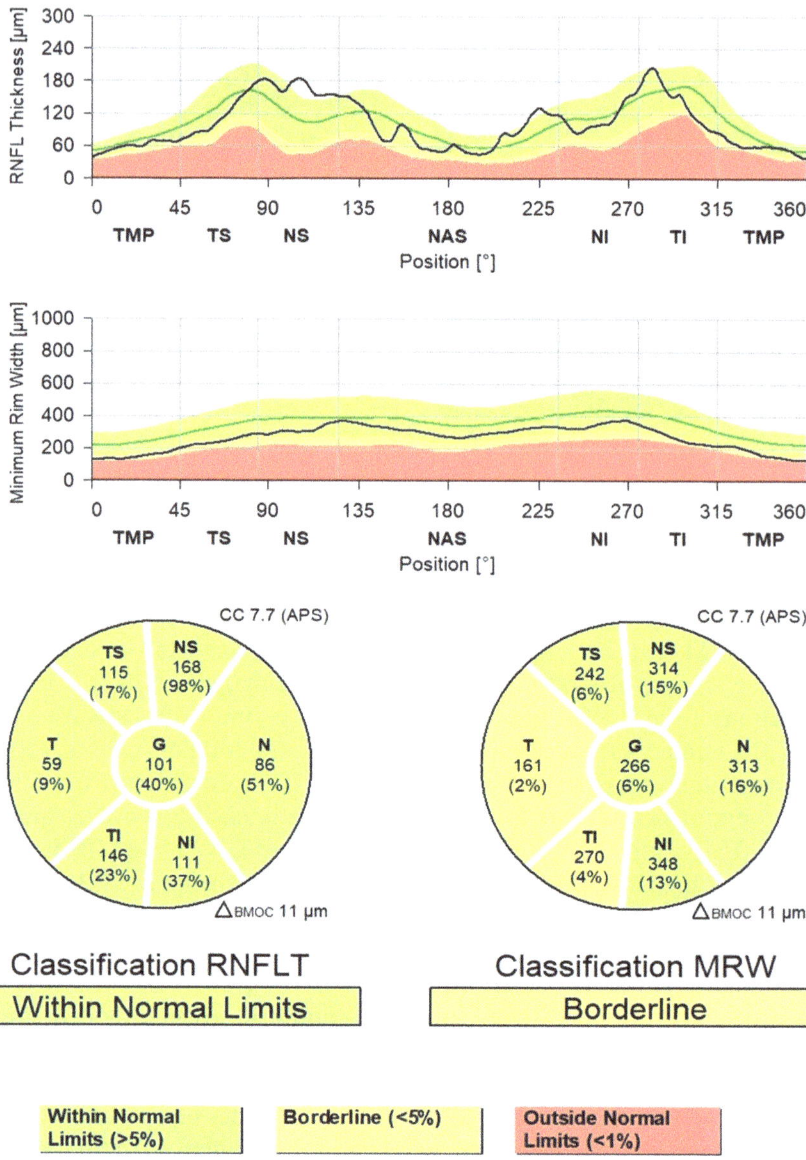

Figure 10.11. Classification of RNFL thickness and MRW (TSNIT)

The green line and green shadow (>5%) indicate the thickness distribution benchmark taken from the European descent database collected in 2014. The yellow shadow indicates the thickness within the borderline limit (<5%), and the red shadow indicates the thickness outside the standard boundary (<1%).

The black line indicates the thickness distribution of the patient, which is compared with the benchmark. Figure 10.11(B) is the pie-chart representation of figure 10.11(A), which shows the numeric value of thinning.

A software-based approach that can segment four separate layers of the retina has been proposed by Juan Xu *et al* [38] and compared to the traditional circumpapillary nerve fibre layer (cpNFL) to identify glaucoma. Macular nerve fibre layer (mNFL), a combination of ganglion cell layer-inner plexiform layer-nucleus layer, outer plexiform layer, and a combination of the outer nuclear layer-photoreceptor layer were the four distinct layers identified in this study. They used 96 subjects, 44 of whom were healthy, and 89 of whom had glaucoma. The experiment revealed that glaucoma patients' average retinal thickness was substantially higher than healthy individuals. Vermeer *et al* [38] segmented six separate sub-retinal layers for both healthy and glaucomatous eyes. With an RMS error of 4–6 m, they could produce the thickness map of many layers. A study by Jakub *et al* [40] has examined the relationship between IOP, vascular density (VD), and the retinal nerve fibre layer in patients who have never received glaucoma treatment. Although it is well known that IOP > 22 mmHG is directly related to a reduced field of vision, this article also provided information on how VD and RNFL depend on IOP of 20 mmHg. This relationship may be helpful for the early identification of normal tension glaucoma.

The Hood Glaucoma Analysis from OCT: A TSNIT RNFL profile is typically used to depict RNFL thickness data (figure 10.11(A)). This profile begins at the disc's most temporal point and moves clockwise through the quadrants of the Temporal (T), Superior (S), Nasal (N), and Inferior (I). Thus, the temporal half of the disc's RNFL thickness data is presented on the right and left of the RNFL profile, while the nasal half of the disc's RNFL thickness data is shown in the middle of the RNFL profile.

The TSNIT profile's drawback is that it can be challenging to compare RNFL damage to a visual field impairment because most retinal ganglion cells are located in the macula. An NSTIN RNFL profile is used to convey the RNFL data in the Hood Glaucoma Report (figure 10.12). It starts at the disc's nasal most point and moves counter clockwise to this profile. As a result, the RNFL profile rearranges the information on RNFL thickness related to the temporal half of the disc given to the centre. By enhancing its visual presentation in relation to the visual field outcomes, the NSTIN plot increases the usefulness of the RNFL information. In addition, the large, high-quality OCT image on the report confirms the details of the structural damage and the accuracy of the segmentation. Figures 10.11(A) and 10.12 are plotted as per TSNIT and NSTIN for the same patient and seem similar in detecting the affected visual zone. But Hood glaucoma report measures thickness between ILM and RNFL rather than between ILM and BM. This approach gives the thickness map of RNFL and ganglion cell layer (GCL) and is used to plot 24–2 and 10–2 visual points. The Hood report is the most significant achievement that may replace the need for visual field analysis to some extent. The visual field plotted for the same patient from the Hood glaucoma report in figure 10.12 is given in figure 10.13.

Impression: It is found that the OCT image seems more promising compared to the Fundus image for the diagnosis of glaucoma. But a fundus image is required to locate the physiological change and focus the OCT camera at a correct distance

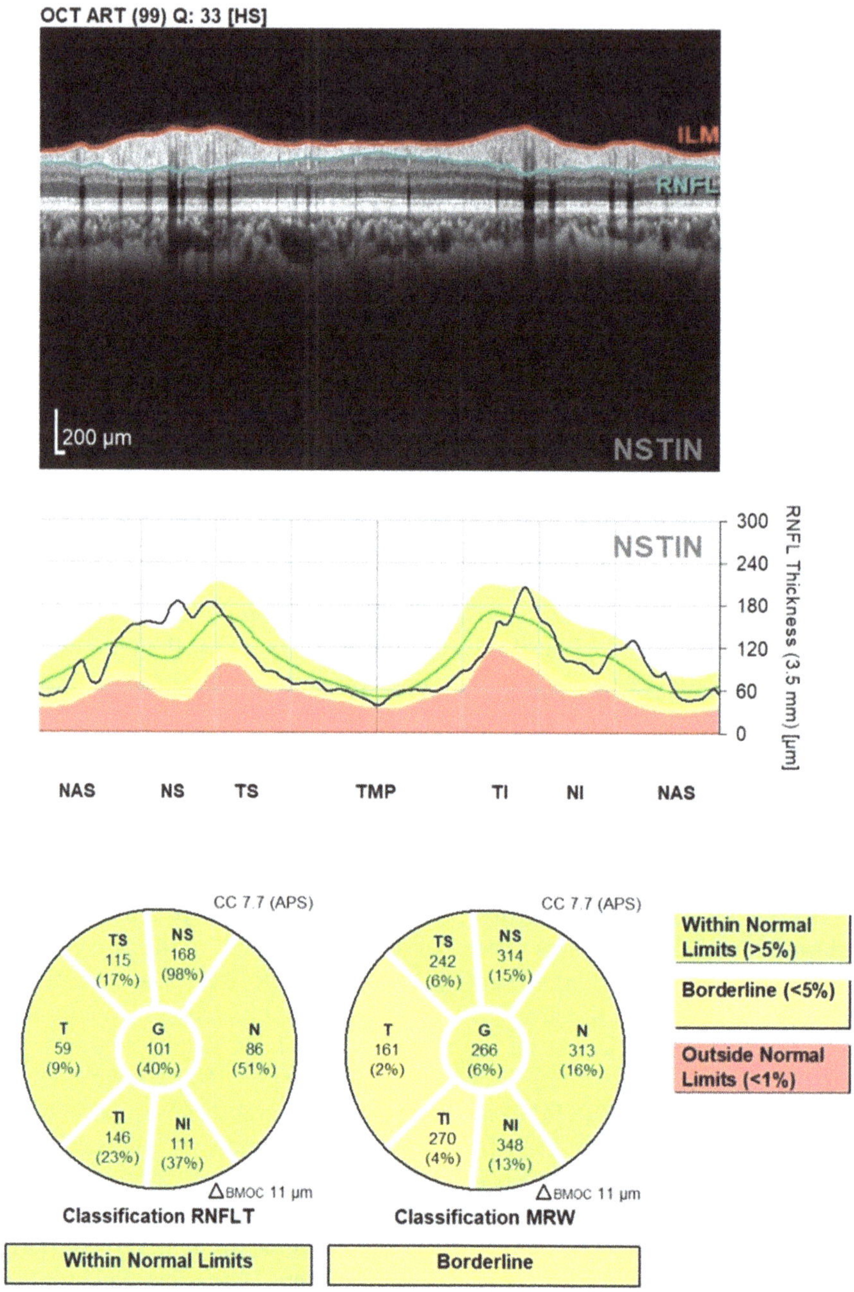

Figure 10.12. Hood glaucoma report (NSTIN).

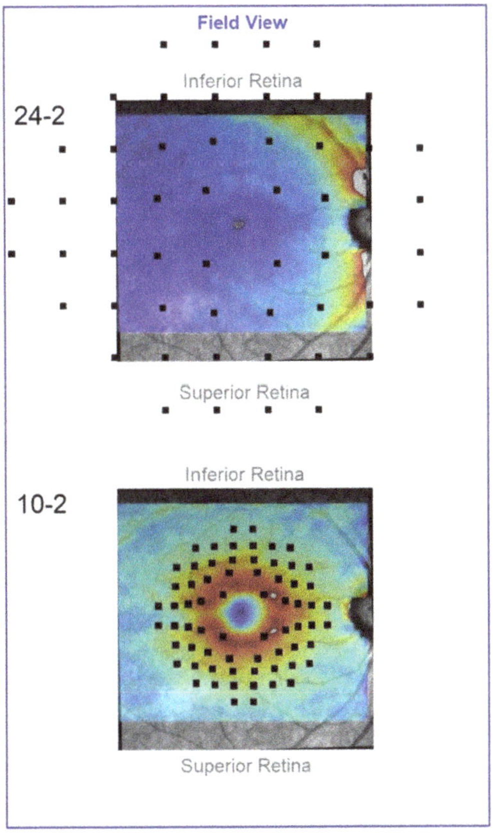

Figure 10.13. Field view generated from Hood glaucoma report.

from ONH. Hence both Fundus and OCT are equally important, with the latter having more scope for detailed examination.

10.5.3 Macular oedema screening

The macula is a portion of the retina found behind the eye. Our centre vision, most of our colour vision, and the fine detail we see are all controlled by this little (5mm) area of the retina. The photoreceptor cells, or cells that sense light, are concentrated in significant numbers in the macula. The brain interprets the data they send as images. The remaining retinal tissue processes our side vision. The macula is susceptible to several different illnesses. Diabetic macular oedema (DME), age-related macular degeneration (AMD), cystoid macular oedema (CME), central serous chorioretinopathy (CSCR) are a few of them. Oedema is a medical term for swelling. Thus, macular oedema refers to any swelling of the retina. The formation of cysts between the sub-retinal layer, an accumulation of retinal fluid, the development of hard exudates owing to protracted diabetes, and other factors are the leading causes of macular oedema. Macular oedema can take many different forms, including diabetic macular oedema (DME), cystoid macular oedema (CME), and clinically significant macular oedema (CSME). DME, as its name implies, is the term for the swelling of the macular area in diabetes patients who have both hard and soft exudates. Patients with this condition are more likely to develop new leaky blood vessels since they are in the proliferative stage of the disease. This condition is known as CME when a cyst develops between sub-retinal layers. When it grows in the fovea region of the eye and impairs vision, it is known as clinically significant macular oedema.

Roychowdhury *et al* [41] localised a cyst in an OCT image with the help of a high-pass filter and also identified six distinct layers. The algorithm the author has suggested can define a cystoid area by segmenting the boundaries of a cyst. The author created a 3D thickness map of a DME patient's eye in a different article. Segmentation and some pre-processing were used to separate the retina into six layers. Additionally, they discovered that the DME retinal layers were 1.5 times thicker than typical retinal layers.

Three different pathologies must be examined to determine the type of oedema. The following describes these pathologies:

10.5.3.1 Intraretinal fluid (IRF)

IRF refers to fluids that build up between an internal limiting membrane (ILM) and an external limiting membrane (ELM). They have fluid-filled, hypo-reflective voids. The IRF reflects circumstances like DME and CME.

10.5.3.2 Subretinal fluid (SRF)

SRF is made up of clear or lipid-rich exudates that build up between the ELM and the retinal pigment epithelium (RPE). Conditions like central serous retinopathy (CSR) are indicated by the SRF.

10.5.3.3 Pigment epithelium detachment (PED)

PED stands for RPE detachment brought on by fluid or drusen buildup, etc. There are two types of AMD-specific conditions: moist AMD and dry AMD. An example of all these pathologies is shown in figure 10.14. In this figure, red indicates IRF, blue indicates PED, and green indicates SRF.

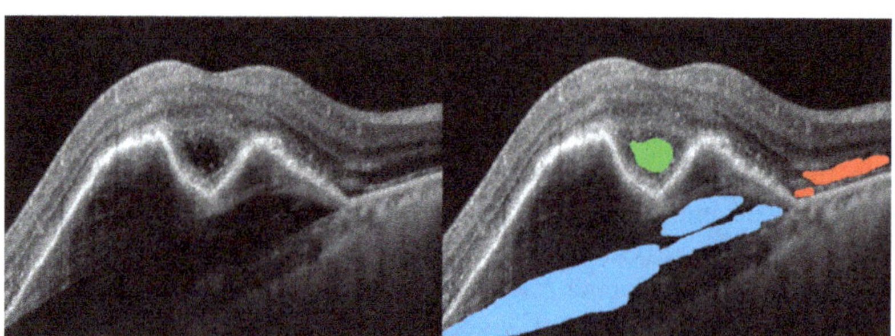

Figure 10.14. OCT image indicating of IRF, SRF and PED

10.5.4 Macular degeneration screening

One of the most prevalent macular illnesses in adults over 50 years of age is ARMD, which can sometimes also occur in young adults. This disease's severity has three stages: early, moderate, and late. The condition is typically only discovered in its late stages when the patient begins to lose all or part of their vision, and their central vision begins to deteriorate.

10.5.4.1 Dry AMD

It is also known as atrophic AMD. The macula thins out at this point as we age. There are three stages of dry AMD: early, middle, and late. In most cases, it develops gradually over several years. Although there is no cure for late dry AMD, there are strategies you can use to maximise your remaining vision.

10.5.4.2 Wet AMD

This condition, which is less frequent but quickly results in visual loss, is also known as advanced neovascular AMD. Wet AMD can develop at any stage of dry AMD; however, wet AMD is always the late stage. The macula is harmed when aberrant blood vessels grow behind the eye.

10.6 Biomarkers associated with fundus and OCT

Clinical interpretation refers to the terminology needed to describe the anatomy of a fundus and an OCT image, whereas a biomarker is the standard used by clinicians to detect various retinal illnesses. The retina has long been recognised as an extension of the brain, and its position allows for extensive anatomical and physiological investigations of neural systems that support the brain's basic information processing. As a result, it's no wonder that retinal imaging, particularly OCT, has risen fast in the medical field as a means of diagnosing, treating, and studying disorders. Its application extends far beyond ophthalmology, as there are ocular signs for a wide range of systemic illnesses, including hypertension, metastatic cancer, autoimmune disorders, immunodeficiency syndrome, and hyperglycemia.

Unsurprisingly, as retinal imaging technology has advanced, so has our ability to recognise retinal biomarkers. This is facilitated further by the use of artificial intelligence (AI), which has allowed for the development of new biomarkers as well as the redefining of previous biomarkers to improve their applicability in treatment. Separating fluid into its constituent parts has been proven to be an essential indicator of disease severity, with intraretinal cystoid spaces having a more deleterious influence on eyesight than subretinal fluid. With the increasing resolution of retinal imaging, retinal layer abnormalities are now easier to identify. As newer OCT models increase scanning speed and depth of penetration while decreasing signal-to-noise ratio, more of the retina will be visible on imaging, including the choroid, which was previously inaccessible. As the complexity of retinal imaging modalities grows, the linkages between clinical features and imaging markers demand analytics tools such as AI to identify positive correlations. AI can currently assist in the

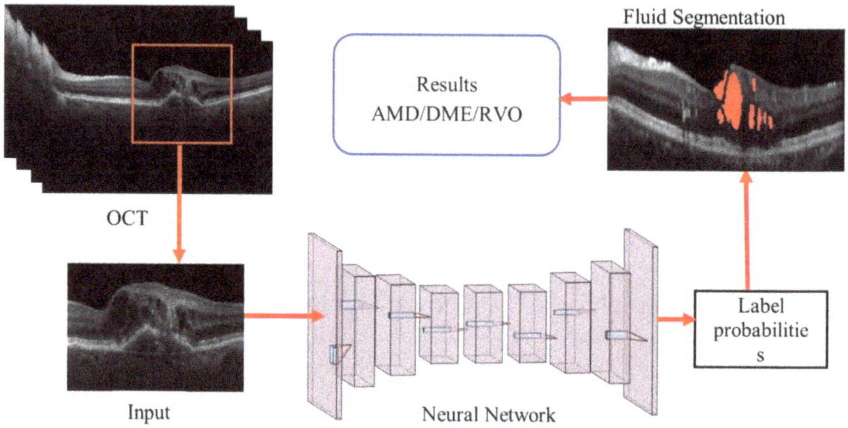

Figure 10.15. General procedure of AI-based disease classification

Table 10.2. Fundus and OCT of various retinal condition.

Retina/disease	Fundus	OCT
Normal retina	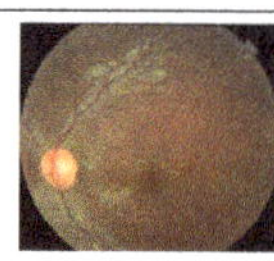	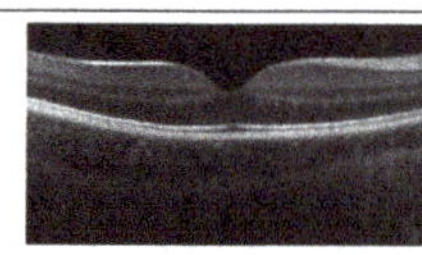
Maculopathy	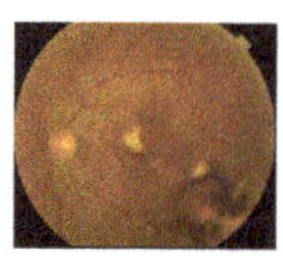	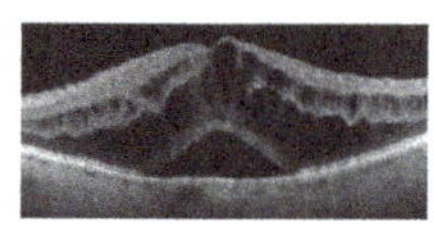
ARMD	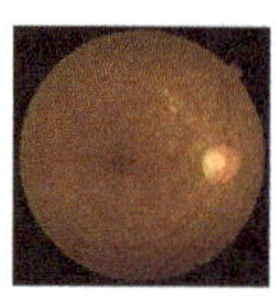	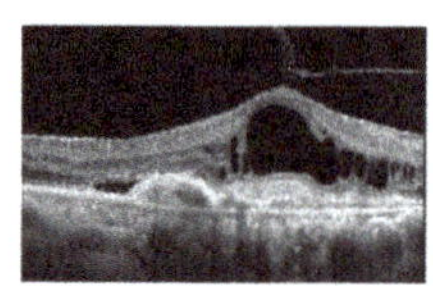
Diabetic retinopathy	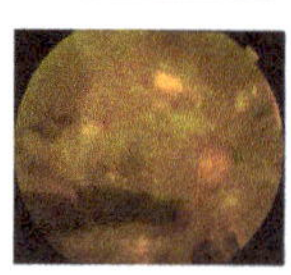	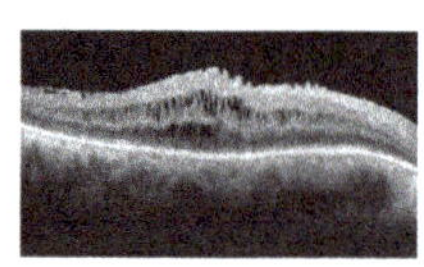

analysis of OCT pictures, fundus photos, and other imaging modalities, and has showed accurate clinical application as well as building patterning of the retina. A general procedure of disease classification using AI is given in figure 10.15.

Various examples of eye disease with corresponding fundus and OCT are given in table 10.2. It will provide a view of visual differences between different pathology associated with the eye.

10.7 Details on publicly available retinal and OCT image datasets for research

Recent advancement in machine learning and the deep learning approach for disease classification has increased the requirement of a significant number of data. Training and testing the newly developed algorithm needs publically available data. Due to a lack of labelled retinal images, many algorithms are never tested for medical use. Hence it is necessary to know about various publically available datasets. Chalakkal *et al* [42] compared the list of publically available fundus images under various categories such as image resolution, features, FOV, etc. Due to requirement of patient approval, most OCT research uses privately collected, non-public photos, which restricts the field of study for OCT image processing. Most of the datasets utilised in the literature are either unclassified or not accessible to the general public. The following list includes several publicly accessible datasets:

Retouch Dataset: This dataset was produced as part of a 2017 MICCAI challenge. It consists of 112 macula-centred OCT volumes, with 128 B-scan images in each volume. Zeiss, Spectralis, and Topcon are three separate vendors from which the photos were obtained.

Duke Dataset: The University of Michigan, Duke University, and Harvard University collaborated to compile this publicly accessible dataset. Forty-five people submitted the 3000 Oct photos in this dataset. There are 15 individuals with dry age-related macular degeneration, 15 patients with diabetic macular oedema, and 15 healthy people.

Labelled OCT and Chest x-ray: There are 68 000 OCT pictures in this dataset. An independent patient testing set and a training set of photos are created. Images are categorised into four directories: CNV, DME, DRUSEN, and NORMAL, and are labelled as train test and validate. The most significant number of relevant datasets is likely this one.

Annotated Retinal OCT Images database—AROI database: It is a limited dataset only provided for research purposes upon request. This database contains 3200 B-Scans obtained from 25 people. The image was captured using the Zeiss Cirrus HD OCT 4000 instrument. Each OCT volume comprised 128 B-scans with a 1024 × 512 pixel resolution. Pathologies such as pigment epithelial detachment (PED), subretinal fluid and subretinal hyperreflective material (designated jointly as SRF), and intra-retinal fluid are labelled on all images.

The University of Waterloo Database: It is a publicly accessible collection containing over 500 high-resolution photos classified into various disease disorders. Normal (NO), Macular Hole (MH), Age-related Macular Degeneration (AMD), Central Serous Retinopathy (CSR), and Diabetic Retinopathy (DR) are the picture classes. These photos were captured using a raster scan procedure with a scan length of 2 mm and a resolution of 512 × 1024.

Optima Dataset: The OPTIMA dataset is a part of the OPTIMA Cyst Segmentation Challenge of MICCAI 2015. This challenge segments IRF from OCT scans of vendors Cirrus, Spectralis, Nidek, and Topcon. The dataset consists of 15 scans for training, eight scans for stage 1 testing, seven scans for stage 2 testing, and the pixel-level annotation masks provided by two different graders.

University of Auckland Dataset: This is a private dataset gathered from two Nepalese eye hospitals: Dristi Eyecare and Nepal Eye Hospital. Volumetric scans from two manufacturers, Topcon and Spectralis, with resolutions of 320×992 and 1008×596, respectively, are used. Topcon images consist of a volumetric scan of 270 patients, with each volumetric scan containing an average of 260 B-scans. Similarly, Spectralis images include a volumetric scan of 112 people, with each volumetric scan containing 98 B-scans from both eyes. All patient information has been deleted and anonymised. Mr Ravi Shankar Chaudhary of Drishti Eye Care and Mrs Manju Yadav of Nepal Eye Hospital collaborated on the image capture and disease classification. As the first observer, Dr Neyaz (MD, Retinal Specialist) of Nepal Eye Hospital classified various diseases. Dr Bhairaja Shrestha (MD, Ophthalmologist) afterwards confirmed it as the second observer. This dataset is available upon request (w.abdulla@auckland.ac.nz) and will be made public in the future.

10.8 Recent advancements in commercially available fundus and OCT cameras

There are numerous manufacturers of the fundus and OCT imaging devices. Some of the widely used devices are listed below:

- TOPCON Fundus Camera

Model	Description	FOV	Image resolution	Features	Photo
SIGNAL camera	Non-mydriatic retinal examination covering the macula and optic disc	50×40 degree	2368×1776	USB and WLAN connectivity, DICOM compatible.	
TRC-50DX	Mydriatic Multimodal fundus imaging	50/35/20 degree	NA	IMAGEnet 6, DICOM, Ez capture	
TRC-NW400	Non-mydriatic fundus and anterior segment	45/30 degree	NA	IMAGEnet 6, DICOM, Ez capture, Teleophthalmology	

- Topcon OCT devices

Model	Description	Scan speed	Report generation	Features	Photo
Maestro2	High-resolution OCT and Color fundus photography	50 000 A-scan per second	3D macula, retina comparision, glaucoma report, Hood'd report	Cataract mode, anterior segment, auto-align and focus	
Triton	Multimodal swept-source OCT	100 000 A-scan per second	3D macula, retina comparision, glaucoma report, Hood'd report	Deep penetration, imaging through opacities	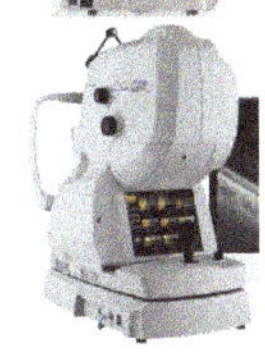

- Nidek OCT and fundus camera

Model	Description	FOV	Image Resolution	Features	Photo
AFC-330	Non-mydriatic Auto Fundus Camera	45/30 degree	NA	Panorama and stereo imaging	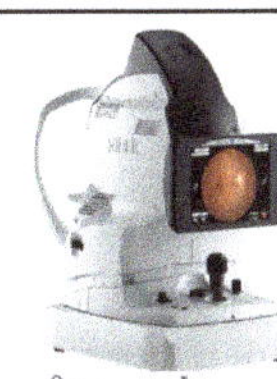
Optical Coherence Tomography RS-330	Duo optical coherence tomography and fundus camera	50/35/20 degree	NA	Denoising technique with deep learning, Fundus autofluorescence (FAF)	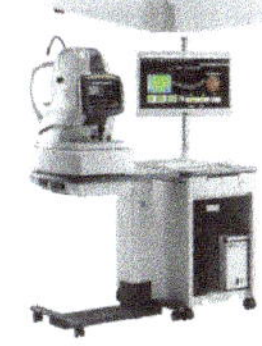

10.9 Conclusion

This chapter presents a comprehensive review of fundus imaging, optical coherence tomography, and the associated biomarkers. If not treated on time, several retinal diseases, such as diabetic retinopathy, macular oedema, and glaucoma, can cause permanent blindness. This chapter gives a clinical interpretation of various reports associated with OCT, such as the macula report, RNFL and MRW report, glaucoma report, and Hood's glaucoma report. Such an interpretation is expected to benefit both clinicians and researchers. We have seen that different OCT and Fundus device manufacturers have their own imaging modalities and features. Some have high resolution, some have different FOV, and some have different approaches to disease-specific report generation. Fundus-based glaucoma report generation relies on the cup-to-disc ratio, whereas OCT-based reports rely on Bruch's membrane opening area.

Hence, it is necessary to understand the recent advancement in imaging devices and features associated with it. It helps to understand the data acquisition process and gives the technical properties of image such as patient information, underlying diseases, age, descent, FOV, resolution etc. Though state-of-art has several research articles on single imaging modality of disease identification, this chapter has given a deep insight into the usablity of fundus and/or OCT for identifying specific eye disease. Most importantly, this chapter has listed all available OCT and fundus image databases that can be referred for research work.

References

[1] Zheng Y, He M and Congdon N 2012 The worldwide epidemic of diabetic retinopathy *Indian J. Ophthalmol.* **60** 428–31

[2] Teo Z L *et al* 2021 Global prevalence of diabetic retinopathy and projection of burden through 2045: systematic review and meta-analysis *Ophthalmology* **128** 1580–91

[3] Tsukikawa M and Stacey A W 2020 A review of hypertensive retinopathy and chorioretinopathy *Clin. Optom.* **12** 67–73

[4] Chalakkal R J, Abdulla W H and Hong S C 2020 *Fundus Retinal Image Analyses for Screening and Diagnosing Diabetic Retinopathy, Macular Edema, and Glaucoma Disorders* (Elsevier)

[5] Hafiz F *et al* 2022 A new approach to non-mydriatic portable fundus imaging *Expert Rev. Med. Devices* **19** 303–14

[6] Mamtora S, Sandinha M T, Ajith A, Song A and Steel D H W 2018 Smart phone ophthalmoscopy: a potential replacement for the direct ophthalmoscope *Eye* **32** 1766–71

[7] Dhaivat Shah A P, Dewan L, Singh A, Jain D, Damani T, Pandit R, Champalal Porwal A, Bhatnagar S and Shrishrimal M 2017 Utility of a smartphone assisted direct ophthalmoscope camera for a general practitioner in screening of diabetic retinopathy at a primary health care center *BMC Ophthalmol.* **17** 1

[8] Chalam K V, Chamchikh J and Gasparian S 2022 Optics and utility of low-cost smartphone-based portable digital fundus camera system for screening of retinal diseases *Diagnostics* **12** 1499

[9] Walsh L, Hong S C, Chalakkal R J and Ogbuehi K C 2021 A systematic review of current teleophthalmology services in New Zealand compared to the four comparable countries of the United Kingdom, Australia, United States of America (USA) and Canada *Clin. Ophthalmol.* **15** 4015–27

[10] Arthur B, Robert Giles C, Renoh C and O'Keeffe 2022 New Zealand's first practical demonstration of the telemedicine system specific to ophthalmology: MedicMind tele-ophthalmology platform *N. Z. Med. J.* **135** 133–5

[11] de Ribot F M, de Ribot A M, Ogbuehi K and Large R 2021 Teleophthalmology in the post-coronavirus era *N. Z. Med. J.* **134** 139–43

[12] Hu R, Chalakkal R J, Linde G and Dhupia J S 2022 Multi-image stitching for smartphone-based retinal fundus stitching *IEEE/ASME Int. Conf. Adv. Intell. Mechatronics, AIM* vol 2022 pp 179–84

[13] Feng X, Cai G, Gou X, Yun Z, Wang W and Yang W 2020 Retinal mosaicking with vascular bifurcations detected on vessel mask by a convolutional network *J. Healthc. Eng.* **2020** 7156408

[14] Xi J, Teng P and Wang J 2019 Multi-retinal images stitching based on the maximum fusion and correction ratio of gray average *ACM Int. Conf. Proc. Ser.* 64–70

[15] Popescu D P *et al* 2011 Optical coherence tomography: Fundamental principles, instrumental designs and biomedical applications *Biophys. Rev.* **3** 155–69

[16] Aumann S, Donner S, Fischer J and Müller F 2019 Optical coherence tomography (OCT): principle and technical realization *High Resolution Imaging in Microscopy and Ophthalmology* ed J F Bille (Cham: Springer International Publishing) pp 59–85

[17] Karn P K, Biswal B and Samantaray S R 2019 Robust retinal blood vessel segmentation using hybrid active contour model *IET Image Process.* **13** 440–50

[18] Chalakkal R J and Abdulla W H 2019 Improved vessel segmentation using curvelet transform and line operators *2018 Asia-Pacific Signal Inf. Process. Assoc. Annu. Summit Conf. APSIPA ASC 2018—Proc* pp 2041–6

[19] Biswal B, T G P P P and karn P K 2021 Robust segmentation of exudates from retinal surface using M-CapsNet via EM routing *Biomed. Signal Process. Control* **68** 102770

[20] Walter T, Klein J C, Massin P and Erginay A 2002 A contribution of image processing to the diagnosis of diabetic retinopathy—detection of exudates in color fundus images of the human retina *IEEE Trans. Med. Imaging* **21** 1236–43

[21] Harangi B and Hajdu A 2014 Automatic exudate detection by fusing multiple active contours and regionwise classification *Comput. Biol. Med.* **54** 156–71

[22] Narang A, Narang G and Singh S 2013 Detection of hard exudates in colored retinal fundus images using the support vector machine classifier *Proc. 2013 6th Int. Congr. Image Signal Process. CISP 2013* vol 2 pp 964–8

[23] Chen X, Bu W E I, Wu X, Dai B and Teng Y A N 2012 A novel method for automatic ha rd exudates detection in color retinal images *Proc. 2012 Int. Conf. Mach. Learn. Cybern* pp 15–7

[24] Shelhamer E, Long J and Darrell T 2017 Fully convolutional networks for semantic segmentation *IEEE Trans. Pattern Anal. Mach. Intell.* **39** 640–51

[25] Hussein S, Kandel P, Bolan C W, Wallace M B and Bagci U 2019 Lung and pancreatic tumor characterization in the deep learning era: novel supervised and unsupervised learning approaches *IEEE Trans. Med. Imaging* **38** 1777–87

[26] Dong Y N and Liang G S 2019 Research and discussion on image recognition and classification algorithm based on deep learning *Proc.—2019 Int. Conf. Mach. Learn. Big Data Bus. Intell. MLBDBI 2019* pp 274–8

[27] Zhang J, Xie Y, Xia Y and Shen C 2019 Attention residual learning for skin lesion classification *IEEE Trans. Med. Imaging* **38** 2092–103

[28] Mahmud M, Kaiser M S, Hussain A and Vassanelli S 2018 Applications of deep learning and reinforcement learning to biological data *IEEE Trans. Neural Networks Learn. Syst* **29** 2063–79

[29] Olaf Ronneberger T B and Fischer P 2015 U-Net: convolutional networks for biomedical image segmentation *Medical Image Computing and Computer-Assisted Intervention—MICCAI 2015. MICCAI 2015.* (Lecture notes in Computer Science vol 9341) ed N Navab, J Hornegger, W Wells and A Frangi (Cham: Springer) pp 234–41

[30] Cai J, Lu L, Zhang Z, Xing F, Yang L and Yin Q 2016 3D U-net learning dense volumetric segmentation from sparse annotation *Med. Image Comput. Comput. Interv.* 424–32

[31] Zhou Z, Siddiquee M M R, Tajbakhsh N and Liang J 2018 *UNet++: A Nested U-Net Architecture for Medical Image Segmentation* **vol 11045** (Springer International Publishing)

[32] Guo X, Lu X, Liu Q and Che X 2019 EMFN: enhanced multi-feature fusion network for hard exudate detection in fundus images *IEEE Access* **7** 176912–20

[33] LaLonde R, Xu Z, Irmakci I, Jain S and Bagci U 2021 Capsules for biomedical image segmentation *Med. Image Anal.* **68** 1–19

[34] Benuwa B B, Zhan Y, Ghansah B, Wornyo D K and Kataka F B 2016 A review of deep machine learning *Int. J. Eng. Res. Africa* **24** 124–36

[35] Sakthi Sree Devi M, Ramkumar S, Vinuraj Kumar S and Sasi G 2021 Detection of diabetic retinopathy using OCT image *Mater. Today Proc.* **47** 185–90

[36] Ghazal M, Ali S S, Mahmoud A H, Shalaby A M and El-Baz A 2020 Accurate Detection of non-proliferative diabetic retinopathy in optical coherence tomography images using convolutional neural networks *IEEE Access* **8** 34387–97

[37] Shen S Y *et al* 2008 The prevalence and types of glaucoma in Malay people: The Singapore Malay eye study *Investig. Ophthalmol. Vis. Sci.* **49** 3846–51

[38] Biswal B, Vyshnavi E, Sairam M V S and Rout P K 2020 Robust retinal optic disc and optic cup segmentation via stationary wavelet transform and maximum vessel pixel sum *IET Image Process.* **14** 592–602

[39] en Chan E W *et al* 2013 Diagnostic performance of the ISNT rule for glaucoma based on the heidelberg retinal tomograph *Transl. Vis. Sci. Technol* **2** 2

[40] Kral J, Lestak J and Nutterova E 2022 OCT angiography, RNFL and the visual fielat different values of intraocular pressure *Biomed. Reports* **16** 1–5

[41] Roychowdhury S, Koozekanani D D, Radwan S and Parhi K K 2013 Automated localization of cysts in diabetic macular edema using optical coherence tomography images *Proc. Annu. Int. Conf. IEEE Eng. Med. Biol. Soc. EMBS* pp 1426–9

[42] Abdulla W and Chalakkal R J 2018 *University of Auckland Diabetic Retinopathy (UoA-DR) Database-End User Licence Agreement* no. 1

Chapter 11

Optical coherence tomography and optical coherence tomography angiography biomarkers of diabetic macular edema

Hossein Nazari, Amber Piazza and Sandra Montezuma

11.1 Introduction

Diabetic retinopathy (DR) is a progressive microvascular complication of diabetes and the leading cause of visual impairment in middle-aged individuals. The World Health Organization estimates that the number of patients with diabetes mellitus (DM) will reach 400 million in the world by 2030, thus, hundreds of millions are at risk of losing vision due to diabetes [1, 2]. Better diabetes control, and early diagnosis, proper treatment, and timely follow up are crucial for preventing vision loss from DR.

The pathogenesis of DR is multifactorial and involves multiple complex biologic pathways. Hyperglycemia and resulting metabolic changes alter the retinal vasculature resulting in decreased perfusion to the retinal tissue. Retinal ischemia results in vascular remodeling, neovascularization, and disruption of the blood-retina barrier (BRB). The disruption of BRB is associated with extravasation and accumulation of fluid in the retina leading to diabetic macular edema (DME), which is the most common cause of vision loss in diabetes. While both outer BRB (composed of retinal pigmented epithelium) and inner BRB (retinal vascular endothelium) contribute to the maintenance of retinal homeostasis, the development of DME is mostly due to the inner BRB breakdown. Hypoxia and hyperglycemia may downregulate proteins essential to the integrity of the endothelial tight junctions and activate vascular endothelial growth factor (VEGF) and placental growth factor (PGF) pathways, further impairing cell junction function and promoting permeability and growth of abnormal new vessels (figure 11.1) [3, 4]. Vascular endothelial dysfunction can increase paracellular and transendothelial

doi:10.1088/978-0-7503-2052-8ch11

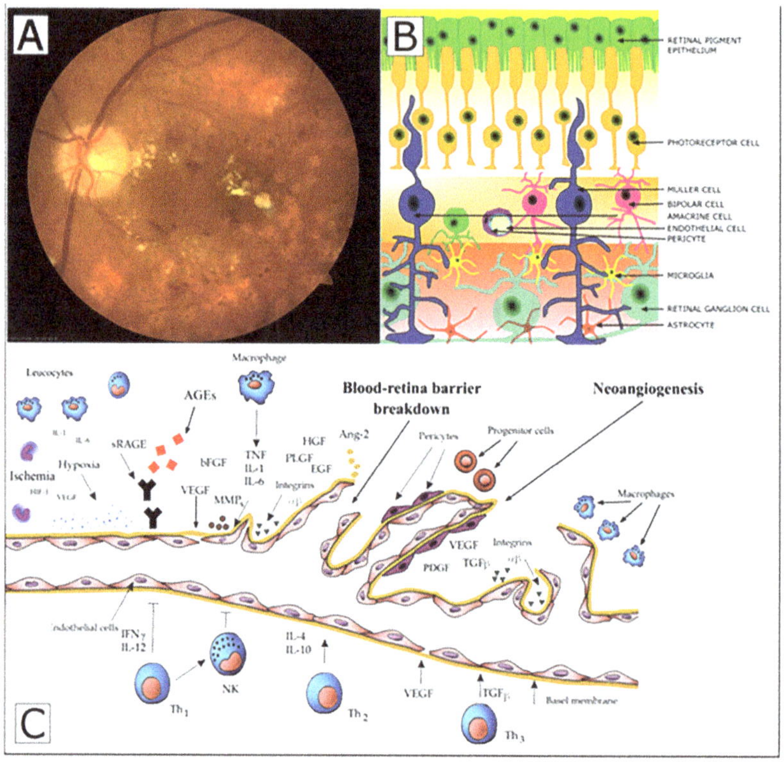

Figure 11.1. Possible biologic pathways involved in the development and progression of diabetic retinopathy. (A) retinal photo of an eye with diabetic retinopathy accompanied by macular oedema. (B) The neurovascular unit of the retina (retinal ganglion cells, Müller cells, microglia, astrocytes, endothelial cells, pericytes, and other). (C) Proposed factors involved in the development and progression of diabetic retinopathy. AGEs—advanced glycation end products, RAGEs—receptor for advanced glycation end products, VEGF—vascular endothelial growth factor, IGF-I—insulin like growth factor, PLGF—placental growth factor, HGF—hepatocyte growth factor, PEDF—pigment epithelium derived factor, bFGF—basic fibroblast growth factor, TGF-beta—transforming growth factor beta, MMPs—metalloproteinases, PDGF—platelet-derived growth factor, EGF—epidermal growth factor, Ang-2—Angiopoietin-2. Reproduced from [4] with permission under an open access Creative Common CC BY license.

transport of intravascular fluid into the neurosensory retina, exceeding the outer BRB's ability to transport fluid out of the retina, leading to retinal edema [5]. The development and progression of DR is known to be associated with the type and duration of diabetes, poor blood sugar control, and cardiovascular risk factors such as hypertension and hypercholesterolemia. In addition, progression of DR is linked to other organ system complications of DM such as diabetic nephropathy, diabetic cardiomyopathy, and diabetic neuropathy. However, none of these non-ocular associations can reliably predict the severity and extent of DR.

Diabetic retinopathy occurs in stages. The earliest stage of DR, non-proliferative diabetic retinopathy (NPDR), is characterized by microaneurysms (MA), retinal hemorrhages, and BRB breakdown leading to retinal edema and exudation. If NPDR is not treated, it often progresses to the more advanced form, proliferative diabetic retinopathy (PDR). PDR is characterized by neovascularization of the retina, optic nerve, and anterior segment of the eye that can result in vitreous hemorrhage, tractional retinal detachment, and neovascular glaucoma, if left untreated. Diabetic macular edema can develop in any stage of DR. Advanced complicated PDR is difficult to treat and typically requires an aggressive medical and surgical intervention in order to stop the proliferative process and preserve vision.

The diagnosis, treatment, and prognosis of DR are mostly determined by retinal examination findings and imaging biomarkers of the disease. Imaging biomarkers of DR and DME are the cornerstone of predicting the risk of development and progression of DR as well as planning for the screening, monitoring, and treatment of DR.

Optical coherence tomography (OCT) and OCT angiography (OCTA) are noninvasive, high-resolution (described as '*in vivo* biopsy'), and rapidly advancing systems for imaging retinal layers and capillary networks. While diagnosis, staging, and treatment decision making depended on examination and findings of fluorescein angiography until recently, OCT has essentially transformed the diagnosis and treatment of retinal diseases, including DR, in the last two decades. OCT is a non-contact optical imaging modality that generates high-resolution cross-sectional retinal images with axial and transverse resolution of 4–8 μm. OCT utilizes scanning near-infrared light that penetrates through the retina and choroid, and an interferometer processes the reflected light to reconstruct the thickness and volume map of the retina. OCTA is the volumetric rendering of retinal vasculature based on OCT imaging. Although fluorescein angiography is the gold standard for visualizing the retinal vasculature, it requires the intravenous injection of fluorescein dye. OCTA uses motion contrast imaging of OCT to generate high-resolution volumetric blood flow map without a need for a contrast agent.

Cross-sectional and en-face OCT and OCTA scans reveal multiple valuable diagnostic and prognostic biomarkers of early and advanced DR [6]. In this chapter, we will review established and emerging OCT and OCTA biomarkers of DR and DME. We will not discuss other imaging modalities such as fluorescein angiography and autofluorescence imaging.

11.2 OCT biomarkers of diabetic macular edema

OCT plays a key role in the management of DME and eligibility to clinical trials. Multiple retinal structural and functional parameters studied with OCT are biomarkers for the development of DR and the evaluation of its progression (figure 11.2).

11.2.1 Retinal thickness

Diabetic macular edema is identified as localized or diffuse retinal thickening in macula as a result of the accumulation of intraretinal and subretinal fluid (figure 11.1). Retinal thickening is due to a combination of diffuse and cystoid interstitial edema with fluid accumulation between neuronal cell body and axonal elements of the retina. The presence of intraretinal cystoid edema at fovea often

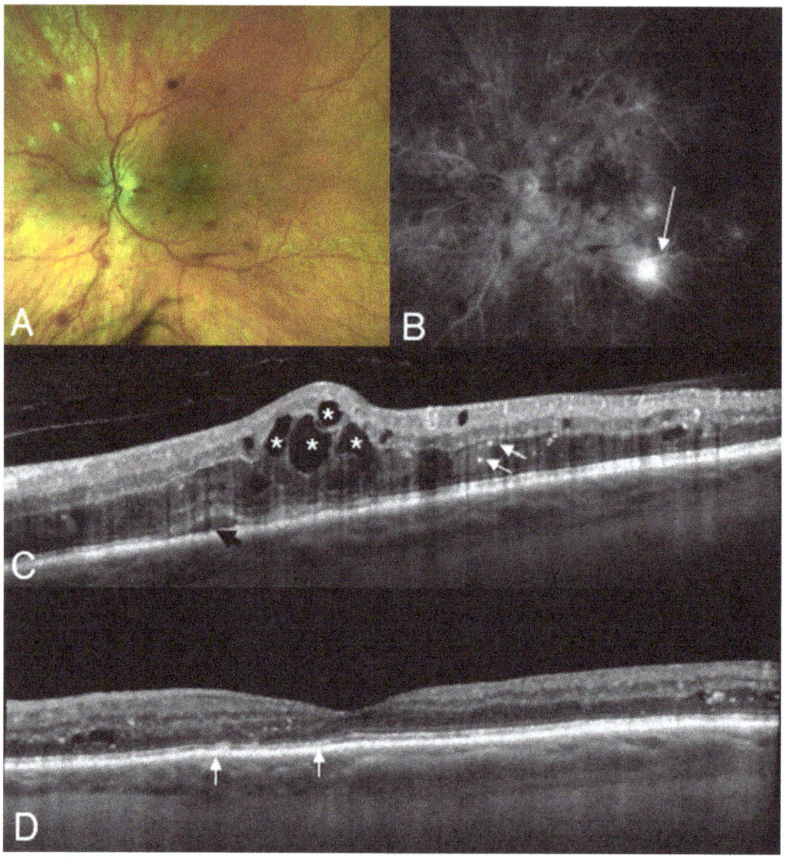

Figure 11.2. (A) Fundus photography from the left of a 55-year-old male with poorly controlled type 2 diabetes mellitus who presented with bilateral vision loss. Visual acuity 20/100 in the left eye. Note superficial and deep retinal hemorrhages, cotton wool spots, sub hyaloid hemorrhage, and retinal neovascularization. (B) Fluorescein angiography confirmed extensive retinal ischemia with capillary drop out areas, leakage from neovascularization elsewhere (NVE – arrow), and petaloid shaped pooling of fluorescein in cystoid edematous spaces at fovea. (C) OCT scan at presentation showed significant intraretinal fluid (asterisk), subretinal fluid (black arrow), and hyperreflective foci (white arrow). (D) Vision improved to 20/30 after multiple monthly injection of anti-vascular endothelial growth factor and panretinal photocoagulation. OCT scan showed normalization of foveal contour. The patient notices parafoveal scotoma, likely due to the residual disruption of outer retina (between white arrows).

significantly impacts vision because of disrupting neuronal and interneuronal structure and organization. Such cystoid spaces are usually located within the inner and outer plexiform layers as a potential space between photoreceptor and bipolar cell axons allow expansion and accumulation of extravasated fluid in the form of cystoid spaces. Cystoid macular edema is sometimes called 'cystic edema', however, 'cystic edema' is a misnomer because 'cyst' by definition is an epithelium-lined space.

It is important to treat center-involving DME (CI DME) that is clinically significant and aim at keeping fovea and parafoveal area 'dry' as much as possible.

Lobo *et al* evaluated 3-year progression of early macular edema in 90 type II DM patients with NPDR [7]. Using OCT imaging, they differentiated subclinical versus clinical CI DME by measuring central retinal thickness. Variation in extracellular fluid accumulation was also measured and tracked via OCT. The presence of fluctuating extracellular fluid accumulation was associated with worse visual outcomes in patients with either subclinical or clinical CI DME [7].

The Early Treatment of Diabetic Retinopathy Study (ETDRS) defined the three clinical criteria for 'clinically significant macular edema'. However, more recently, central subfield thickness (CST), which is an OCT measurement of the retina thickness underneath the fovea, became a more objective measurement of DME and has been found easier to detect and follow. Central subfield portends a circular area 1 mm in diameter centered around the center point of the fovea. In addition to its widespread use to quantify macular edema when following a patient with DME, CST has also been used in retinal clinical trials as one of the eligibility criteria, outcome measure, and a criterion for retreatment. CST has only a moderate correlation with visual acuity (VA) and response to treatment in DME and may not consistently correlate with impairment in visual acuity [8]. Permanent retinal cell damage can accompany a long standing DME involving fovea and result in permanent vision loss. It is important to know that a combination of multiple biomarkers reflect DR status, prognosis, and treatment outcomes and one cannot rely on CST as the sole indicator of DR severity and need for treatment.

11.2.2 Retinal volume

Retinal volume scan is another objective biomarker provided by OCT software that correlates with DME severity and response to treatment.

11.2.3 Intraretinal fluid (IRF)

Size of DME cystoid spaces may be classified as either small (<100 μm), large (101–200μm), or giant (>200μm), which has a direct correlation with retinal ischemia and macular disorganization and therefore, visual disturbance [9]. Additionally, location of cystoid spaces is important in determining need for treatment and predicting vision outcomes, as sub- and parafoveal edema and cystoid edema involving the outer retinal layers and associated with the disruption of the inner segment-outer segment junction (IS/OS) in OCT may likely cause more severe vision loss. In a study of 26 patients with clinically significant macular edema, Deak *et al* found that patients with outer nuclear layer edema may be less likely to benefit from treatment with biologics compared to patients with DME due to diffuse retinal thickening [10].

11.2.4 Subretinal fluid (SRF)

Subfoveal fluid collection is usually seen in acute and severe DME in poorly controlled diabetic patients. The presence of subfoveal fluid is usually associated with significant decline in visual acuity and requires aggressive treatment, with monthly anti-VEGF injections possibly combined with focal laser or panretinal photocoagulation, depending on the presence of proliferative diabetic retinopathy or the presence of localized microaneurysms from which retinal fluid originates.

11.2.5 Integrity of the photoreceptor inner segment/outer segment (IS/OS) junction

The photoreceptor inner segment/outer segment junction, also known as the ellipsoid zone (EZ band), is an indicator of the photoreceptor health. The IS/OS junction represents the anatomic-functional correlation between the integrity of photoreceptors and visual acuity and its disruption is likely responsible for the lack of linear correlation between CST and visual acuity in DME. Subfield IS/OS integrity can be quantified as normal, questionably abnormal, definitely abnormal (patchy defects), and definitely abnormal (absent) [11]. Alternatively, some studies have measured the extent of IS/OS junction disruption or the thickness of the IS/OS-retinal pigment epithelium [12] and showed a correlation and visual acuity and response to treatment in DME and inflammatory macular edemas (figure 11.3).

11.2.6 Disorganization of retinal inner layers

In addition to CST, location and distribution of intraretinal cystoid spaces, and IS/OS junction integrity, other factors also contribute to visual function in DME. Disorganization of retinal inner layers (DRIL) is a potentially important biomarker of reduced vision, because it possibly represents disruption of inner retinal layers composed of ganglion cells, bipolars, and amacrine and horizontal cells. DRIL is defined as inability to distinguish between the boundaries of the ganglion cell–inner plexiform layer complex, inner nuclear layer, and outer plexiform layer (figure 11.3) [13]. In one study, DRIL was better associated with presenting vision and visual outcomes after DME treatment as compared to the previously mentioned biomarkers [13]. DRIL is measured on OCT as the horizontal extent of such disorganization in microns in the central 1 mm of retina centered at fovea. A DRIL larger than 500 μm or affecting >50% of the central 1-mm of the fovea is associated

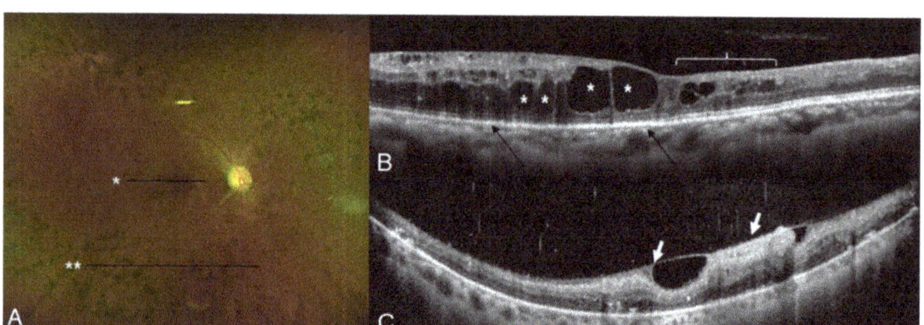

Figure 11.3. 54-year-old male with long standing type II diabetes mellitus with proliferative diabetic retinopathy and diabetic macular edema (A). (B) 30° OCT scan showed significant intraretinal fluid (asterisk), inner segment/outer segment (IS/OS) disruption (black arrow), and disorganization of inner retinal layers (DRIL - bracket), (arrow). (C) 55° OCT scan from the area of neovascularization and focal traction (white arrow).

with poorer VA. Early increases in DRIL predicts worse visual outcomes during DME treatment [13, 14]. Additionally, DRIL has also been shown as a potential marker of further damage to the outer retinal layer as well as worsening of diabetic retinopathy [15]. DRIL is not specific to DME but rather a marker of retinal disorganization due to multiple stressors.

11.2.7 Hyperreflective foci

Hyperreflective foci (HRF) within retinal layers are considered to be lipoproteins extravasated as a result of BRB breakdown (figures 11.2 and 11.3). Larger hyperreflective foci are visualized in clinical examination as hard exudates. It is noted that exudates smaller than 30 μm in diameter are not visualized in examination but can be seen in OCT line scans as hyperreflective foci [16]. Zur *et al* evaluated 284 patients with DME treated with a dexamethasone implant and found that those without HRF responded better to treatment than those with HRF, indicating a potentially valuable prognostic biomarker to be noninvasively tracked via OCT imaging [17].

11.2.8 Vitreoretinal traction

In a poorly controlled DM, vitreous humor may undergo cross-linking of the collagen fibers of the cortical vitreous, breakdown of hyaluronic acid, and extension of neovascularization to the vitreous cavity. A taut posterior hyaloid occurs when the posterior vitreous cortex forms a denser layer along the inner surface of the posterior pole of the retina. As a result of the anterior-posterior and tangential traction applied on the retinal surface, the retinal structure may distort leading to disorganization and dysfunction of retinal neuronal elements (figure 11.4). Such mechanical disturbances are readily visible on OCT and may guide follow up and treatment choices. A partial posterior vitreous detachment, where more adherent posterior hyaloid remains attached to fovea, may enhance anterior-posterior traction applied on the fovea and lead to treatment-refractory macular edema. Such macular edema may need vitrectomy surgery and removal of the taut posterior hyaloid. Vitreoretinal traction may result in tractional retinal detachment that needs surgical removal of the membranes if the detachment involves or threatens the macula (figure 11.5). OCT scanning is instrumental for the diagnosis and follow up of such complications after surgical intervention (figure 11.5).

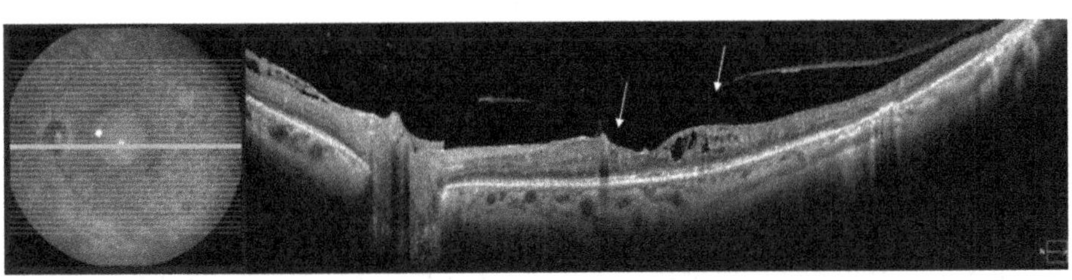

Figure 11.4. Vitreomacular traction (white arrow) contributing to diabetic macular edema.

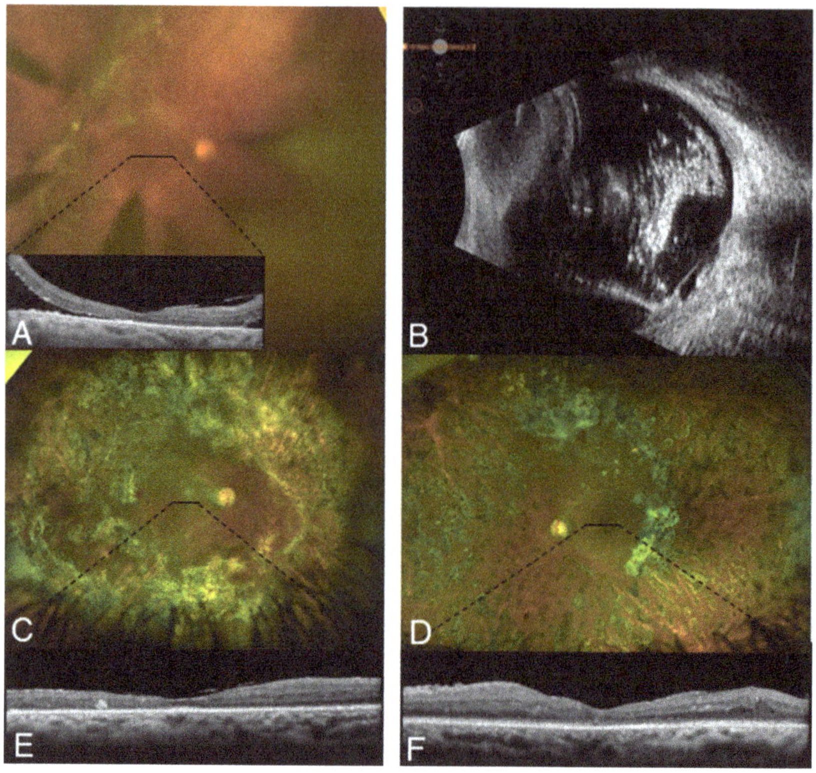

Figure 11.5. 56-year-old male with advance type 2 diabetes mellitus presented after tractional retinal detachment involved macula of the right eye (A). Line scan from foveal center (inset) shows tractional detachment involving fovea. Left eye vision was hand motion for more than 2 years (B) due to vitreous hemorrhage, tractional retinal detachment, and mature cataract. After pars plana vitrectomy and cataract extraction in both eyes, retina was flat (C and D, right and left eye) and foveal anatomy improved (E and F, right and left eye). Retinal line scan passing through the fovea in C and D is shown in E and F. Vision improved to 20/30 and 20/40.

11.2.9 Choroidal thickness, vascularity, and reflectivity

Alterations of choroidal vessel structure and blood flow induced by diabetes have been previously studied via histopathologic examination and fluorescein angiography. However, OCT offers a noninvasive early-stage visualization and tracking of choroidal thickness, vascularity, and reflectivity. It is shown that choroidal thickness significantly increases with the progression of DR from non-proliferative to proliferative stages of DR and choroidal thickness decreased after panretinal photocoagulation [18]. Moreover, individuals with a higher subfoveal choroidal thickness usually show a better response to anti-VEGF treatment compared to those with thinner choroid thickness [19].

Choroidal vascularity can be quantified in more sophisticated approaches than only measuring the thickness. Choroidal vascularity index (CVI), defined as the ratio of choroidal luminal area to total choroidal area, has been found to be a quantifiable

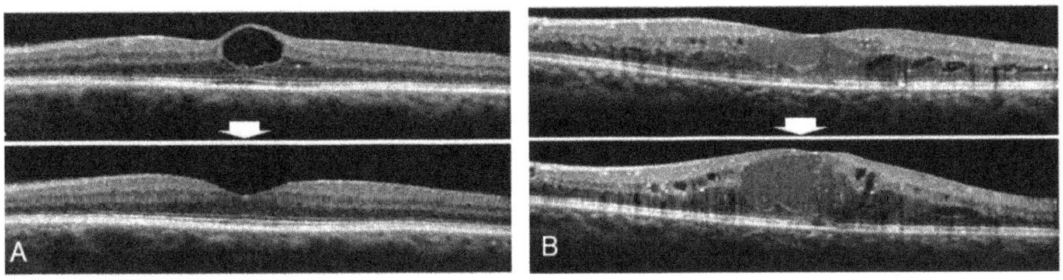

Figure 11.6. (A) Representative optical coherence tomography from a patient with diabetic macular edema before and after antivascular endothelial growth factor injection. Note complete resolution of the low-reflectivity intraretinal fluid. (B) OCT scan from a patient with long-standing poorly controlled diabetic macular edema. This macular edema containing hyperreflectivity fluid was refractory to multiple anti-VEGF injections.

indicator of choroidal vasculature in DR. As reported by Gupta *et al* in a study comparing 82 eyes with DME and 86 healthy control eyes, higher CVI was associated with worsening DR and a higher occurrence of PDR [20].

Hyperreflective choroidal foci (HCF), as seen by enhanced depth OCT (ED OCT), are considered lipofuscin deposits within the choroid. The presence of HCF has been found to be associated with PDR and poor visual acuity [21]. Both vascularity and reflectivity of the choroid have the potential to add valuable information aiding in the management of DR.

11.2.10 Optical reflectivity of intraretinal fluid

It is hypothesized that intraretinal or subretinal fluid with a more reflective signal on OCT likely signifies a thicker fluid (i.e. exudative fluid as compared to transudative fluid) with a higher density of lipids or macroscopic particles [22]. In the setting of DR, an exudative fluid with possibly a higher OCT signal reflectivity is likely to resolve more slowly and recur faster, because it may be associated with more severely disrupted BRB. Our preliminary study (pilot—not published) suggested that eyes with persistent DME at the end of the study period presented with a more reflective IRF compared to the eyes with completely resolved DME (figure 11.6). Whether the optical density of intraretinal fluid in DME correlates with visual acuity at presentation and after treatment, the severity of macular edema measured as central subfield thickness, and the number of intravitreal injections needed to treat the edema need to be studied.

11.3 OCTA biomarkers of diabetes mellitus

11.3.1 Retinal vascular density and blood flow

Until recently, the study of retinal vessel anatomy and function was only possible in two-dimensional images of fluorescein angiography (FA). While FA is still the gold standard of such studies, optical coherence tomography angiography (OCTA) offers a noninvasive, depth-resolved visualization and quantification of the retinal microvascular network. Low coherence interferometry, as used in OCT, is the basic

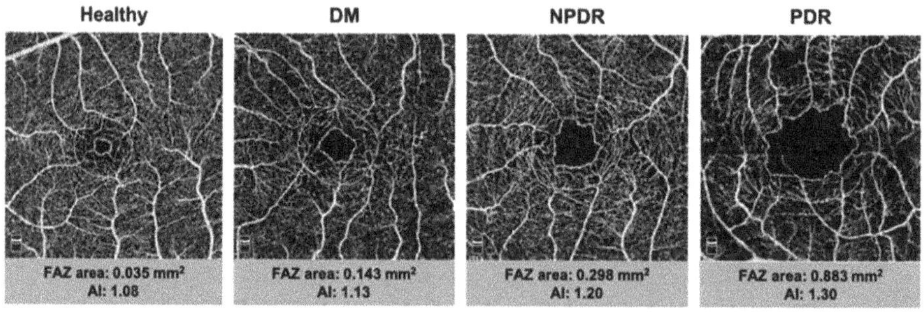

Figure 11.7. Representative quantitative evaluation of foveal avascular zone (FAZ) area and acircularity index (AI) with worsening diabetic retinopathy (DR) severity in optical coherence tomography angiography (OCTA). Foveal avascular zone (FAZ) area and AI measurements under each image were generated from built-in Optovue RTVue XR Avanti software with AngioAnalytics (Optovue Inc., Fremont, CA, USA, Version 2018.1.8.60). As DR severity worsens, the area of the FAZ enlarges and the FAZ becomes increasingly acircular. Reporduced from [27] with permission under an open access Creative Common CC BY license.

principle of OCTA that uses particle motion within the vessels as intrinsic contrast. Retinal vascular map is reconstituted by comparing motion-related differences between repeated OCT B-scans, taken exactly at the same location. OCTA is a relatively fast and dyeless imaging technique that provides high-resolution images of the retinal microvasculature at different depths, allowing visualization and delineation of superficial, intermediate, and deep retinal vascular plexus in DR (figure 11.7). OCTA delineates retinal capillary nonperfusion (FAZ and capillary drop out caused by ischemia) with a better resolution compared to FA [26]. However, it should be noted that the nonperfusion areas, as seen on OCTA, may represent either a complete occlusion or loss of capillaries or an extremely slow flow within an existing retinal capillary bed. In addition, OCTA can display the choroidal vascular network at distinct depths, separating the choriocapillaris, middle and outer choroidal vessels. Many of the classic vascular changes visualized in examination and FA imaging in DR, including microaneurysms, intraretinal microvascular abnormalities (IRMA), retinal neovascularization, and retinal nonperfusion regions, have been comprehensively described using OCTA. Indices of vascular density and the size of the foveal avascular zone (FAZ) as shown on OCTA en-face images are promising biomarkers to be incorporated in the care of DR [23, 24]. For example, microaneurysms located in deep vascular plexus are shown to be strongly associated with the development of intraretinal fluid at 12 months [25].

Systemic cardiovascular diseases may impact retinal microvascular structure and density and should be considered when interpreting OCT and OCTA results.

11.3.2 Peripheral findings in wide field OCT and OCTA as biomarkers of DR

While widefield FA remains the gold-standard imaging tool for detecting peripheral retinal nonperfusion and neovascularization, widefield OCTA has shown comparable diagnostic performance [28].

11.4 Systemic implications of local biomarkers

DR biomarkers as visualized on OCT and OCTA imaging may have significant predictive value for an individual's systemic health. Total retinal thickness and inner retinal thickness in the central subfield have been correlated with HbA1c and fasting blood glucose in youth patients with Type 2 DM [29]. Additionally, a better correlation between DR OCT biomarkers was noted amongst patients with type I DM with more severe systemic and laboratory disease parameters that warrants further investigation into the value of interlacing OCT and systemic biomarkers of DM and other cardiovascular diseases. [30–32]

11.5 Future applications

Noninvasive OCT and OCTA biomarkers of diabetic retinopathy are intertwined with the management of DR. Such biomarkers may reliably predict need for treatment, response to treatment, and visual outcomes. However, there is a subset of patients who do not confine to known patterns of disease and response to treatment. Novel biomarkers such as DRIL, intraretinal fluid reflectivity, cystoid space volume and localization, and peripheral retinal nonperfusion may perform better if integrated within a matrix of biomarkers to predict the prognosis and response to treatment.

Removing access barriers to high quality diagnostic technology for eye care will benefit patients across both rural and urban living environments [33]. Portable OCT is a promising technological advancement that may enable millions of patients who did not previously have access to equitable eye care benefit from early diagnosis and treatment. An ideal portable OCT system should be low-cost and easy to operate while offering the same resolution of established office-based OCT machines [34]. Such technologies can be integrated with automated analysis algorithms. At costs multiple times less than that of commercial systems, the availability of portable OCT devices of comparable system performance is becoming a reality. If made widely accessible, home-health and e-diagnostic care would not only expand access to earlier diagnosis and more accurate prognosis but would also broaden the range of available follow-up modalities. For example, home-health care using a portable OCT device would promote continued monitoring of disease status in the setting of a patient's own home, an option invaluable to rural communities, the elderly, those at high-risk of communicable diseases such as COVID-19 infection. Alternatively, patients could undergo follow-up imaging at local 'satellite' centers out of a pharmacy or outpatient primary care practice that is not necessarily ophthalmology-specific. Such practices could drastically change the landscape of ophthalmologic care in communities facing high DME burden with no easy or affordable access to specialty care clinics.

Automation of ophthalmic imaging is an inevitable development in the future of the field. OCT imaging is subject to segmentation error and subjectivity of interpretation while also requiring the refined skill set of ophthalmologists, retina specialists, and trained evaluators who currently exist in limited numbers. The use of artificial intelligence (AI) in the field of ophthalmology is ideal given the robust use of OCT imaging in a variety of ophthalmic diseases as well as the large number of images that can be fed to the program to feed developing algorithms. Several studies have assessed the performance of automated image assessment systems using artificial intelligence (AI), machine learning (ML), or deep learning (DL) in the diagnosis of DR, delivering promising findings [35]. Specific to automated analysis of fundus photos, Abràmoff *et al* examined the Messidor-2 data set with DL technology, showing an area under the curve (AUC) of 0.98 with sensitivity 98.8 and specificity of 87% in detecting moderate NPDR or worse, which included DME [36]. FDA approval has been given to a system developed by Abràmoff *et al* for its analysis of fundus photography in DR diagnosis [37]. It is likely that AI use in OCT analysis will shape the future landscape of ophthalmic imaging, diagnosis, and prognostication.

References

[1] *National Diabetes Statistics Report* | Diabetes | CDC (https://cdc.gov/diabetes/data/statistics-report/index.html) (accessed 27 December 2022)

[2] *Diabetes and Vision Loss* | Diabetes | CDC. (https://cdc.gov/diabetes/managing/diabetes-vision-loss.html) (accessed 27 December 2022)

[3] Daruich A, Matet A and Moulin A *et al* 2018 Mechanisms of macular edema: beyond the surface *Prog. Retin. Eye Res.* **63** 20–68

[4] Mrugacz M, Bryl A and Zorena K 2021 Retinal vascular endothelial cell dysfunction and neuroretinal degeneration in diabetic patients *J. Clin. Med.* **10** 458

[5] Bhagat N, Grigorian R A, Tutela A and Zarbin M A 2009 Diabetic macular edema: pathogenesis and treatment *Surv. Ophthalmol.* **54** 1–32

[6] Tang F Y, Chan E O and Sun Z *et al* 2020 Clinically relevant factors associated with quantitative optical coherence tomography angiography metrics in deep capillary plexus in patients with diabetes *Eye Vis.* **7** 1–11

[7] Lobo C, Santos T and Marques I P *et al* 2022 Characterisation of progression of macular oedema in the initial stages of diabetic retinopathy: a 3-year longitudinal study *Eye* **2022** 1–7

[8] Sivaprasad S, Crosby-Nwaobi R and Esposti S *et al* 2013 Structural and functional measures of efficacy in response to bevacizumab monotherapy in diabetic macular oedema: exploratory analyses of the BOLT study (report 4) *PLoS One* **8** 1371

[9] Yalçın N G and Özdek Ş 2019 The relationship between macular cyst formation and ischemia in diabetic macular edema *Turk. J. Ophthalmol.* **49** 194

[10] Deák G G, Bolz M, Ritter M, Prager S, Benesch T and Schmidt-Erfurth U 2010 A systematic correlation between morphology and functional alterations in diabetic macular edema *Invest. Ophthalmol. Vis. Sci.* **51** 6710–4

[11] Ciulla T A, Kapik B, Grewal D S and Ip M S 2021 Visual acuity in retinal vein occlusion, diabetic, and uveitic macular edema: central subfield thickness and ellipsoid zone analysis *Ophthalmol. Retina* **5** 633–47

[12] Ehlers J P, Uchida A and Hu M *et al* 2019 Higher-order assessment of OCT in diabetic macular edema from the VISTA study: ellipsoid zone dynamics and the retinal fluid index *Ophthalmology Retina* **3** 1056–66

[13] Sun J K, Lin M M and Lammer J *et al* 2014 Disorganization of the retinal inner layers as a predictor of visual acuity in eyes with center-involved diabetic macular edema *JAMA Ophthalmol.* **132** 1309–16

[14] Fickweiler W, Schauwvlieghe ASME, Schlingemann R O, Maria Hooymans J M, Los L I and Verbraak F D 2018 Predictive value of optical coherence tomographic features in the Bevacizumab and ranibiZumab in patients with diabetic macular edema (brdme) study *Retina* **38** 812–9

[15] Das R, Spence G, Hogg R E, Stevenson M and Chakravarthy U 2018 Disorganization of inner retina and outer retinal morphology in diabetic macular edema *JAMA Ophthalmol.* **136** 202–8

[16] Midena E, Torresin T, Velotta E, Pilotto E, Parrozzani R and Frizziero L 2021 OCT hyperreflective retinal foci in diabetic retinopathy: a semi-automatic detection comparative study *Front. Immunol.* **12** 1

[17] Zur D, Iglicki M and Busch C *et al* 2018 OCT biomarkers as functional outcome predictors in diabetic macular edema treated with dexamethasone implant *Ophthalmology* **125** 267–75

[18] Kim J T, Lee D H, Joe S G, Kim J G and Yoon Y H 2013 Changes in choroidal thickness in relation to the severity of retinopathy and macular edema in type 2 diabetic patients *Investigative Ophthalmol. Vis. Sci.* **54** 3378–84

[19] Rayess N, Rahimy E and Ying G S *et al* 2015 Baseline choroidal thickness as a predictor for response to anti-vascular endothelial growth factor therapy in diabetic macular edema *Am. J. Ophthalmol.* **159** 85–91

[20] Gupta C, Tan R and Mishra C *et al* 2018 Choroidal structural analysis in eyes with diabetic retinopathy and diabetic macular edema—a novel OCT based imaging biomarker *PLoS One* **13** e0207435

[21] Roy R, Saurabh K, Shah D, Chowdhury M and Goel S 2019 Choroidal hyperreflective foci: a novel spectral domain optical coherence tomography biomarker in eyes with diabetic macular edema *Asia-Pac. J. Ophthalmol.* **8** 314–8

[22] Kashani A H, Cheung A Y, Robinson J and Williams G A 2015 Longitudinal optical density analysis of subretinal fluid after surgical repair of rhegmatogenous retinal detachment *Retina* **35** 149–56

[23] Mirshahi R, Riazi-Esfahani H and Khalili Pour E *et al* 2021 Differentiating features of OCT angiography in diabetic macular edema *Sci. Rep.* **11** 1–7

[24] AttaAllah H R, Mohamed AAM and Ali M A 2018 Macular vessels density in diabetic retinopathy: quantitative assessment using optical coherence tomography angiography *Int. Ophthalmol.* **39** 1845–59

[25] Parravano M, De Geronimo D and Scarinci F *et al* 2019 Progression of diabetic microaneurysms according to the internal reflectivity on structural optical coherence tomography and visibility on optical coherence tomography angiography *Am. J. Ophthalmol.* **198** 8–16

[26] Couturier A, Mané V and Bonnin S *et al* 2015 Capillary plexus anomalies in diabetic retinopathy on optical coherence tomography angiography *Retina* **35** 2384–91

[27] Moir J, Khanna S and Skondra D *et al* 2021 Review of OCT angiography findings in diabetic retinopathy: insights and perspectives *Int. J. Transl. Med.* **1** 286–305

[28] Couturier A, Rey P A and Erginay A *et al* 2019 Widefield OCT-angiography and fluorescein angiography assessments of nonperfusion in diabetic retinopathy and edema treated with anti-vascular endothelial growth factor *Ophthalmology* **126** 1685–94

[29] Mititelu M, Uschner D and Doherty L *et al* 2022 Retinal thickness and morphology changes on OCT in youth with type 2 diabetes: findings from the TODAY study *Ophthalmol. Sci.* **2** 100191

[30] Wong T Y, Kamineni A and Klein R *et al* 2006 Quantitative retinal venular caliber and risk of cardiovascular disease in older persons: the cardiovascular health study *Arch. Intern. Med.* **166** 2388–94

[31] Kawasaki R, Xie J and Cheung N *et al* 2012 Retinal microvascular signs and risk of stroke: the Multi-Ethnic Study of Atherosclerosis (MESA) *Stroke* **43** 3245–51

[32] López-Cuenca I, Salobrar-García E and Sánchez-Puebla L *et al* 2022 Retinal vascular study using OCTA in subjects at high genetic risk of developing Alzheimer's disease and cardiovascular risk factors *J. Clin. Med.* **11** 3248

[33] Elam A R, Tseng V L and Rodriguez T M *et al* 2022 Disparities in vision health and eye care *Ophthalmology* **129** e89–e113

[34] Kim S, Crose M and Eldridge W J *et al* 2018 Design and implementation of a low-cost, portable OCT system *Biomed. Opt. Express* **9** 1232–43

[35] Moraru A D, Costin D, Moraru R L and Branisteanu D C 2020 Artificial intelligence and deep learning in ophthalmology—present and future (Review) *Exp. Therap. Med.* **20** 3469–73

[36] Abràmoff M D, Lou Y and Erginay A *et al* 2016 Improved automated detection of diabetic retinopathy on a publicly available dataset through integration of deep learning *Invest. Ophthalmol. Vis. Sci.* **57** 5200–6

[37] Abràmoff M D, Lavin P T, Birch M, Shah N and Folk J C 2018 Pivotal trial of an autonomous AI-based diagnostic system for detection of diabetic retinopathy in primary care offices *npj Digit. Med.* **1** 1–8

IOP Publishing

Photo Acoustic and Optical Coherence Tomography Imaging, Volume 1
Diabetic retinopathy
Ayman El-Baz and Jasjit S Suri

Chapter 12

Early identification of diabetic retinopathy through a computer-assisted diagnostic system and a higher-order spatial appearance model of 3D-OCT

Mohamed Elsharkawy, Ahmed Sharafeldeen, Ahmed Soliman, Fahmi Khalifa, Ali Mahmoud, Ahmed El-Baz, Mohammed Ghazal, Harpal Singh Sandhu and Ayman El-Baz

The timely detection of diabetic retinopathy (DR) is crucial to prevent serious harm to the optic nerve and potential visualization damage. This research proposes a computer-aided diagnosis (CAD) procedure to recognize DR in the earlier phase using 3D-OCT images, by analyzing these structural 3D retinal scans. An adaptive, appearance-based approach is employed in this system to automatically segment all retinal layers within the 3D-OCT scans, utilizing prior knowledge of the shape of the layers. After the segmentation of the retinal layers, distinctive texture features are extracted from each segmented layer of the OCT B-scans volume. These features are then utilized for the identification of DR. In order to obtain the second-order reflectivity for each individual retina layer, a Markov-Gibbs random field (MGRF) method is utilized. For representing these image-derived features, cumulative distribution function (CDF) values are used. The construction of the CDF is described using the percentages of the 10th through the 90th CDF percentiles, with a 10% growth. To perform layer-wise classification in the 3D volume, the system feeds the obtained feature for each layer into an artificial neural network (ANN). After performing layer-wise classification using the ANN for each of the twelve retinal layers, the system combines the classification outputs using a majority voting schema to arrive at a global subject diagnosis. This approach ensures that the system considers the information from all layers to make an accurate diagnosis. The performing of this method is calculated with a cohort of 188 3D-OCT subjects.

doi:10.1088/978-0-7503-2052-8ch12

Various k-fold methods and metrics of validation are used to evaluate the system's performance. This helps to ensure that the system's performance is consistent across different validation techniques and metrics.

12.1 Introduction

Ocular pathology poses a significant medical concern, given that the eye serves as a vital sensory organ. The possibility of blindness exists in ocular conditions like age-related macular degeneration (AMD) and diabetic retinopathy (DR). Unfortunately, these conditions may remain asymptomatic during their early stages and require regular eye exams for early diagnosis. The current chapter describes a study illustrating DR, a severe neurodegenerative and vascular disease that damages retinal cells yet may not cause noticeable visual impairment at first. DR often results from uncontrolled blood sugar levels, leading to progressive damage to the visual damage and retina. Thus, early detection of DR is crucial in preventing retinal damage that could eventually lead to blindness. As stated by [1], there were roughly 463 million people worldwide with DR in 2019. The United States only has 31 million cases. These numbers are predicted to increase to 578 million globally (34.4 million in the U.S.) by 2030 and to 700 million (367 million in the U.S.) by 2045. Furthermore, the Centers for Disease Control and Prevention (CDC) [2] reported an incidence of 4.1 million cases of DR in the United States. The economic cost of diabetes-related blindness amounts to roughly 500 million per year in the U.S. [2]. Notably, an estimated 899 000 Americans experience visual impairments. The timely identification of diabetic retinopathy (DR) is essential in the prevention of permanent blindness since it can advance without obvious symptoms and cause structural complications like neovascular glaucoma, proliferative diabetic retinopathy (PDR), and tractional retinal detachment.

The clinical management of DR begins with the improvement of systemic blood pressure and blood glucose control. Apart from regulating the body's systems, three standard methods for treating DR are vitreoretinal surgery, intravitreal anti-VEGF injections, and laser photocoagulation. In most cases, the first two treatments are sufficient and can be performed in an outpatient clinic setting. Intravitreal anti-VEGF injections and laser photocoagulation are the mainstays of DR therapy due to their ease of use and routine application. The most severe cases of proliferative vitreoretinopathy, marked by dense vitreous hemorrhages and/or tractional retinal detachment, are treated with vitrectomy. To monitor the progression of DR, clinicians use different imaging modalities such as Fundus Photography (FP), Optical Coherence Tomography (OCT), OCT Angiography (OCTA), and Fluorescein Angiography (FA). Each imaging modality has its own strengths in visualizing the retina. In recent years, researchers have increasingly employed retinal imaging techniques to identify eye diseases that can lead to vision loss. Nevertheless, some of these studies still rely on FP and FA, which necessitate a considerable amount of time for manual screening to detect morphological changes of the optic disk and exudates.

OCT has emerged as a valuable imaging modality for the early detection of various eye diseases, including DR. Interferometry, a non-invasive technique, produces cross-sectional images that offer a volumetric perspective of the retina, enabling visualization of minor changes like macular edema. In the next paragraph, we provide a summary of the literature on computer-aided diagnosis (CAD) systems that utilize OCT images for the diagnosis of ocular diseases.

12.1.1 Related work

Researchers have developed CAD systems to detect the disease (DR) using FP and FA. As an example, in their study [3], the researchers assessed the effectiveness of a support vector machine (SVM) for precisely DR grading in FA. To enhance the quality of images in their system, they utilized morphological procedures and median filters. Following that, they created a gray-level co-occurrence matrix to obtain statistical characteristics like correlation, energy, contrast, and homogeneity. Finally, they employed an SVM to categorize these features, and they stated an accuracy of 82.35%. Another system [4] uses fuzzy image preprocessing along with algorithms of ML to diagnose DR. This system utilizes color fundus images.

Additionally, OCT has displayed substantial potential in identifying retinal ailments such as DR, AMD, and diabetic macular edema (DME). Sandhu *et al* [5] created a computer-aided detection (CAD) framework that employed both OCTA and OCT techniques to identify and classify DR, integrating patient demographic information with characteristics obtained from OCTA and OCT images. Another study by [6] also utilized OCT images to grade DR in their CAD system, extracting features from OCT histograms and using SVM to classify DR. Although they achieved a correct classification rate of 66.7% based on leave-one-subject-out validation, their results were less accurate when compared to other studies. Maetschke and colleagues (Maetschke *et al* [7]) introduced an alternative CAD system that utilizes OCT images for detecting glaucoma. This method employs a CNN to extract features from OCT images and a SoftMax layer to classify them, achieving an area under the curve (AUC) of 94%. Additional research that employs OCT for the diagnosis of DR can be located in the references [8–14].

Non-invasive imaging techniques such as OCT and OCTA have been utilized in various studies to detect retinal disorders. For instance, Alam *et al* [15] designed an approach for DR grading and detection using OCTA scans. They used a classification approach built on SVM and obtained six features from the OCTA images, including blood vessel density, vessel perimeter index, and blood vessel caliber. The fusion of all features resulted in the best diagnostic performance, with an accuracy of 94.41% and 92.96% for normal vs. DR and normal vs. different grades of DR, respectively. In another study [16], a CAD system was developed for DR diagnosis using 3D-OCT volumes. The system utilized the histogram of oriented gradients and local binary pattern features, with dimensionality reduction via principal component analysis (PCA) and separate classification using several classifiers. The optimal classifier employed in this study was SVM with kernel of linear, achieving specificity and sensitivity of 87.5%. However, this study had some limitations, including absence of segmentation of OCT layers and bad performance. Ibrahim and colleagues (Ibrahim *et al* [17]) introduced a CAD system that utilizes a pre-trained

VGG16 for detecting DME, choroidal neovascularization (CNV), and drusenoid disorders in three-dimensional optical coherence tomography (3D-OCT) volumes. The deep neural network extracted features, which were combined with hand-crafted features from the region of interest to achieve a precise diagnosis. This system achieved 98.8%, 99.4%, and 98.2% for accuracy, sensitivity, and specificity, respectively.

Ghazal *et al* [18] have demonstrated the use of CNN CAD to detect DR in OCT B-scans. The B-scan is divided into five regions, including nasal, distal nasal, central, distal temporal, and temporal. Seven CNNs were trained, each based on one of these regions, along with two transfers based only on the nasal and temporal regions. The results from the seven CNNs were combined with the two regions and transfer learning to obtain an overall diagnosis. The system achieved a reported performance of 94%. In another study [5], OCT and OCTA were integrated to diagnose DR grades using a random forest classifier along with clinical and demographic data. The reported accuracies of detecting DR were 98.2% and 98.7%, respectively, and the grades assigned to the DR were reported. There have also been additional research paper that used OCTA and OCT to diagnose various retinal disorders [6–10, 14, 19–33].

The CAD system being proposed in this study includes a segmentation method that is used to segment 3D-OCT volume into twelve layers for each B-scan. An adaptive structure prior learning is used in the segmentation process. Subsequently, a novel texture feature is developed from a Markov-Gibbs random field (MGRF) shape that exploits the second-order characteristics of image gray quantities. The 2D image in the OCT volume is regarded as a singular occurrence of an MGRF. The Gibbs energy values are used to establish a cumulative distribution function (CDF) for that layer, which is then divided into nine deciles. The decile features for each layer are combined, and an ANN is used for the purpose of training and testing. Various validation techniques are used to evaluate and conduct the proposed system, utilizing a majority voting representation.

The work involved in this chapter represents an advancement to the previous work on segmenting 12 layers in 3D-OCT images [34], with the following contributions: First, the OCT is analyzed on a per-layer basis, enabling local and individualized analysis. Second, a 3D-MGRF model is incorporated instead of using a first-order reflectivity. Third, to enhance the descriptive power of the extracted features, a statistical approach based on CDF percentiles is utilized. Finally, the ANN takes the CDF percentile values as input to generate the ultimate diagnosis of the 3D-OCT images.

12.2 Computer aided diagnosis system

In this chapter, we describe a new 3D-OCT CAD system that utilizes a spatial appearance model of higher order, depicted in figure 12.1. This CAD method developed consists of three main stages. The initial phase involves 3D-OCT segmentation of twelve OCT layers using a segmentation method we developed [34], which uses an appearance-based technique. In the following stage, distinct

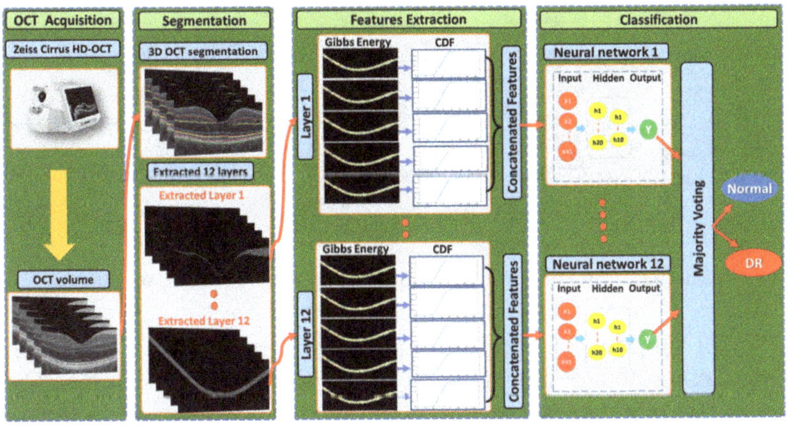

Figure 12.1. The proposed framework for DR diagnosis using 3D-OCT images.

higher-order reflectivity features are obtained. In the third and final stage, the individual layers are classified based on the features extracted, and a majority voting approach is used to determine the global diagnosis from the classification outputs of the layers. The upcoming subsections will elaborate on the 3D-OCT segmentation method and the unique feature derived from segmented retinal layers.

12.2.1 OCT volume segmentation

In this method, the macular OCT is segmented using the initial cross-sectional image passing through the center of the fovea, where the twelve OCT layers encircling the fovea are identified using MGRF modeling. To help convey the qualified contrast, anticipated shape, and homogeneity information, an atlas of the retina is utilized. To segment a 3D-OCT volume, the information acquired from the central slice is transferred to the inferior and superior B-scans in relation to the central slice. The segmentation technique involves a series of steps aimed at accurately identifying and isolating the retinal layers. The first step is to match the OCT 2D-scans to a model database that includes segmentation done manually for the foveal B-scans of both diseased retinas and normal one. This phase helps to establish a reference for the segmentation algorithm to follow. Secondly, the area among the choroid and vitreous is divided into twelve individual sectors created on a model that takes into account shape, intensity, and spatial interactions. These segments are carefully defined to ensure that each layer is properly identified and delineated. Thirdly, the B-scans that have not undergone processing are employed as shape models to segment each B-scan that has already been segmented. This phase involves using the information obtained in the previous two steps to refine the segmentation further. Finally, the initial shape models are iteratively processed to achieve the optimal segmentation. This refinement process involves the use of statistical shape models and level-set methods to adjust the boundaries of the segmented layers. The statistical shape models are constructed from a training set of labeled data, and they help to capture the variability in the shape and position of the retinal layers

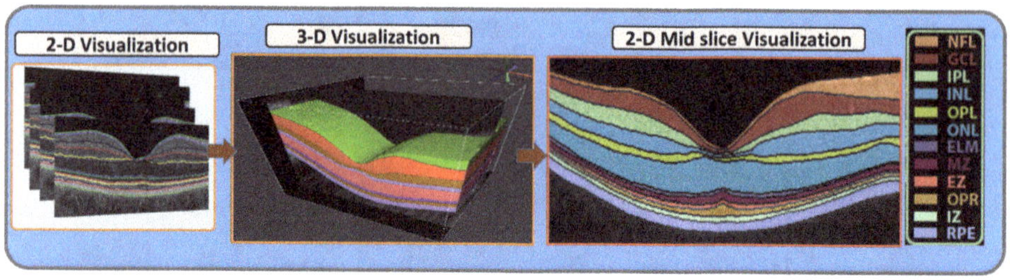

Figure 12.2. An Illustrative figure for the layer segmentation of the retina.

across different subjects. The level-set method is used to evolve the shape models towards the actual boundaries of the layers in the 3D-OCT volume. The refined shape models are then used to generate the final segmentation of the retinal layers.

In figure 12.2, the layer segmentation of the retina can be visualized using 3D-OCT.

The segmentation includes twelve layers, with the first layer being the nerve fiber layer (NFL), and the twelfth layer being the retinal pigment epithelium (RPE). The segmentation of these layers is crucial for accurate diagnosis, as each layer provides important information about the health of the retina. The segmentation method described here is designed to be reliable, efficient, and accurate, ensuring that the resulting 3D-OCT volume can be analyzed effectively to aid in the diagnosis of various retinal diseases.

The algorithm below presents the Prior Shape Propagation Algorithm, which is used for segmenting the 12 retinal layers of macular OCT images. The algorithm takes as input a set of K B-scans and the segmentation of B-scan labeling twelve OCT layers. The algorithm works as follows:

1. Iterate over the B-scans in the range [n−1, K]:
   (a) For each B-scan, warp it using non-rigid registration to align it with the previous (for B-scans n − 1 to 1) or next (for B-scans n + 1 to K) B-scan.
   (b) For each pixel v in the B-scan:
      i. Find the equivalent pixel v′ in the warped B-scan using the non-rigid mapping.
      ii. A window of size $K_{1i} \times K_{2i}$ is specified over the voxel v′.
      iii. Pixels in the window w whose gray-level intensities drop inside a threshold τ of that of the voxel v are specified.
      iv. If there are no such pixels found in step (2), the window w is expanded until in-range gray quantities are observed, or the maximum window size is reached.
      v. We compute the probability of v belonging to a particular retinal layer using the label frequency of that layer among the pixels in the window w that meet the threshold requirement.
      vi. Finally, the voxel v is assigned the label of the segment that has the highest corresponding probability calculated in the previous step.

The algorithm for prior shape propagation involves using non-rigid registration to align the B-scans and then considering the intensity of pixels in a window around each pixel to calculate the probability of belonging to a retinal layer. The algorithm starts with the initial segmentation of a B-scan and then propagates the past shape patterns to the neighboring B-scans to improve the segmentation accuracy for each B-scan. This is done by warping each 2D-OCT scan to align with its neighboring 2D-OCT utilizing non-rigid registration. Then, for each pixel in the 2D-OCT scan, a window is centered on the corresponding pixel in the neighboring B-scan, and pixels in the window with gray-level intensities surrounded by a particular threshold of the corresponding pixel are identified. The posterior probability of belonging to a layer is calculated based on the occurrence of that layer's label with the pixels in the window that satisfy the threshold criterion. The algorithm repeats this process for all B-scans in both directions to refine the past shape patterns and get the optimal segmentation.

A novel technique has been developed to segment the whole 3D-OCT images using only one slice. This method incorporates an adaptive shape prior that considers both reflectivity values and voxel locations represented to the reference slice. In contrast to prior shape models that solely rely on shape information, this approach considers optical presence when producing probabilistic maps. Specifically, the adaptive shape-intensity model of first order predominantly spreads segment labels to nearby B-scans that display similar OCT reflectivity as the labeled region in the reference slice. This process is aimed at maintaining consistency in appearance while propagating the segment labels. The segmentation propagation process is delineated in the algorithm. Figure 12.2 depicts the segmenting development of the complete 3D-OCT images in both directions, commencing from the middle slice. Using our 2D segmentation method, we developed a segmentation approach for the macular OCT by segmenting the fovea B-scan, followed by expanding it to segmentation of 3D-OCT volume for the leftover four slices. In this approach, we used the patient-specific atlas of center B-scan for the segmentation propagation. We demonstrated the segmentation process using a 3D-OCT image volume comprising 5 B-scans. The segmentation was carried out on 12 layers using a joint-MGRF algorithm in the third slice, which intersected the fovea. The segmentation of the 12 layers in slice 3 acted as a patient-specific atlas to propagate the segmentation to slices 2 and 4. Finally, the segmentation of slices 2 and 4 was propagated to slices 1 and 5, guided by the patient-specific atlas.

12.2.2 Feature extraction

The second phase of our approach involves the analysis of higher-order reflectivity features from the segmented retina layers, with the goal of capturing the second-order structure of image gray levels. To accomplish this, we consider each B-scan in the OCT volume as a Markov random field (MGRF) instance, where the interactions among pixels in a fully connected neighborhood or clique are specified. Our MGRF system is two-dimensional, and its translation-invariant system includes four neighbors: $N = (0, 1), (0, -1), (1, 0), (-1, 0)$. However, we do not consider pixel interactions between B-scans since the slice spacing is much larger than the pixel dimensions within a B-scan.

The notation used in this study is defined as follows:

The image g in grayscale is defined on a discrete domain R, which is a subset of the Cartesian product of the integers Z, i.e., $R \subset Z \times Z$. The pixel values of g lie in the set $Q = 0, 1, \ldots, Q - 1$, where Q is the maximum grayscale intensity. We can represent g as a function that maps each pixel coordinate in R to its corresponding grayscale intensity value in Q, i.e., g: $R \rightarrow Q$. The neighborhood system is represented by a set of (x, y)-offsets, $N = (\xi i, \eta i)$, $i = 1, \ldots, n$, which specifies pairwise interactions. The graph of pixel interactions on R is denoted by C, and the neighborhood system for pixel $(x, y) \in R$ is given by $C(x, y) = (x, y), (x + \xi, y + \eta), (\xi, \eta) \in N$. The Gibbs potential associated with neighborhood $(\xi i, \eta i)$ is represented as $Vi : Q \times Q \rightarrow R$.

The Gibbs probability distribution function for the second-order MGRF on R can be defined as follows:

$$P(q) = \frac{1}{Z} \exp\left(|R| \sum_{i-1}^{n} \sum_{q=0} V_i(q, q') \right) Q - 1, \tag{12.1}$$

The MGRF model can be defined by its Gibbs probability distribution function, which includes the partition function Z, the cardinality of R denoted by $|R|$, and the bivariate empirical probability distribution F_i over the neighborhood family i. To estimate the potentials Vi in the Gibbs probability function (equation (12.1)), We apply the analytic maximum likelihood approximator to estimate the second-order potentials of Gibbs proposed in [35]. This estimator is given by the following equation:

$$Vi(q, s) = \rho i(Fi(q, s) - fi(q)fi(s)) \tag{12.2}$$

where ρi is a scaling parameter, and $fi(q)$ and $fi(s)$ are the marginal probability distributions of gray levels q and s, respectively, and where $f_i(q) = \frac{1}{Q} \sum_{q'=0}^{Q-1} F_i(q, q')$ is the gray-level marginal distribution. To calculate the potentials of Gibbs, empirical gray-level distributions are first calculated within each B-scan. Then, these distributions are averaged across B-scans to obtain the final Fi and fi for every layer in the OCT volumes. This information is then used to determine the potentials of Gibbs for each layer during the OCT images. Specifically, for every single layer l, a vector of Gibbs energies is calculated to describe that layer's texture.

12.2.3 Classification system

The third phase of our methodology involves constructing a diagnostic approach for DR using an artificial neural network (ANN)-based classification system. To obtain a conclusive diagnosis of DR, we employed a classifier consisting of two stages. Initially, we trained 12 ANNs to classify the 3D-OCT volume on a layer-by-layer basis. Different features for each layer across all five scans of the 3D-OCT were combined into a single vector or array and then fed into the ANNs as input. In the second stage, we utilized a majority voting (MV) method to determine the global diagnosis for the subject based on the output of the ANNs. To give more significance to the layer representation, we generated a cumulative distribution function (CDF) of the feature values. The process for training the ANN is described in Algorithm 1.

1. The initial step in training an artificial neural network is to assign random weights to the network's connections. These random values are typically generated from a normal distribution. This process is known as initialization.
2. In the feed-forward step, the neurons in the hidden layers and output layer are activated by calculating the weighted sum of their inputs and applying a non-linear activation function. The outputs are then passed to the next layer and this process continues until the output is produced. This is called a feed-forward step because the activation flows from the input to the output layer without any feedback.
3. Backpropagation requires computation of the slope of the loss function in connection with the network's weights using the chain rule and then using the stochastic gradient descent optimization technique to update the weights in the opposite direction of the gradient. This process iteratively adjusts the weights to minimize the difference between the actual output of the network and the desired output.
4. The second and third step are repeated iteratively until the weights of ANN reach convergence or modify significantly less than a pre-defined threshold.

The network architecture was optimized through hyperparameter tuning, which involved adjusting the number and size of hidden layers. The optimal configuration for the ANN was found to be two hidden layers, with one layer consisting of 20 neurons and the other with 10 neurons. The activation function for the hidden layer was set to a hyperbolic tangent, while the output layer used a SoftMax activation function. The activation function was selected from a set of four functions, namely tangent, SoftMax, sigmoid, and rectified linear. The optimal function was chosen based on performance.

12.3 Experimental results

Our proposed method was tested on 188 OCT volumes, with 100 volumes from normal retinas and 88 volumes from DR patients. The OCT data utilized in this investigation were acquired from the Zeiss Cirrus HD-OCT 5000 device [36] and were gathered at the University of Louisville Hospital. The study was authorized by the University Institutional Review Board (IRB), and all subjects provided informed consent. The Zeiss Cirrus equipment captured 5 OCT image scans of both eyes, with dimensions of 1024 ×1024×5 voxels, 8.80 ×1.96 ×500.00 mm^3, and 9.01 ×2.00 ×2.50 mm^3. The subjects' ages varied amongst 21 and 84 years old. Manual segmentation was performed by medical retina specialists. To reduce the likelihood of human error, each image was independently reviewed by two specialists who then reached a consensus on the segmentation for that image. The 3D-OCT scans of the macula revealed twelve layers separated by thirteen boundary surfaces. These scans were included in diabetic studies regardless of their ocular background, with a minimum signal strength of 7/10, in a clinical setting. The clinical grade and severity of disease were evaluated by physicians specializing in retinal disease. Dilated fundus examinations were also conducted on each eye to determine the presence of significant retinopathy.

To evaluate the performance of our system on each class of the dataset, which is composed of DR and normal cases, we employed three evaluation metrics: accuracy, sensitivity, and specificity. To calculate these metrics, we computed the true positive (TP), false positive (FP), true negative (TN), and false negative (FN) rates for each class.

Then, the three evaluation metrics were computed as in figure 12.3:

We employed K-fold cross-validation to evaluate the performance of our proposed system. The validation experiments were carried out to test the system's quantitative performance using various scenarios, including 4-fold, 5-fold, and 10-fold cross-validation. The system's accuracy, sensitivity, and specificity were evaluated as performance metrics. In addition, the dataset was divided into a training set and a test set. The training set was utilized to train and fine-tune the model parameters to obtain the optimal hyperparameters for the ANN. Furthermore, the dataset was split into a 70% training set and a 30% test set for additional evaluation. The accuracy results for the layer-wise categorization utilizing 10-folds cross-validation. The established method achieved high accuracies for most layers, ranging from 74.38% for ONL to 94.61% for NFL. The accuracy values for

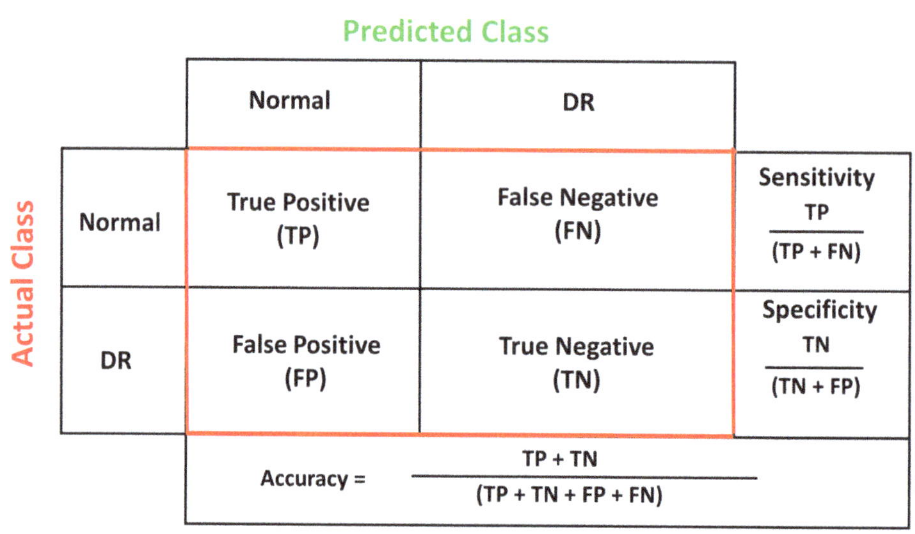

Figure 12.3. An illustrative figure for how to find three evaluation metrics: accuracy, sensitivity, and specificity.

GCL, IPL, INL, OPL, EZ, IZ, and RPE were between 85.14% to 88.21%. Moreover, the majority voting method resulted in the best accuracy value of 95.7% for the developed system, surpassing the 4-folds and 5-folds cross-validation methods, which obtained accuracy values of 89.40% and 91.50%, respectively.

As the existing literature presents 2D systems that utilize CNN and/or FP to classify DR, it is inappropriate to compare them to the proposed system since the proposed system differs greatly from them. We assessed the effectiveness of the proposed method by comparing it with several other machine learning (ML) algorithms, including a support vector machine with kernels of radial and linear, Naive Bayes, logistic regression, k-nearest neighbors, decision tree, and random forest. The findings outcomes of these reasonable conduct experiments are produced. Our proposed method achieved the highest accuracy among all the compared methods. As an example, we can consider the results obtained from 10-fold cross-validation, where we compared the accuracies of various ML algorithms such as SVM with linear and radial basis function kernels, logistic regression, Naive Bayes, K-nearest neighbors (KNN), random forest (RF), and decision tree (DT). The obtained accuracies for these algorithms were as follows: 76.60%, 76.06%, 81.91%, 71.28%, 76.06%, 77.70%, and 81.91%, respectively. Evaluating every one of the ML procedures along with our proposed approach, it achieved the greatest accuracy of 95.70% using ten-fold cross-validation. Additionally, we conducted a comparison of our proposed method with established deep learning techniques such as AlexNet, ResNet50, and Inceptionv3 [37–39]. Many ML and deep learning algorithms used before in some medical image diagnosis and prove the efficiency of using it [40–51]. To obtain reasonable outcomes with the proposed method, we used and optimized the hyperparameters for these pre-trained CNNs. We conducted a grid search on the optimizer using Adam and Stochastic Gradient Descent with Momentum (SGDM) as search spaces to identify the optimal hyperparameters designed for the entire pre-trained networks. Moreover, we reduced the learning rate by a factor of 10 from 0.001 after every five iterations. We set a maximum number of epochs of 500 and monitored the validation loss to determine the best-fit training model. Our developed method achieved higher accuracy compared to well-known pre-trained CNNs. Our method achieved an accuracy of 95.70% using 10-fold cross-validation, while the pre-trained CNNs AlexNet, ResNet50, and Inceptionv3 achieved accuracies of 91.40%, 94.90%, and 94.51%, respectively. Furthermore, in order to demonstrate the robustness of our proposed system, we constructed receiver operating characteristic (ROC) curves for every single retina layer. The calculated area under the curve (AUC) values for all retina layers reveal that our suggested method outperforms the other ML methods such as SVM with linear kernel, KNN, random forest, and Naive Bayes utilizing the ten-fold cross-validation. Our proposed system demonstrated excellent performance in terms of AUC values for all retina layers, with scores of 0.99, 0.95, 0.95, 0.92, 0.96, 0.95, 0.95, 0.82, 0.85, 0.86, 0.87, and 0.94 for NFL, GCL, IPL, INL, OPL, ONL, ELM, MZ, EZ, OPR, IZ, and RPE, respectively.

12.4 Conclusions and future work

This paper presents a novel OCT CAD system for detecting DR at an earlier stage, which enables early intervention to prevent vision loss by utilizing more informative 3D-OCT volumes. The system focuses on mild and moderate grades of non-proliferative diabetic retinopathy (NPDR) and does not include cases with macula edema or severe NPDR and proliferative diabetic retinopathy. The proposed approach employs the concept of MGRF to estimate a spatial appearance model of higher order from the 3D-OCT retina layers that have been segmented automatically and achieved a higher accuracy of 95.70% compared to state-of-the-art deep learning networks and ML algorithms. However, to further improve the accuracy, the authors plan to incorporate additional features and analyze a larger dataset. They also intend to explore the ability of combining the suggested method through another clinical biomarkers and image modalities in upcoming studies. Moreover, this diagnostic system holds promising potential for expansion to identify diverse ocular conditions that negatively impact the retinal layers and result in impaired vision.

In addition to the retina [52–55], this work could also be applied to various other applications in medical imaging, such as the prostate [56–59], the kidney [60–77], the heart [78–95], the lung [96–143], the brain [144–165], the vascular system [166–176], the bladder [177–181], the head and neck [182, 183], and injury prediction [184], as well as several non-medical applications [185–191].

References

[1] Saeedi P *et al* 2019 Global and regional diabetes prevalence estimates for 2019 and projections for 2030 and 2045: results from the international diabetes federation diabetes atlas, 9th edition *Diabetes Res. Clin. Pract* **157** 107843

[2] *Centers for Disease Control and Prevention* (https://cdc.gov/visionhealth/pdf/factsheet.pdf) (accessed 12 January 2021)

[3] Foeady A Z, Novitasari D C R, Asyhar A H and Firmansjah M 2018 Automated diagnosis system of diabetic retinopathy using glcm method and svm classifier *2018 5th Int. Conf. on Electrical Engineering, Computer Science and Informatics (EECSI)* pp 154–60

[4] Rahim S S, Palade V, Shuttleworth J and Jayne C 2016 Automatic screening and classification of diabetic retinopathy and maculopathy using fuzzy image processing *Brain Inform.* **3** 249–67

[5] Sandhu H S, Elmogy M, Sharafeldeen A T, Elsharkawy M, Eladawi N, Eltanboly A, Shalaby A, Keynton R and El-Baz A 2020 Automated diagnosis of diabetic retinopathy using clinical biomarkers, optical coherence tomography, and optical coherence tomography angiography *Am. J. Ophthalmol.* **216** 201–6

[6] Bernardes R, Serranho P, Santos T, Gonçalves V and Cunha-Vaz J 2012 Optical coherence tomography: automatic retina classification through support vector machines *Eur. Ophthalmic Rev.* **6** 200–3

[7] Maetschke S, Antony B, Ishikawa H, Wollstein G, Schuman J and Garnavi R 2019 A feature agnostic approach for glaucoma detection in OCT volumes *PLoS One* **14** e0219126

[8] Ko C-E, Chen P-H, Liao W-M, Lu C-K, Lin C-H and Liang J-W 2019 Using a cropping technique or not: impacts on svm-based amd detection on OCT images *2019 IEEE Int. Conf. on Artificial Intelligence Circuits and Systems* pp 199–200

[9] Serener A and Serte S 2019 Dry and wet age-related macular degeneration classification using OCT images and deep learning *2019 Scientific Meeting on Electrical-Electronics Biomedical Engineering and Computer Science (EBBT)* pp 1–4

[10] Pekala M, Joshi N, Alvin Liu T Y, Bressler N M, Cabrera DeBruc D and Bulina P *et al* 2019 Deep learning based retinal OCT segmentation *Comput. Biol. Med.* **114** 103445

[11] Mohammed S, Li T, Chen X D, Warner E, Shankar A, Abalem M F, Jayasundera T, Gardner T W and Rao A 2020 Density-based classification in diabetic retinopathy through thickness of retinal layers from optical coherence tomography *Sci. Rep.* **10** 1–13

[12] Haggag S, Elnakib A, Sharafeldeen A, Elsharkawy M, Khalifa F, Farag R K, Mohamed M A, Sandhu H S, Mansoor W and Sewelam A *et al* 2022 A computer-aided diagnostic system for diabetic retinopathy based on local and global extracted features *Appl. Sci.* **12** 8326

[13] Elsharkawy M, Elrazzaz M, Sharafeldeen A, Alhalabi M, Khalifa F, Soliman A, Elnakib A, Mahmoud A, Ghazal M and El-Daydamony E *et al* 2022 The role of different retinal imaging modalities in predicting progression of diabetic retinopathy: a survey *Sensors* **22** 3490

[14] Elsharkawy M, Sharafeldeen A, Soliman A, Khalifa F, Ghazal M, El-Daydamony E, Atwan A, Sandhu H S and El-Baz A 2022 Diabetic retinopathy diagnostic CAD system using 3D-OCT higher order spatial appearance model *2022 IEEE 19th Int. Symp. on Biomedical Imaging (ISBI)* (Piscataway, NJ: IEEE) pp 1–4

[15] Alam M, Zhang Y, Lim J, Chan R, Yang M and Yao X 2018 Quantitative optical coherence tomography angiography features for objective classification and staging of diabetic retinopathy *Retina* 1

[16] Alsaih K, Lemaitre G, Rastgoo M, Massich J and Sidibe D 2017 Ma-´ chine learning techniques for diabetic macular edema (dme) classification on SD-OCT images *Biomed. Eng. Online* **16** 68

[17] Ibrahim M, Fathalla K and Youssef S 2020 Hycad-OCT: a hybrid computer-aided diagnosis of retinopathy by optical coherence tomography integrating machine learning and feature maps localization *Appl. Sci.* **10** 4716

[18] Ghazal M, Ali S S, Mahmoud A H, Shalaby A M and El-Baz A 2020 Accurate detection of non-proliferative diabetic retinopathy in optical coherence tomography images using convolutional neural networks *IEEE Access* **8** 34387–97

[19] Banerjee I, Sisternes L, Hallak J, Leng T, Osborne A, Durbin M and Rubin D 2019 A deep-learning approach for prognosis of age-related macular degeneration disease using SD-OCT imaging biomarkers ArXiv:1902.10700

[21] An G, Omodaka K, Hashimoto K, Tsuda S, Shiga Y, Takada N, Kikawa T, Yokota H, Akiba M and Nakazawa T 2019 Glaucoma diagnosis with machine learning based on optical coherence tomography and color fundus images *J. Healthcare Eng.* **2019** 4061313

[22] Elsharkawy M, Sharafeldeen A, Soliman A, Khalifa F, Ghazal M, El-Daydamony E, Atwan A, Sandhu H S and El-Baz A 2022 A novel computer-aided diagnostic system for early detection of diabetic retinopathy using 3D-OCT higher-order spatial appearance model *Diagnostics* **12** 461

[23] Mateen M, Wen J, Hassan M, Nasrullah N, Sun S and Hayat S 2020 Automatic detection of diabetic retinopathy: a review on datasets, methods and evaluation metrics *IEEE Access* **8** 48784–811

[24] Shankar K, Wahab Sait A R, Gupta D, Lakshmanaprabu S K, Khanna A and Mohan H 2020 Automated detection and classification of fundus diabetic retinopathy images using synergic deep learning model *Pattern Recognit. Lett.* **133** 210–6

[25] Cao K, Xu J and Zhao W-Q 2019 Artificial intelligence on diabetic retinopathy diagnosis: an automatic classification method based on grey level co-occurrence matrix and Naive Bayesian model *Int. J. Ophthalmol.* **12** 1158–62

[26] Ng W S, Mahmud W M H W, Huong A K C, Kairuddin W N H W, Gan H S and Izaham R M A R 2019 Computer aided diagnosis of eye disease for diabetic retinopathy *J. Phys.: Conf. Ser.* **1372** 012030

[27] Bannigidad P and Deshpande A 2018 Exudates detection from digital fundus images using glcm features with decision tree classifier *Int. Conf. on Recent Trends in Image Processing and Pattern Recognition (Berlin)* (Springer) pp 245–57

[28] Rashed N, Ali S and Dawood A 2018 Diagnosis retinopathy disease using GLCM and ANN *J. Theor. Appl. Inform. Technol.* **96** 6028–40

[29] Giraddi S, Pujari J and Seeri S 2015 Role of GLCM features in identifying abnormalities in the retinal images *Int. J. Image, Graph. Signal Process.* **7** 45–51

[30] Elsharkawy M, Elrazzaz M, Ghazal M, Alhalabi M, Soliman A, Mahmoud A, El-Daydamony E, Atwan A, Thanos A and Sandhu H S *et al* 2021 Role of optical coherence tomography imaging in predicting progression of age-related macular disease: a survey *Diagnostics* **11** 2313

[31] Le D, Alam M, Yao C, Lim J, Chan R, Toslak D and Yao X 2020 Transfer learning for automated OCT a detection of diabetic retinopathy *Transl. Vis. Sci. Technol.* **9** 7

[32] Heisler M, Karst S, Lo J, Mammo Z, Yu T, Warner S, Maberley D, Beg M F, Navajas E and Sarunic M 2020 Ensemble deep learning for diabetic retinopathy detection using optical coherence tomography angiography *Transl. Vision Sci. Technol.* **9** 20

[33] Sharafeldeen A, Elsharkawy M, Khalifa F, Soliman A, Ghazal M, AlHalabi M, Yaghi M, Alrahmawy M, Elmougy S and Sandhu H *et al* 2021 Precise higher-order reflectivity and morphology models for early diagnosis of diabetic retinopathy using OCT images *Sci. Rep.* **11** 1–16

[34] Sleman A A, Soliman A, Elsharkawy M, Giridharan G, Ghazal M, Sandhu H, Schaal S, Keynton R, Elmaghraby A and El-Baz A 2021 A novel 3D segmentation approach for extracting retinal layers from optical coherence tomography images *Med. Phys.* **48** 1584–95

[35] El-Baz A *et al* 2016 *Stochastic Modeling for Medical Image Analysis* (Boca Raton: CRC Press)

[36] ZEISS *Cirrus HD-OCT 5000* 2020 (https://zeiss.com/meditec/us/customercare/customer-care-for-ophthalmology-optometry/quick-help-for-cirrushd-oct-5000.html) (accessed 25 October 2020)

[37] Krizhevsky A *et al* 2012 Imagenet classification with deep convolutional neural networks *Adv. Neural Inform. Process. Syst.* **25** 1097–105

[38] He K *et al* 2016 Deep residual learning for image recognition *Proc. of the IEEE Conf. on Computer Vision and Pattern Recognition* pp 770–8

[39] Szegedy C *et al* 2016 Rethinking the inception architecture for computer vision *Proc. of the IEEE Conf. on Computer Vision and Pattern Recognition* pp 2818–26

[40] Abdel Razek A A K, Alksas A, Shehata M, AbdelKhalek A, Abdel Baky K, El-Baz A and Helmy E 2021 Clinical applications of artificial intelligence and radiomics in neuro-oncology imaging *Insights Imaging* **12** 1–17

[41] Alksas A, Shehata M, Saleh G A, Shaffie A, Soliman A, Ghazal M, Khelifi A, Khalifeh H A, Razek A A and Giridharan G A *et al* 2021 A novel computer-aided diagnostic system for accurate detection and grading of liver tumors *Sci. Rep.* **11** 1–18

[42] Alksas A, Shehata M, Saleh G A, Shaffie A, Soliman A, Ghazal M, Khalifeh H A, Razek A A and El-Baz A 2021 A novel computer-aided diagnostic system for early assessment of hepatocellular carcinoma *2020 25th Int. Conf. on Pattern Recognition (ICPR)* (Piscataway, NJ: IEEE) pp 10375–82

[43] Ayyad S M, Badawy M A, Shehata M, Alksas A, Mahmoud A, Abou El-Ghar M, Ghazal M, El-Melegy M, Abdel-Hamid N B and Labib L M *et al* 2022 A new framework for precise identification of prostatic adenocarcinoma *Sensors* **22** 1848

[44] Shehata M, Alksas A, Abouelkheir R T, Elmahdy A, Shaffie A, Soliman A, Ghazal M, Abu Khalifeh H, Salim R and Abdel Razek A A K *et al* 2021 A comprehensive computer-assisted diagnosis system for early assessment of renal cancer tumors *Sensors* **21** 4928

[45] Farahat I S, Sharafeldeen A, Elsharkawy M, Soliman A, Mahmoud A, Ghazal M, Taher F, Bilal M, Abdel Razek A A K and Aladrousy W *et al* 2022 The role of 3D CT imaging in the accurate diagnosis of lung function in coronavirus patients *Diagnostics* **12** 696

[46] Elsharkawy M, Sharafeldeen A, Taher F, Shalaby A, Soliman A, Mahmoud A, Ghazal M, Khalil A, Alghamdi N S and Razek A A K A *et al* 2021 Early assessment of lung function in coronavirus patients using invariant markers from chest x-rays images *Sci. Rep.* **11** 12095

[47] Sharafeldeen A, Elsharkawy M, Khaled R, Shaffie A, Khalifa F, Soliman A, Abdel Razek A A k, Hussein M M, Taman S and Naglah A *et al* 2022 Texture and shape analysis of diffusion-weighted imaging for thyroid nodules classification using machine learning *Med. Phys.* **49** 988–99

[48] Sharafeldeen A, Elsharkawy M, Shaffie A, Khalifa F, Soliman A, Naglah A, Khaled R, Hussein M, Alrahmawy M and Elmougy S *et al* 2022 Thyroid cancer diagnostic system using magnetic resonance imaging *2022 26th Int. Conf. on Pattern Recognition (ICPR)* (Piscataway, NJ: IEEE) pp 4365–70

[49] Elsharkawy M, Sharafeldeen A, Soliman A, Khalifa F, Widjajahakim R, Switala A, Elnakib A, Schaal S, Sandhu H S and Seddon J M *et al* 2021 Automated diagnosis and grading of dry age-related macular degeneration using optical coherence tomography imaging *Invest. Ophthalmol. Vis. Sci.* **62** 107–7

[50] Alksas A, Shehata M, Atef H, Sherif F, Alghamdi N S, Ghazal M, Abdel Fattah S, El-Serougy L G and El-Baz A 2022 A novel system for precise grading of glioma *Bioengineering* **9** 532

[51] Shalata A T, Shehata M, Van Bogaert E, Ali K M, Alksas A, Mahmoud A, El-Gendy E M, Mohamed M A, Giridharan G A and Contractor S *et al* 2022 Predicting recurrence of non-muscle-invasive bladder cancer: current techniques and future trends *Cancers* **14** 019

[52] Sleman A A, Soliman A, Ghazal M, Sandhu H, Schaal S, Elmaghraby A and El-Baz A 2019 Retinal layers OCT scans 3-D segmentation *2019 IEEE Int. Conf. on Imaging Systems and Techniques (IST)* (Piscataway, NJ: IEEE) pp 1–6

[53] Eladawi N, Elmogy M, Ghazal M, Helmy O, Aboelfetouh A, Riad A, Schaal S and El-Baz A 2018 Classification of retinal diseases based on OCT images *Front. Biosci. (Landmark Ed)* **23** 247–64

[54] ElTanboly A, Ismail M, Shalaby A, Switala A, El-Baz A, Schaal S, Gimel'farb G and El-Azab M 2017 A computer-aided diagnostic system for detecting diabetic retinopathy in optical coherence tomography images *Med. Phys.* **44** 914–23

[55] Sandhu H S, El-Baz A and Seddon J M 2018 Progress in automated deep learning for macular degeneration *JAMA Ophthalmol.* **136** 1366–7

[56] Reda I, Ghazal M, Shalaby A, Elmogy M, AbouEl-Fetouh A, Ayinde B O, AbouEl-Ghar M, Elmaghraby A, Keynton R and El-Baz A 2018 A novel ADCS-based CNN classification system for precise diagnosis of prostate cancer *2018 24th Int. Conf. on Pattern Recognition (ICPR)* (Piscataway, NJ: IEEE) pp 3923–8

[57] Reda I, Khalil A, Elmogy M, Abou El-Fetouh A, Shalaby A, Abou El-Ghar M, Elmaghraby A, Ghazal M and El-Baz A 2018 Deep learning role in early diagnosis of prostate cancer *Technol. Cancer Res. Treat.* **17** 1533034618775530

[58] Reda I, Ayinde B O, Elmogy M, Shalaby A, El-Melegy M, El-Ghar M A, El-fetouh A A, Ghazal M and El-Baz A 2018 A new CNN-based system for early diagnosis of prostate cancer *2018 IEEE 15th Int. Symp. on Biomedical Imaging (ISBI 2018)* (Piscataway, NJ: IEEE) pp 207–10

[59] Hammouda K, Khalifa F, El-Melegy M, Ghazal M, Darwish H E, El-Ghar M A and El-Baz A 2021 A deep learning pipeline for grade groups classification using digitized prostate biopsy specimens *Sensors* **21** 6708

[60] Shehata M, Shalaby A, Switala A E, El-Baz M, Ghazal M, Fraiwan L, Khalil A, El-Ghar M A, Badawy M and Bakr A M *et al* 2020 A multimodal computer-aided diagnostic system for precise identification of renal allograft rejection: preliminary results *Med. Phys.* **47** 2427–40

[61] Shehata M, Khalifa F, Soliman A, Ghazal M, Taher F, Abou El-Ghar M, Dwyer A C, Gimel'farb G, Keynton R S and El- Baz A 2018 Computer-aided diagnostic system for early detection of acute renal transplant rejection using diffusion-weighted MRI *IEEE Trans. Biomed. Eng.* **66** 539–52

[62] Hollis E, Shehata M, Abou El-Ghar M, Ghazal M, El-Diasty T, Merchant M, Switala A E and El-Baz A 2017 Statistical analysis of ADCS and clinical biomarkers in detecting acute renal transplant rejection *Br. J. Radiol.* **90** 20170125

[63] Khalifa F, Beache G M, El-Ghar M A, El-Diasty T, Gimel'farb G, Kong M and El-Baz A 2013 Dynamic contrast-enhanced MRI- based early detection of acute renal transplant rejection *IEEE Trans. Med. Imaging* **32** 1910–27

[64] Khalifa F, El-Ghar M A, Abdollahi B, Frieboes H, El-Diasty T and El-Baz A 2013 A comprehensive non-invasive framework for automated evaluation of acute renal transplant rejection using DCE-MRI *NMR Biomed.* **26** 1460–70

[65] Khalifa F, Elnakib A, Beache G M, Gimel'farb G, El-Ghar M A, Sokhadze G, Manning S, McClure P and El-Baz A 2011 3D kidney segmentation from CT images using a level set approach guided by a novel stochastic speed function *Proc. of Int. Conf. Medical Image Computing and Computer-Assisted Intervention, (MICCAI'11) (Toronto, Canada, September 18–22)* pp 587–94

[66] Shehata M, Khalifa F, Hollis E, Soliman A, Hosseini-Asl E, El-Ghar M A, El-Baz M, Dwyer A C, El-Baz A and Keynton R 2016 A new non-invasive approach for early classification of renal rejection types using diffusion-weighted MRI *IEEE Int. Conf. on Image Processing (ICIP), 2016* (Piscataway, NJ: IEEE) pp 136–40

[67] Khalifa F, Soliman A, Takieldeen A, Shehata M, Mostapha M, Shaffie A, Ouseph R, Elmaghraby A and El-Baz A 2016 Kidney segmentation from CT images using a 3D NMF-guided active contour model *IEEE 13th Int. Symp. on Biomedical Imaging (ISBI), 2016* (Piscataway, NJ: IEEE) pp 432–5

[68] Shehata M, Khalifa F, Soliman A, Takieldeen A, El-Ghar M A, Shaffie A, Dwyer A C, Ouseph R, El-Baz A and Keynton R 2016 3D diffusion MRI-based CAD system for early diagnosis of acute renal rejection *Biomedical Imaging (ISBI), 2016 IEEE 13th Int. Symp. on* (Piscataway, NJ: IEEE) pp 1177–80

[69] Shehata M, Khalifa F, Soliman A, Alrefai R, El-Ghar M A, Dwyer A C, Ouseph R and El-Baz A 2015 A level set-based framework for 3D kidney segmentation from diffusion Mr images *IEEE Int. Conf. on Image Processing (ICIP), 2015* (Piscataway, NJ: IEEE) pp 4441–5

[70] Shehata M, Khalifa F, Soliman A, El-Ghar M A, Dwyer A C, Gimel'farb G, Keynton R and El-Baz A 2016 A promising non- invasive CAD system for kidney function assessment *Int. Conf. on Medical Image Computing and Computer-Assisted Intervention* (Springer) pp 613–21

[71] Khalifa F, Soliman A, Elmaghraby A, Gimel'farb G and El-Baz A 2017 3D kidney segmentation from abdominal images using spatial-appearance models *Comput. Math. Methods Med.* **2017** pp 1–10

[72] Hollis E, Shehata M, Khalifa F, El-Ghar M A, El-Diasty T and El-Baz A 2016 Towards non-invasive diagnostic techniques for early detection of acute renal transplant rejection: a review *Egypt. J. Radiol. Nucl. Med.* **48** 257–69

[73] Shehata M, Khalifa F, Soliman A, El-Ghar M A, Dwyer A C and El-Baz A 2017 Assessment of renal transplant using image and clinical-based biomarkers *Proc. of 13th Annual Scientific Meeting of American Society for Diagnostics and Interventional Nephrology (ASDIN'17) (New Orleans, LA, February 10–12, 2017)*

[74] 2016 Early assessment of acute renal rejection *Proceedings of 12th Annual Scientific Meeting of American Society for Diagnostics and Interventional Nephrology (ASDIN'16) (Pheonix, AZ, 19–21 February 2016)*

[75] Eltanboly A, Ghazal M, Hajjdiab H, Shalaby A, Switala A, Mahmoud A, Sahoo P, El-Azab M and El-Baz A 2019 Level sets-based image segmentation approach using statistical shape priors *Appl. Math. Comput.* **340** 164–79

[76] Shehata M, Mahmoud A, Soliman A, Khalifa F, Ghazal M, El-Ghar M A, El-Melegy M and El-Baz A 2018 3D kidney segmentation from abdominal diffusion MRI using an appearance-guided deformable boundary *PLoS One* **13** e0200082

[77] Abdeltawab H, Shehata M, Shalaby A, Khalifa F, Mahmoud A, El-Ghar M A, Dwyer A C, Ghazal M, Hajjdiab H and Keynton R *et al* 2019 A novel CNN-based CAD system for early assessment of transplanted kidney dysfunction *Sci. Rep.* **9** 5948

[78] Hammouda K, Khalifa F, Abdeltawab H, Elnakib A, Giridharan G, Zhu M, Ng C, Dassanayaka S, Kong M and Darwish H *et al* 2020 A new framework for performing cardiac strain analysis from cine MRI imaging in mice *Sci. Rep.* **10** 1–15

[79] Abdeltawab H, Khalifa F, Hammouda K, Miller J M, Meki M M, Ou Q, El-Baz A and Mohamed T 2021 Artificial intelligence based framework to quantify the cardiomyocyte structural integrity in heart slices *Cardiovasc. Eng. Technol.* pp 1–11

[80] Khalifa F, Beache G M, Elnakib A, Sliman H, Gimel'farb G, Welch K C and El-Baz A 2013 A new shape-based framework for the left ventricle wall segmentation from cardiac first-pass perfusion MRI *Proc. of IEEE Int. Symp. on Biomedical Imaging: From Nano to Macro, (ISBI'13) (San Francisco, CA)* pp 41–4

[81] 2012 A new nonrigid registration framework for improved visualization of transmural perfusion gradients on cardiac first–pass perfusion MRI *Proceedings of IEEE Int. Symp. on Biomedical Imaging: From Nano to Macro (ISBI'12)(May 2–5) (Barcelona)* pp 828–31

[82] Khalifa F, Beache G M, Firjani A, Welch K C, Gimel'farb G and El-Baz A 2012 A new nonrigid registration approach for motion correction of cardiac first-pass perfusion MRI *Proc. of IEEE Int. Conf. on Image Processing, (ICIP'12) (Lake Buena Vista, FL, September 30–October 3)* pp 1665–8
[83] Khalifa F, Beache G M, Gimel'farb G and El-Baz A 2012 A novel CAD system for analyzing cardiac first-pass MR images *Proc. of IAPR Int. Conf. on Pattern Recognition (ICPR'12) (Tsukuba Science City, Japan)* pp 77–80
[84] A novel approach for accurate estimation of left ventricle global indexes from short-axis cine MRI *Proceedings of IEEE Int. Conf. on Image Processing, (ICIP'11) (Brussels 11–14 September 2011)* pp 2645–9
[85] Khalifa F, Beache G M, Gimel'farb G, Giridharan G A and El-Baz A 2011 A new image-based framework for analyzing cine images *Handbook of Multi Modality State-of-the-Art Medical Image Segmentation and Registration Methodologies* ed A El-Baz, U R Acharya, M Mirmedhdi and J S Suri (New York: Springer) Vol 2 pp 69–98
[86] Accurate automatic analysis of cardiac cine images *IEEE Trans. Biomed. Eng* **59** 445 455 2012
[87] Khalifa F, Beache G M, Nitzken M, Gimel'farb G, Giridharan G A and El-Baz A 2011 Automatic analysis of left ventricle wall thickness using short-axis cine CMR images *Proc. of IEEE Int. Symp. on Biomedical Imaging: From Nano to Macro, (ISBI'11) (Chicago, IL, March 30–April 2)* pp 1306–9
[88] Nitzken M, Beache G, Elnakib A, Khalifa F, Gimel'farb G and El-Baz A 2012 Accurate modeling of tagged cmr 3D image appearance characteristics to improve cardiac cycle strain estimation *Image Processing (ICIP), 2012 19th IEEE Int. Conf. on (Orlando, FL)* (Piscataway, NJ: IEEE) pp 521–4
[89] Improving full-cardiac cycle strain estimation from tagged cmr by accurate modeling of 3D image appearance characteristics *2012 9th IEEE Int. Symp. on Biomedical Imaging (ISBI) (May 2012* Barcelona) (Piscataway, NJ: IEEE) pp 462–5 (selected for oral presentation)
[90] Nitzken M J, El-Baz A S and Beache G M 2012 Markov-gibbs random field model for improved full-cardiac cycle strain estimation from tagged cmr *J. Cardiovasc. Magn. Resonan.* **14** 1–2
[91] Sliman H, Elnakib A, Beache G, Elmaghraby A and El-Baz A 2014 Assessment of myocardial function from cine cardiac MRI using a novel 4D tracking approach *J. Comput. Sci. Syst. Biol.* **7** 169–73
[92] Sliman H, Elnakib A, Beache G M, Soliman A, Khalifa F, Gimel'farb G, Elmaghraby A and El-Baz A 2014 A novel 4D PDE-based approach for accurate assessment of myocardium function using cine cardiac magnetic resonance images *Proc. of IEEE Int. Conf. on Image Processing (ICIP'14) (Paris)* pp 3537–41
[93] Sliman H, Khalifa F, Elnakib A, Beache G M, Elmaghraby A and El-Baz A 2013 A new segmentation-based tracking framework for extracting the left ventricle cavity from cine cardiac MRI *Proc. of IEEE Int. Conf. on Image Processing, (ICIP'13) (Melbourne)* pp 685–9
[94] Sliman H, Khalifa F, Elnakib A, Soliman A, Beache G M, Elmaghraby A, Gimel'farb G and El-Baz A 2013 Myocardial borders segmentation from cine MR images using bi-directional coupled parametric deformable models *Med. Phys.* **40** 1–13
[95] Sliman H, Khalifa F, Elnakib A, Soliman A, Beache G M, Gimel'farb G, Emam A, Elmaghraby A and El-Baz A 2013 Accurate segmentation framework for the left ventricle wall from cardiac cine MRI *Proc. of Int. Symp. on Computational Models for Life Science, (CMLS'13)* vol 1559 *(Sydney)* pp 287–96

[96] Abdollahi B, Civelek A C, Li X-F, Suri J and El-Baz A 2014 PET/CT nodule segmentation and diagnosis: a survey *Multi Detector CT Imaging* ed L Saba and J S Suri (London: Taylor and Francis) ch 30 pp 639–51
[97] Abdollahi B, El-Baz A and Amini A A 2011 A multi-scale non-linear vessel enhancement technique *Engineering in Medicine and Biology Society, EMBC, 2011 Annual Int. Conf. of the IEEE* (Piscataway, NJ: IEEE) pp 3925–9
[98] Abdollahi B, Soliman A, Civelek A, Li X-F, Gimel'farb G and El-Baz A 2012 A novel gaussian scale space-based joint MGRF framework for precise lung segmentation *Proc. of IEEE Int. Conf. on Image Processing, (ICIP'12)* (Piscataway, NJ: IEEE) pp 2029–32
[99] Abdollahi B, Soliman A, Civelek A C, Li X-F, Gimel'farb G and El-Baz A 2012 A novel 3D joint MGRF framework for precise lung segmentation *Machine Learning in Medical Imaging* (Springer) pp 86–93
[100] Ali A M, El-Baz A S and Farag A A 2007 A novel framework for accurate lung segmentation using graph cuts *Proc. of IEEE Int. Symp. on Biomedical Imaging: From Nano to Macro, (ISBI'07)* (Piscataway, NJ: IEEE) pp 908–11
[101] El-Baz A, Beache G M, Gimel'farb G, Suzuki K and Okada K 2013 Lung imaging data analysis *Int. J. Biomed. Imaging* **2013** 1–2
[102] El-Baz A, Beache G M, Gimel'farb G, Suzuki K, Okada K and Elnakib A 2013 Computer-aided diagnosis systems for lung cancer: challenges and methodologies *Int. J. Biomed. Imaging* **2013** 1–46
[103] El-Baz A, Elnakib A, Abou El-Ghar M, Gimel'farb G, Falk R and Farag A 2013 Automatic detection of 2D and 3D lung nodules in chest spiral CT scans *Int. J. Biomed. Imaging* **2013** 1–11
[104] El-Baz A, Farag A A, Falk R and La Rocca R 2003 A unified approach for detection, visualization, and identification of lung abnormalities in chest spiral CT scans *Int. Congress Series* **1256** 998–1004
[105] El-Baz A, Farag A A, Falk R and La Rocca R 2002 Detection, visualization and identification of lung abnormalities in chest spiral CT scan: phase-I *Proc. of Int. Conf. on Biomedical Engineering (Cairo) vol 12*
[106] El-Baz A, Farag A, Gimel'farb G, Falk R, El-Ghar M A and Eldiasty T 2006 A framework for automatic segmentation of lung nodules from low dose chest CT scans *Proc. of Int. Conf. on Pattern Recognition, (ICPR'06)* vol 3 (Piscataway, NJ: IEEE) pp 611–4
[107] El-Baz A, Farag A, Gimel'farb G, Falk R and El-Ghar M A 2011 A novel level set-based computer-aided detection system for automatic detection of lung nodules in low dose chest computed tomography scans *Lung Imaging and Computer Aided Diagnosis* (Boca Raton, Fl: CRC Press) vol 10 pp 221–38
[108] El-Baz A, Gimel'farb G, Abou El-Ghar M and Falk R 2012 Appearance-based diagnostic system for early assessment of malignant lung nodules *Proc. of IEEE Int. Conf. on Image Processing, (ICIP'12)* (Piscataway, NJ: IEEE) pp 533–6
[109] El-Baz A, Gimel'farb G and Falk R 2011 A novel 3D framework for automatic lung segmentation from low dose CT images *Lung Imaging and Computer Aided Diagnosis* ed A El-Baz and J S Suri (London: Taylor and Francis) ch 1 pp 1–16
[110] El-Baz A, Gimel'farb G, Falk R and El-Ghar M 2010 Appearance analysis for diagnosing malignant lung nodules *Proc. of IEEE Int. Symp. on Biomedical Imaging: From Nano to Macro (ISBI'10)* (Piscataway, NJ: IEEE) pp 193–6

[111] El-Baz A, Gimel'farb G, Falk R and El-Ghar M A 2011 A novel level set-based CAD system for automatic detection of lung nodules in low dose chest CT scans *Lung Imaging and Computer Aided Diagnosis* ed A El-Baz and J S Suri (London: Taylor and Francis) vol 1 pp 221–38

[112] A new approach for automatic analysis of 3D low dose CT images for accurate monitoring the detected lung nodules *Proc. of Int. Conf. on Pattern Recognition, (ICPR'08)* (Piscataway, NJ: IEEE) 2008 pp 1–4

[113] El-Baz A, Gimel'farb G, Falk R and El-Ghar M A 2007 A novel approach for automatic follow-up of detected lung nodules *Proc. of IEEE Int. Conf. on Image Processing, (ICIP'07)* vol 5 (Piscataway, NJ: IEEE) pp V–501

[114] El-Baz A, Gimel'farb G, Falk R and El-Ghar M A 2007 A new CAD system for early diagnosis of detected lung nodules *IEEE Int. Conf. on Image Processing, 2007. ICIP 2007* vol 2 (Piscataway, NJ: IEEE) pp II–461

[115] El-Baz A, Gimel'farb G, Falk R, El-Ghar M A and Refaie H 2008 Promising results for early diagnosis of lung cancer *Proc. of IEEE Int. Symp. on Biomedical Imaging: From Nano to Macro, (ISBI'08)* (Piscataway, NJ: IEEE) pp 1151–4

[116] El-Baz A, Gimel'farb G L, Falk R, Abou El-Ghar M, Holland T and Shaffer T 2008 A new stochastic framework for accurate lung segmentation *Proc. of Medical Image Computing and Computer-Assisted Intervention, (MICCAI'08)* pp 322–30

[117] El-Baz A, Gimel'farb G L, Falk R, Heredis D and Abou M 2008 El-Ghar, A novel approach for accurate estimation of the growth rate of the detected lung nodules *Proc. of Int. Workshop on Pulmonary Image Analysis* pp 33–42

[118] El-Baz A, Gimel'farb G L, Falk R, Holland T and Shaffer T 2008 A framework for unsupervised segmentation of lung tissues from low dose computed tomography images *Proc. of British Machine Vision, (BMVC'08)* pp 1–10

[119] El-Baz A, Gimel'farb G, Falk R and El-Ghar M A 2011 3D MGRF-based appearance modeling for robust segmentation of pulmonary nodules in 3D LDCT chest images *Lung Imaging and Computer Aided Diagnosis* (Boca Raton, FL: CRC Press) ch 3 51–63

[120] El-Baz A, Gimel'farb G, Falk R and El-Ghar M A 2009 Automatic analysis of 3D low dose CT images for early diagnosis of lung cancer *Pattern Recogn* **42** 1041–51

[121] El-Baz A, Gimel'farb G, Falk R, El-Ghar M A, Rainey S, Heredia D and Shaffer T 2009 Toward early diagnosis of lung cancer *Proc. of Medical Image Computing and Computer-Assisted Intervention, (MICCAI'09)* (Berlin: Springer) pp 682–9

[122] El-Baz A, Gimel'farb G, Falk R, El-Ghar M A and Suri J 2011 Appearance analysis for the early assessment of detected lung nodules *Lung Imaging and Computer Aided Diagnosis* (Boca Raton, FL: CRC Press) ch 17 pp 395–404

[123] El-Baz A, Khalifa F, Elnakib A, Nitkzen M, Soliman A, McClure P, Gimel'farb G and El-Ghar M A 2012 A novel approach for global lung registration using 3D Markov Gibbs appearance model *Proc. of Int. Conf. Medical Image Computing and Computer-Assisted Intervention, (MICCAI'12) (Nice)* pp 114–21

[124] El-Baz A, Nitzken M, Elnakib A, Khalifa F, Gimel'farb G, Falk R and El-Ghar M A 2011 3D shape analysis for early diagnosis of malignant lung nodules *Proc. of Int. Conf. Medical Image Computing and Computer-Assisted Intervention, (MICCAI'11) (Toronto)* pp 175–82

[125] El-Baz A, Nitzken M, Gimel'farb G, Van Bogaert E, Falk R, El-Ghar M A and Suri J 2011 Three-dimensional shape analysis using spherical harmonics for early assessment of detected lung nodules *Lung Imaging and Computer Aided Diagnosis.* (Boca Raton, FL: CRC Press) ch 19 421–38

[126] El-Baz A, Nitzken M, Khalifa F, Elnakib A, Gimel'farb G, Falk R and El-Ghar M A 2011 3D shape analysis for early diagnosis of malignant lung nodules *Proc. of Int. Conf. on Information Processing in Medical Imaging, (IPMI'11) (Monastery Irsee, Germany (Bavaria))* pp 772–83

[127] El-Baz A, Nitzken M, Vanbogaert E, Gimel'Farb G, Falk R and Abo M 2011 A novel shape-based diagnostic approach for early diagnosis of lung nodules *2011 IEEE Int. Symp. on Biomedical Imaging: From Nano to Macro* (Piscataway, NJ: IEEE) pp 137–40

[128] El-Baz A, Sethu P, Gimel'farb G, Khalifa F, Elnakib A, Falk R and El-Ghar M A 2011 Elastic phantoms generated by microfluidics technology: validation of an imaged-based approach for accurate measurement of the growth rate of lung nodules *Biotechnol. J.* **6** 195–203

[129] 2010 A new validation approach for the growth rate measurement using elastic phantoms generated by state-of-the-art microfluidics technology *Proc. of IEEE Int. Conf. on Image Processing, (ICIP'10) (Hong Kong, September 26–29)* pp 4381–3

[130] El-Baz A, Sethu P, Gimel'farb G, Khalifa F, Elnakib A, Falk R and Suri J S 2011 Validation of a new imaged-based approach for the accurate estimating of the growth rate of detected lung nodules using real CT images and elastic phantoms generated by state-of-the-art microfluidics technology *Handbook of Lung Imaging and Computer Aided Diagnosis* ed A El-Baz and J S Suri (New York: Taylor and Francis) vol 1 pp 405–20

[131] El-Baz A, Soliman A, McClure P, Gimel'farb G, El-Ghar M A and Falk R 2012 Early assessment of malignant lung nodules based on the spatial analysis of detected lung nodules *Proc. of IEEE Int. Symp. on Biomedical Imaging: From Nano to Macro (ISBI'12)* (Piscataway, NJ: IEEE) pp 1463–6

[132] El-Baz A, Yuksel S E, Elshazly S and Farag A A 2005 Non-rigid registration techniques for automatic follow-up of lung nodules *Proc. of Computer Assisted Radiology and Surgery, (CARS'05)* vol 1281 (Amsterdam: Elsevier) pp 1115–20

[133] El-Baz A S and Suri J S 2011 *Lung Imaging and Computer Aided Diagnosis* (Boca Raton, FL: CRC Press)

[134] Soliman A, Khalifa F, Dunlap N, Wang B, El-Ghar M and El-Baz A 2016 An iso-surfaces based local deformation handling framework of lung tissues *2016 IEEE 13th Int. Symp. on Biomedical Imaging (ISBI)* (Piscataway, NJ: IEEE) pp 1253–9

[135] Soliman A, Khalifa F, Shaffie A, Dunlap N, Wang B, Elmaghraby A and El-Baz A 2016 Detection of lung injury using 4D-CT chest images *2016 IEEE 13th Int. Symp. on Biomedical Imaging (ISBI)* (Piscataway, NJ: IEEE) pp 1274–7

[136] Soliman A, Khalifa F, Shaffie A, Dunlap N, Wang B, Elmaghraby A, Gimel'farb G, Ghazal M and El-Baz A 2017 A comprehensive framework for early assessment of lung injury *2017 IEEE Int. Conf. on Image Processing (ICIP)* (Piscataway, NJ: IEEE) pp 3275–9

[137] Shaffie A, Soliman A, Ghazal M, Taher F, Dunlap N, Wang B, Elmaghraby A, Gimel'farb G and El-Baz A 2017 A new framework for incorporating appearance and shape features of lung nodules for precise diagnosis of lung cancer *2017 IEEE Int. Conf. on Image Processing (ICIP)* (Piscataway, NJ: IEEE) pp 1372–6

[138] Soliman A, Khalifa F, Shaffie A, Liu N, Dunlap N, Wang B, Elmaghraby A, Gimel'farb G and El-Baz A 2016 Image-based CAD system for accurate identification of lung injury *2016 IEEE Int. Conf. on Image Processing (ICIP)* (Piscataway, NJ: IEEE) pp 121–5

[139] Soliman A, Shaffie A, Ghazal M, Gimel'farb G, Keynton R and El-Baz A 2018 A novel CNN segmentation framework based on using new shape and appearance features *2018 25th IEEE Int. Conf. on Image Processing (ICIP)* (Piscataway, NJ: IEEE) pp 3488–92

[140] Shaffie A, Soliman A, Khalifeh H A, Ghazal M, Taher F, Keynton R, Elmaghraby A and El-Baz A 2018 On the integration of ct- derived features for accurate detection of lung cancer *2018 IEEE Int. Symp. on Signal Processing and Information Technology (ISSPIT)* (Piscataway, NJ: IEEE) pp 435–40

[141] Shaffie A, Soliman A, Khalifeh H A, Ghazal M, Taher F, Elmaghraby A, Keynton R and El-Baz A 2019 Radiomic-based framework for early diagnosis of lung cancer *2019 IEEE 16th Int. Symp. on Biomedical Imaging (ISBI 2019)* (Piscataway, NJ: IEEE) pp 1293–7

[142] Shaffie A, Soliman A, Ghazal M, Taher F, Dunlap N, Wang B, Van Berkel V, Gimelfarb G, Elmaghraby A and El-Baz A 2018 A novel autoencoder-based diagnostic system for early assessment of lung cancer *2018 25th IEEE Int. Conf. on Image Processing (ICIP)* (Piscataway, NJ: IEEE) pp 1393–7

[143] Shaffie A, Soliman A, Fraiwan L, Ghazal M, Taher F, Dunlap N, Wang B, van Berkel V, Keynton R and Elmaghraby A *et al* 2018 A generalized deep learning-based diagnostic system for early diagnosis of various types of pulmonary nodules *Technol. Cancer Res. Treat.* **17** 1533033818798800

[144] ElNakieb Y, Ali M T, Dekhil O, Khalefa M E, Soliman A, Shalaby A, Mahmoud A, Ghazal M, Hajjdiab H and Elmaghraby A *et al* 2018 Towards accurate personalized autism diagnosis using different imaging modalities: SMRI, FMRI, and DTI *2018 IEEE Int. Symp. on Signal Processing and Information Technology (ISSPIT)* (Piscataway, NJ: IEEE) pp 447–52

[145] ElNakieb Y, Soliman A, Mahmoud A, Dekhil O, Shalaby A, Ghazal M, Khalil A, Switala A, Keynton R S and Barnes G N *et al* 2019 Autism spectrum disorder diagnosis framework using diffusion tensor imaging *2019 IEEE Int. Conf. on Imaging Systems and Techniques (IST)* (Piscataway, NJ: IEEE) pp 1–5

[146] Haweel R, Dekhil O, Shalaby A, Mahmoud A, Ghazal M, Keynton R, Barnes G and El-Baz A 2019 A machine learning approach for grading autism severity levels using task-based functional MRI *2019 IEEE Int. Conf. on Imaging Systems and Techniques (IST)* (Piscataway, NJ: IEEE) pp 1–5

[147] Dekhil O, Ali M, Haweel R, Elnakib Y, Ghazal M, Hajjdiab H, Fraiwan L, Shalaby A, Soliman A and Mahmoud A *et al* 2020 A comprehensive framework for differentiating autism spectrum disorder from neurotypicals by fusing structural MRI and resting state functional MRI *Seminars in Pediatric Neurology* (Amsterdam: Elsevier) p 100805

[148] Haweel R, Dekhil O, Shalaby A, Mahmoud A, Ghazal M, Khalil A, Keynton R, Barnes G and El-Baz A 2020 A novel framework for grading autism severity using task-based FMRI *2020 IEEE 17th Int. Symp. on Biomedical Imaging (ISBI)* (Piscataway, NJ: IEEE) pp 1404–7

[149] El-Baz A, Elnakib A, Khalifa F, El-Ghar M A, McClure P, Soliman A and Gimel'farb G 2012 Precise segmentation of 3-D magnetic resonance angiography *IEEE Trans. Biomed. Eng.* **59** 2019–29

[150] El-Baz A, Farag A, Elnakib A, Casanova M F, Gimel'farb G, Switala A E, Jordan D and Rainey S 2011 Accurate automated detection of autism related corpus callosum abnormalities *J. Med. Syst.* **35** 929–39

[151] El-Baz A, Gimel'farb G, Falk R, El-Ghar M A, Kumar V and Heredia D 2009 A novel 3D joint Markov-gibbs model for extracting blood vessels from PC–mra images *Medical Image Computing and Computer-Assisted Intervention–MICCAI 2009* vol 5762 (Berlin: Springer) pp 943–50

[152] Elnakib A, El-Baz A, Casanova M F, Gimel'farb G and Switala A E 2010 Image-based detection of corpus callosum variability for more accurate discrimination between dyslexic and normal brains *Proc. IEEE Int. Symp. on Biomedical Imaging: From Nano to Macro (ISBI'2010)* (Piscataway, NJ: IEEE) pp 109–12

[153] Elnakib A, Casanova M F, Gimel'farb G, Switala A E and El-Baz A 2011 Autism diagnostics by centerline-based shape analysis of the corpus callosum *Proc. IEEE Int. Symp. on Biomedical Imaging: From Nano to Macro (ISBI'2011)* (Piscataway, NJ: IEEE) pp 1843–6

[154] Elnakib A, Nitzken M, Casanova M, Park H, Gimel'farb G and El-Baz A 2012 Quantification of age-related brain cortex change using 3D shape analysis *Pattern Recognition (ICPR), 2012 21st Int. Conf. on* (Piscataway, NJ: IEEE) pp 41–4

[155] Nitzken M, Casanova M, Gimel'farb G, Elnakib A, Khalifa F, Switala A and El-Baz A 2011 3D shape analysis of the brain cortex with application to dyslexia *Image Processing (ICIP), 2011 18th IEEE Int. Conf. on (Brussels)*(Piscataway, NJ: IEEE) pp 2657–60 (Selected for oral presentation. Oral acceptance rate is 10 percent and the overall acceptance rate is 35 percent)

[156] El-Gamal F E-Z A, Elmogy M M, Ghazal M, Atwan A, Barnes G N, Casanova M F, Keynton R and El-Baz A S 2017 A novel CAD system for local and global early diagnosis of Alzheimer's disease based on PIB-PET scans *2017 IEEE Int. Conf. on Image Processing (ICIP)* (Piscataway, NJ: IEEE) pp 3270–4

[157] Ismail M M, Keynton R S, Mostapha M M, ElTanboly A H, Casanova M F, Gimel'farb G L and El-Baz A 2016 Studying autism spectrum disorder with structural and diffusion magnetic resonance imaging: a survey *Front. Human Neurosci.* **10** 211

[158] Alansary A, Ismail M, Soliman A, Khalifa F, Nitzken M, Elnakib A, Mostapha M, Black A, Stinebruner K and Casanova M F *et al* 2016 Infant brain extraction in t1-weighted MR images using bet and refinement using LCDG and MGRF models *IEEE J. Biomed. Health Inform.* **20** 925–35

[159] Asl E H, Ghazal M, Mahmoud A, Aslantas A, Shalaby A, Casanova M, Barnes G, Gimel'farb G, Keynton R and El-Baz A 2018 Alzheimer's disease diagnostics by a 3D deeply supervised adaptable convolutional network *Front. Biosci. (Landmark edition)* **23** 584–96

[160] Dekhil O *et al* 2019 A personalized autism diagnosis CAD system using a fusion of structural MRI and resting-state functional MRI data *Front. Psych.* **10** 392

[161] Dekhil O, Shalaby A, Soliman A, Mahmoud A, Kong M, Barnes G, Elmaghraby A and El-Baz A 2021 Identifying brain areas correlated with ados raw scores by studying altered dynamic functional connectivity patterns *Med. Image Anal.* **68** 101899

[162] Elnakieb Y A, Ali M T, Soliman A, Mahmoud A H, Shalaby A M, Alghamdi N S, Ghazal M, Khalil A, Switala A and Keynton R S *et al* 2020 Computer aided autism diagnosis using diffusion tensor imaging *IEEE Access* **8** 191 298–1308

[163] Ali M T, Elnakieb Y A, Shalaby A, Mahmoud A, Switala A, Ghazal M, Khelifi A, Fraiwan L, Barnes G and El-Baz A 2021 Autism classification using SMRI: a recursive features selection based on sampling from multi-level high dimensional spaces *2021 IEEE 18th Int. Symp. on Biomedical Imaging (ISBI)* (Piscataway, NJ: IEEE) pp 267–70

[164] Ali M T, ElNakieb Y, Elnakib A, Shalaby A, Mahmoud A, Ghazal M, Yousaf J, Abu Khalifeh H, Casanova M and Barnes G *et al* 2022 The role of structure MRI in diagnosing autism *Diagnostics* **12** 165

[165] ElNakieb Y, Ali M T, Elnakib A, Shalaby A, Soliman A, Mahmoud A, Ghazal M, Barnes G N and El-Baz A 2021 The role of diffusion tensor MR imaging (DTI) of the brain in diagnosing autism spectrum disorder: promising results *Sensors* **21** 8171

[166] Mahmoud A, El-Barkouky A, Farag H, Graham J and Farag A 2013 A non-invasive method for measuring blood flow rate in superficial veins from a single thermal image *Proc. of the IEEE Conf. on Computer Vision and Pattern Recognition Workshops* pp 354–9

[167] Elsaid N, Saied A, Kandil H, Soliman A, Taher F, Hadi M, Giridharan G, Jennings R, Casanova M and Keynton R *et al* 2021 Impact of stress and hypertension on the cerebrovasculature *Front. Biosci.-Landmark* **26** 1643

[168] Taher F, Kandil H, Gebru Y, Mahmoud A, Shalaby A, El-Mashad S and El-Baz A 2021 A novel mra-based framework for segmenting the cerebrovascular system and correlating cerebral vascular changes to mean arterial pressure *Appl. Sci.* **11** 4022

[169] Kandil H, Soliman A, Taher F, Ghazal M, Khalil A, Giridharan G, Keynton R, Jennings J R and El-Baz A 2020 A novel computer-aided diagnosis system for the early detection of hypertension based on cerebrovascular alterations *NeuroImage: Clin.* **25** 102107

[170] Kandil H, Soliman A, Ghazal M, Mahmoud A, Shalaby A, Keynton R, Elmaghraby A, Giridharan G and El-Baz A 2019 A novel framework for early detection of hypertension using magnetic resonance angiography *Sci. Rep.* **9** 1–12

[171] Gebru Y, Giridharan G, Ghazal M, Mahmoud A, Shalaby A and El-Baz A 2018 Detection of cerebrovascular changes using magnetic resonance angiography *Cardiovascular Imaging and Image Analysis* (Boca Raton, FL: CRC Press) pp 1–22

[172] Mahmoud A, Shalaby A, Taher F, El-Baz M, Suri J S and El-Baz A 2018 Vascular tree segmentation from different image modalities *Cardiovascular Imaging and Image Analysis* (Boca Raton, FL: CRC Press) pp 43–70

[173] Taher F, Mahmoud A, Shalaby A and El-Baz A 2018 A review on the cerebrovascular segmentation methods *2018 IEEE Int. Symp. on Signal Processing and Information Technology (ISSPIT)* (Piscataway, NJ: IEEE) pp 359–64

[174] Kandil H, Soliman A, Fraiwan L, Shalaby A, Mahmoud A, ElTanboly A, Elmaghraby A, Giridharan G and El-Baz A 2018 A novel MRA framework based on integrated global and local analysis for accurate segmentation of the cerebral vascular system *2018 IEEE 15th Int. Symp. on Biomedical Imaging (ISBI 2018)* (Piscataway, NJ: IEEE) pp 1365–8

[175] Taher F, Soliman A, Kandil H, Mahmoud A, Shalaby A, Gimel'farb G and El-Baz A 2020 Accurate segmentation of cerebrovasculature from TOF-MRA images using appearance descriptors *IEEE Access* **8** 96139–49

[176] Taher F *et al* 2020 Precise cerebrovascular segmentation *2020 IEEE Int. Conf. on Image Processing (ICIP)* (IEEE) pp 394–7

[177] Hammouda K, Khalifa F, Soliman A, Ghazal M, Abou El-Ghar M, Haddad A, Elmogy M, Darwish H, Khalil A and Elmaghraby A *et al* 2019 A CNN-based framework for bladder wall segmentation using MRI *2019 Fifth Int. Conf. on Advances in Biomedical Engineering (ICABME)* (Piscataway, NJ: IEEE) pp 1–4

[178] Hammouda K, Khalifa F, Soliman A, Ghazal M, Abou El-Ghar M, Haddad A, Elmogy M, Darwish H, Keynton R and El-Baz A 2019 A deep learning-based approach for accurate segmentation of bladder wall using MR images *2019 IEEE Int. Conf. on Imaging Systems and Techniques (IST)* (Piscataway, NJ: IEEE) pp 1–6

[179] Hammouda K, Khalifa F, Soliman A, Abdeltawab H, Ghazal M, Abou El-Ghar M, Haddad A, Darwish H E, Keynton R and El-Baz A 2020 A 3D CNN with a learnable

adaptive shape prior for accurate segmentation of bladder wall using MR images *2020 IEEE 17th Int. Symp. on Biomedical Imaging (ISBI)* (Piscataway, NJ: IEEE) pp 935–8

[180] Hammouda K, Khalifa F, Soliman A, Ghazal M, Abou El-Ghar M, Badawy M, Darwish H, Khelifi A and El-Baz A 2021 A multiparametric MRI-based CAD system for accurate diagnosis of bladder cancer staging *Comput. Med. Imaging Graph.* **90** 101911

[181] Hammouda K, Khalifa F, Soliman A, Ghazal M, Abou El-Ghar M, Badawy M, Darwish H and El-Baz A 2021 A CAD system for accurate diagnosis of bladder cancer staging using a multiparametric MRI *2021 IEEE 18th Int. Symp. on Biomedical Imaging (ISBI)* (Piscataway, NJ: IEEE) pp 1718–21

[182] Razek A A K A, Khaled R, Helmy E, Naglah A, AbdelKhalek A and El-Baz A 2022 Artificial intelligence and deep learning of head and neck cancer *Magn. Resonan. Imaging Clin.* **30** 81–94

[183] Naglah A, Khalifa F, Khaled R, Abdel Razek A A K, Ghazal M, Giridharan G and El-Baz A 2021 Novel MRI-based cad system for early detection of thyroid cancer using multi-input CNN *Sensors* **21** 3878

[184] Naglah A, Khalifa F, Mahmoud A, Ghazal M, Jones P, Murray T, Elmaghraby A S and El-Baz A 2018 Athlete-customized injury prediction using training load statistical records and machine learning *2018 IEEE Int. Symp. on Signal Processing and Information Technology (ISSPIT)* (Piscataway, NJ: IEEE) pp 459–64

[185] Mahmoud A H 2014 Utilizing radiation for smart robotic applications using visible, thermal, and polarization images *PhD Dissertation* (University of Louisville)

[186] Mahmoud A, El-Barkouky A, Graham J and Farag A 2014 Pedestrian detection using mixed partial derivative based his togram of oriented gradients *2014 IEEE Int. Conf. on Image Processing (ICIP)* (Piscataway, NJ: IEEE) pp 2334–7

[187] El-Barkouky A, Mahmoud A, Graham J and Farag A 2013 An interactive educational drawing system using a humanoid robot and light polarization *2013 IEEE Int. Conf. on Image Processing* (Piscataway, NJ: IEEE) pp 3407–11

[188] Mahmoud A H, El-Melegy M T and Farag A A 2012 Direct method for shape recovery from polarization and shading *2012 19th IEEE Int. Conf. on Image Processing* (Piscataway, NJ: IEEE) pp 1769–72

[189] Ghazal M A, Mahmoud A, Aslantas A, Soliman A, Shalaby A, Benediktsson J A and El-Baz A 2019 Vegetation cover estimation using convolutional neural networks *IEEE Access* **7** 132 563–176

[190] Ghazal M, Mahmoud A, Shalaby A and El-Baz A 2019 Automated framework for accurate segmentation of leaf images for plant health assessment *Environ. Monit. Assess.* **191** 491

[191] Ghazal M, Mahmoud A, Shalaby A, Shaker S, Khelifi A and El-Baz A 2020 Precise statistical approach for leaf segmentation *2020 IEEE Int. Conf. on Image Processing (ICIP)* (Piscataway, NJ: IEEE) pp 2985–9

IOP Publishing

Photo Acoustic and Optical Coherence Tomography Imaging, Volume 1

Diabetic retinopathy

Ayman El-Baz and Jasjit S Suri

Chapter 13

Prevention of age-related macular degeneration disease: current strategies and future directions

Mohamed Elsharkawy, Ahmed Soliman, Ali Mahmoud, Mohammed Ghazal, Marah Alhalabi, Ahmed El-Baz, Aristomenis Thanos, Harpal Singh Sandhu, Guruprasad Giridharan and Ayman El-Baz

Age-related macular degeneration (AMD) is a disorder that concerns the retina and is a major cause of vision loss among older adults in developed nations. Optical coherence tomography (OCT) is widely regarded as the extremely effective technique of detecting AMD in its early stages. This chapter explores the utilization of OCT imaging in diagnosing AMD, with a specific emphasis on assessing and comparing computer-aided diagnostic (CAD) systems that can automatically diagnose and grade AMD. We briefly discuss performance evaluation and lay the foundation for further research in AMD diagnosis. Despite ongoing research, preventing AMD remains the only viable strategy for avoiding this devastating eye disease and its associated sight damage. Moreover, speedy recognition of AMD is crucial to stop patients from progressing to advanced stages of the disease. As such, we delve into the persisting obstacles in constructing automatic CAD for AMD discovery that utilize OCT images and explore possible paths for creating examination and diagnostic systems constructed on telemedicine applications and OCT Images.

13.1 Introduction

Age-related macular degeneration (AMD) is a retinal disorder that predominantly causes sight damage in the macula, the central part of the retina. It is most commonly found in individuals aged 50 years and above in developed countries [1, 2]. The macula has a crucial function in daily activities such as driving, reading, screen use, and other visual tasks [3]. Regular eye examinations are necessary to identify initial markers of the vision disorder, monitoring its development, and initiate treatment when necessary. To aid in this process, various computer-aided

doi:10.1088/978-0-7503-2052-8ch13

diagnosis (CAD) methods have been developed for detecting AMD at initial phases [3–7]. These CAD methods are essential to reduce the specific time frame of ophthalmologists. In elderly patients, AMD typically affects central sight damage due to the accumulation of drusen, which are acellular deposits between Bruch's membrane and the retinal pigment epithelium (RPE) [8, 9]. In the advanced stages of AMD, the loss of RPE tissue can result in geographic atrophy (GA). Alternatively, wet AMD is characterized by the growth of pathological blood vessels, under or into the retina, a condition known as choroidal neovascularization (CNV). Class 3 neovascularization is a subtype of wet AMD additionally referred to as retinal angiomatous proliferation [10]. This can cause fluid leakage into the retina (intraretinal fluid, or IRF) or beneath it (subretinal fluid, or SRF), leading to progressive sight damage.

The development of advanced AMD is influenced via various factors, including age, genetics, and ethnicity. AMD is highly linked to older age [11]. A paper examining individuals aged 43 to 86 years found that the prevalence of AMD is three times higher in those aged 75 years and above compared to those aged 65 to 74 years [12, 13]. In young patients, AMD is rare. Caucasians have the highest prevalence of AMD in the United States, followed by Hispanics and Asians, while African Americans have the lowest prevalence [14]. A family history of the disease increases the risk of developing AMD. Additionally, several other factors increase the risk of AMD, including abdominal obesity [15], hyperlipidemia [16], hyperopia [17], light iris color [18], cardiovascular conditions [11], hormonal changes [19], alcohol intake [20], and a low level of vitamin B and D in the blood [21, 22]. Various imaging methods are currently used for detecting abnormalities in the retina correlated through AMD. One of the highest important imaging tools that has emerged in recent years is spectral-domain optical coherence tomography (SD-OCT). SD-OCT is a cross-sectional technique that has become a key tool for ophthalmologists for AMD diagnosis and monitoring of its development, especially in patients who demand management [23, 24]. SD-OCT is an imaging tool that has greatly aided in the diagnosis and management of AMD. Its high-resolution imaging capabilities enable the *in vivo* imaging of the macula of a human at a level of detail that surpasses the medical assessment of the color fundus photographs. With a resolution of 5–7 microns, SD-OCT can identify subtle changes in the retina, such as the presence of SRF or IRF, cutoffs of the exterior retinal layers, RPE rips, shallow PED, sub-RPE or subretinal tissue formation, and external retinal tubulations. It is capable of also providing accurate measurements of the thickness of the middle macular. Thus, it is essential for both comprehensive ophthalmologists and retina experts to be proficient in interpreting OCT images [25].

Spectral-domain optical coherence tomography (SD-OCT) utilizes low-coherence optical maser light at an infrared occurrence to partially penetrate the retina, generating cross-sectional images of the retina through the interference pattern of backscattered light [26]. The images obtained can identify changes in the retina and detect the presence of fluids, tissue above and below the retinal pigment epithelium, and other abnormalities. This imaging technique aids in distinguishing occult membranes, proliferation of retinal angiomas, classic membranes, and disciform scars affected through the disease, as well as in facilitating the follow-up of VGEF therapy [27–29].

Numerous studies have been conducted on retinal imaging, focusing on the diagnosis of AMD. This chapter presents a complete review of various research that examined the OCT role and the modalities of images used in grading AMD. We provide in the next section a detailed overview of the distinct categories of AMD, which can be categorized into two essentials catgories: (1) wet AMD and (2) dry AMD.

13.2 The different grades of AMD

It is important to clinically classify AMD to predict its progression and develop an appropriate identification of a disease, interventions and strategies used to manage, alleviate, or cure a medical condition. AMD can be categorized into two types: non-exudative (dry) and exudative (wet), with 85%–90% of cases being of the dry type. Dry AMD typically progresses slowly and can lead to RPE atrophy, with a certain percentage of cases progressing to wet AMD over time.

13.2.1 Dry AMD

Dry AMD is characterized by the abnormal leaking of blood vessels, and while patients can experience a vision damage. However, they may also experience limitations such as fluctuating vision, poor night vision, and trouble in the reading due to a reduced mid field of perception. Drusen, which are small yellow deposits found under the macula, are the primary hallmark of dry AMD. When drusen form, they can cause the macula to thin and dry out, leading to a decline in macular function and eventual vision loss. Dry AMD is further categorized to "initial stages", "intermediate stages", and "advanced or late" stages based on the dimension of drusen and AMD pigmentary irregularities [30, 31].

13.2.1.1 Early dry AMD

In the initial stage of AMD, patients might not have any clinical symptoms, but they may experience impaired dark adaptation, such as difficulty seeing in dim light or needing brighter light. Ophthalmoscopy may reveal medium-sized drusen (larger than 63 μm and up to 125 μm), and no pigmentary abnormalities are guaranteed.

The danger of development from early or initial AMD to late or advanced AMD within 5–10 years is minimal [31].

Several studies have investigated the ability of OCT to detect early signs of AMD and to predict the development of the AMD disease from the initial to late stages. OCT features such as disruption of drusenoid RPE detachment, the ellipsoid zone, retinal pigmentary hyperreflective material, and RPE thickening were found to be related with an increased danger of progression to late AMD. Additionally, certain OCT findings such as total retinal thickness, choroidal vessel irregularities, and developing GA features were found to be independently related with an advanced danger of development to advanced AMD. These outcomes are important for clinicians to identify high-risk patients and develop appropriate treatment plans to prevent the progression of the disease.

13.2.1.2 Intermediate dry AMD

Intermediate AMD is characterized by the presence of one or more large drusen (⩾125 μm) and/or RPE disturbances, which can result in perception damage in both eyes. Patients with intermediate AMD may experience signs such as blurry dots in their field of perception, difficulty seeing in minimal light, and reduced contrast sensitivity. Research on AMD development has indicated that inside a 5-year period, 6.3% of patients with intermediate AMD and large drusen in one eye may progress to late AMD, while the danger rises to 26% if large drusen exists in both eyes [31, 32].

13.2.1.3 Advanced dry AMD

GA is a form of late dry AMD that is characterized by sharply defined areas of the outer retina that undergo atrophy, leading to a damage of the photoreceptors, RPE, and the choriocapillaris in both eyes. This condition results in a gradual and irreversible decline in visual function, typically over the course of several years, and often affects the foveal center late. The clinical manifestation of GA is the appearance of blind spots in the parafoveal region that progressively coalesce and enlarge, eventually involving the foveal center and causing critical central perception damage [32].

OCT imaging of GA can reveal a loss of three outer layers, which includes the photoreceptors. This loss causes thinning of the external hyperreflective tissue and attenuation of RPE/Bruch's membrane complexes. OCT scans can also help detect changes that occur before the development of GA. Overall, OCT imaging can be helpful in diagnosing and tracking the progression of GA [30].

The utilization of optical coherence tomography angiography (OCTA) has provided valuable insight into the development and progression of geographic atrophy (GA) by allowing visualization of choriocapillaris atrophy beneath the photoreceptors and RPE. Studies have shown that OCTA is a useful tool in understanding the pathophysiology of GA [30, 31].

13.2.2 Wet AMD

Wet AMD is characterized by blood or fluid leakage under the macula, resulting in dark spots in the central vision. Choroidal neovascularization (CNV) occurring beneath the retina and macula is the leading cause of wet AMD. This abnormal growth of blood vessels can lead to macular swelling, temporary perception damage, and bleeding that can severely damage the photoreceptors and cause permanent perception damage. Peripheral vision is typically unaffected. Wet AMD may progress rapidly, and the development of CNV in one eye increases the risk of CNV in the other eye, necessitating periodic eye exams. There are two types of wet AMD: classic and occult, and CNV can be either active or inactive. Wet AMD is always categorized as late AMD and is mostly headed by dry AMD. Diagnosis is proven by images of OCT, which can predict the success of surgical removal and correlates with response to treatment [32, 33].

13.2.2.1 Inactive wet AMD

The inactive form of CNV in wet AMD is characterized by less leakage and a less demarcated appearance compared to the active form. Visual acuity is also less impaired, typically ranging from 20/80 to 20/200. OCT imaging is useful in detecting subepithelial occult AMD features, exudative reactions, PED presence, and alterations in the RPE band. In OCT scans of occult inactive AMD, the lesion seems flat and ill-defined with a convex plane. These findings have been reported in multiple studies [27, 30, 34].

13.2.2.2 Active wet AMD

The active form of wet AMD may be characterized by a distinct, highly reflective, spindle-shaped thickening between the RPE and Bruch's membrane in the subretinal space, leading to visual acuity ranging from 20/250 to worse than 20/800. However, with the use of OCT imaging, the diagnosis of active wet AMD at initial phases has become possible. The structure of organisms of the lesion can be visualized by OCT as a well-defined, steeply marginated lesion with a concave configuration resembling a crater [33].

13.3 The image modalities used for AMD classification

Figure 13.1 illustrates the structure of the eye, which includes the sclera, pupil, iris, and cornea. Light passes via the anterior chamber and cornea, is refracted by the lens and pupil, and ultimately reaches the retina. Several medical imaging techniques can capture different parts of the eye, and the resulting images can reveal various pathological changes. Different imaging technologies are employed for this purpose.

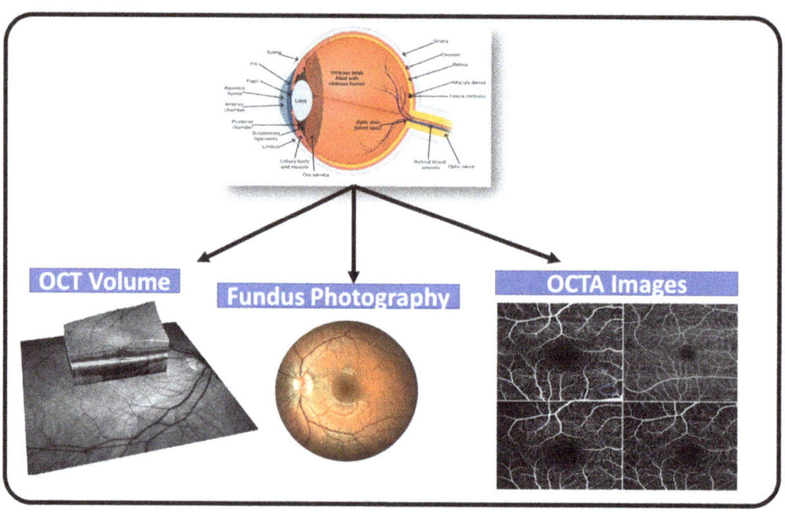

Figure 13.1. An illustrative figure for the structure of the eye.

13.3.1 Fundus image

Fundus photography is an ophthalmic imaging technique used to capture the appearance of the retina using an ophthalmoscope. This imaging modality is commonly used in various eye diseases, including AMD. The classification and staging of AMD can be aided by the use of color fundus photography [35]. Drusen, which are yellow or white deposits that accumulate between Bruch's membrane and RPE, can be observed in the early stages of AMD using fundus photography. There are two types of drusen, hard and soft, which appear as bright spots in color fundus photographs. The ophthalmic record of the patient's retina, called a fundus photograph, is captured using an ophthalmoscope. This photographic evidence is frequently required in the diagnosis of various eye diseases. In the staging and classification of AMD, color fundus photography is particularly useful as it can reveal drusen. Drusen can be hard or soft, with hard drusen being small and well-defined while soft drusen are less distinct and may meet. While drusen occurrence is common in normal aging, an increase in their amount and dimension can increase the threat of AMD and subsequent vision loss. As AMD progresses, pigmentary changes in the RPE can indicate GA, while exudative abnormalities can indicate wet AMD. Although drusen is able be detected manually by sight, computer algorithms have been developed to aid in the detection and differentiation of drusen from other pathological appearances.

13.3.2 Optical coherence tomography (OCT)

OCT is an imaging modality that utilizes interferometry to produce high-resolution cross-sectional images of the retina. Figure 13.2 depicts the various grades of AMD using OCT images. It is similar to ultrasound but employs light waves instead of acoustic waves. OCT allows for visualization of the cellular layers of the retina and measurement of their thickness, aiding in the early detection and diagnosis of retinal diseases. Recent advancements in OCT technology have led to real-time imaging at

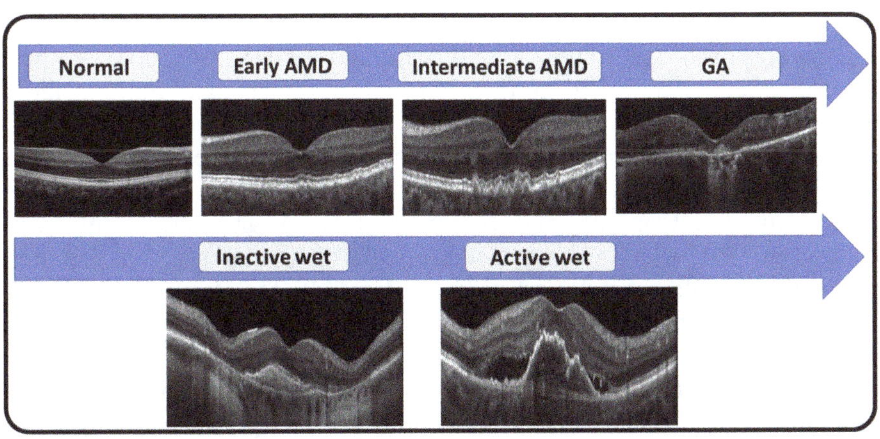

Figure 13.2. An illustrative figure for AMD grading using OCT images.

a rate of several frames per second, and it has been used to analyze developmental biology specimens at the cellular level. Additionally, OCT can provide internal body imaging using catheters, endoscopes, and laparoscopes. Its ability to provide cross-sectional imaging of the retina and visualize microscopic structures is a significant benefit in ophthalmology [36, 37].

OCT imaging is highly sensitive, allowing for the visualization of weakly backscattering features like the vitreoretinal interface, despite the low optical backscattering of the retina. RPE and choroid, on the other hand, appear bright and hyperreflective in OCT images. The retinal nerve fiber layer, which can be visualized using OCT, is thickest near the optic disk and gradually thins towards the fovea. In addition to analyzing the retina's dynamic responses, OCT has been used to study retinal laser injury. Intelligent algorithms can quantitatively analyze OCT images, including measuring the thickness of the retinal nerve fiber layer or retinal nerve tissue, which can provide prognostic information for conditions like glaucoma or diabetic macular edema. As a result, OCT is useful for detecting and diagnosing early stages of disease before the development of clinical symptoms, helping prevent irreversible vision loss [38].

13.4 The pathologies associated with AMD

13.4.1 Drusen

OCT imaging was initially challenged by motion artifacts that could make it difficult to differentiate between drusen and apparent RPE undulation. However, high-speed spectral-domain technologies have made it much easier to determine the size, reflectivity, and form of drusen. Larger drusen are typically identified by a more significant elevation of the RPE, with hyporeflective or moderately reflective material separating the RPE from the underlying Bruch's membrane. On the other hand, small and intermediate-sized drusen can be observed as discrete bulges in the RPE, with nonuniform reflectivity. Large confluent drusen may not have a singular dome-shaped lesion, and an OCT study suggests that they may be indicative of fluid accumulation under the retina in the absence of CNV. This finding could modify the managing approach for some patients who have CNV by allowing them to implement careful care with cautious follow-up rather than more determined therapy with anti-angiogenic. Further research is needed to confirm this feature [39–42]. Drusen deposits in AMD typically form between the RPE and the inner edge of Bruch's membrane, and there are different forms of drusen that can be observed using OCT in different conditions [43]. A grading system was first announced in 1991, which characterized reticular drusen as a complex pattern of broad interwoven ribbons with indistinct borders [44]. The Beaver Dam Eye Study identified these as a major risk factor for advanced AMD [45]. The introduction of SD-OCT has allowed for better characterization of reticular drusen, also known as pseudo-drusen, which appear as granular hyperreflective material in the subretinal space between the RPE and the IS-OS junctions. Some experts have suggested using the term "subretinal drusenoid deposits" for this drusen [46, 47].

When cuticular drusen are examined using OCT, it shows that the RPE is elevated and the IS-OS junction appears rippled [48, 49]. Vitelliform lesions may coexist with cuticular drusen, especially in the early stages of AMD [50, 51]. These lesions appear as hyperreflective material in the subretinal space on OCT and can be mistaken for CNV on FA. Correct identification of these lesions is crucial for effective treatment of AMD since they may have subretinal fluid accumulation caused by incomplete RPE phagocytosis of subretinal material. Multiple researchers have highlighted the prognostic significance of the size of drusen, expressed as area or diameter, in predicting the development of late-stage AMD. Manual evaluation of drusen on color fundus photographs is subject to significant inter-observer variability, and computerized degree and recognition of drusen using SD-OCT are more promising and reliable [52, 53]. Prophylactic interventions for extrafoveal GA using therapies that target complement pathways may be introduced in the future, thanks to these advances in imaging technology.

13.4.2 Geographic atrophy

OCT imaging is useful in detecting the characteristic features of GA in AMD, such as dehydration and calcification of drusen, as well as RPE atrophy and photoreceptor and choriocapillaris degradation [54, 55]. In addition, OCT can reveal markedly hyperreflective choroidal tissue in areas where the RPE is absent, indicating GA [56]. When GA is associated with retinal atrophy, OCT may show thinning or loss of the outer nuclear layer and the absence of ELM and IS–OS junctions. Despite the preserved outer retina, some hypereflective, drusenoid material may be observed at the level of the RPE on OCT. However, variable dynamic changes may indicate imminent atrophy, highlighting the importance of early detection and intervention [45]. Additionally, tapering of the ELM and IS-OS junctions, pigment migration, and alterations in drusen height may also be observed. The significance of junctional zone changes in GA pathogenesis and the qualified tasks of RPE and photoreceptors remain uncertain. Current research has utilized imaging software to register OCT images with fundus photos, allowing for the precise delineation of the boundaries of GA [57, 58]. A study investigating the development of GA found that patients with a rapidly progressing form of the condition displayed partition of the RPE and Bruch's membrane in these border zones. Studies have suggested that OCT fundus images acquired using SD-OCT devices may reveal higher signal intensity in individuals with clinical manifestations of GA.

13.4.3 Neovascular AMD

In order to properly evaluate neovascular AMD using OCT, it is important to understand the underlying pathogenesis of neovascularization. In neovascular AMD, abnormal blood circulation begins in the choroidal circulation, with the proliferation of incompetent vessels in the subretinal or RPE space after passing through Bruch's membrane breaks [59]. These immature vessels lead to fluid exudation and hemorrhage, which can result in the formation of pathologic "compartments" involving the Bruch's membrane (PED) and the neurosensory retina (serous retinal detachment) in cases of severe retinal detachment. The overlying retinal architecture may become significantly disorganized, and there can be a loss of RPE and photoreceptors in disciform scars. The neovascular invasion also results in the significant degradation and remodeling of the extracellular space in the retina, with fibroblasts invading the space and restoring the extracellular space [60–62].

13.5 Methods

Our analysis delves deep into the identification and diagnosis of diabetic retinopathy (DR), incorporating the latest research and developments. We present a comprehensive comparative analysis of various imaging modalities used for the detection and diagnosis of DR.

An abundance of algorithms and databases have been developed for the management of AMD. To assess the performance of image-based methods, a comprehensive review of the major imaging modalities used in CAD applications and research domains has been conducted. This chapter provides insights into the current state of image-based research on AMD disease.

Using the PubMed database, we conducted a comprehensive survey of peer-reviewed articles published up to June 2021 and included all relevant papers to review the state-of-the-art in AMD diagnosis and progression using machine learning (ML) techniques. In this study, we present in this chapter a systematic review of articles that focus on the automated detection of drusen, intraretinal and subretinal fluids, as well as subretinal tissue and sub-RPE tissue in OCT images. Our aim is to provide a comprehensive analysis of the current research in ML-based AMD diagnosis and identify potential gaps for future research.

13.5.1 CAD approaches for AMD grading

In the last century, ocular imaging has made significant progress and has become a crucial element of ophthalmic disease administration and medical treatment. Since the early 1980s, there has been considerable systematic research and development on computer-aided diagnosis (CAD) systems based on radiology and medical images. Retinal image analysis was first reported in 1973, with a focus on vessel segmentation [64]. In 1984, Baudoin and colleagues [65] presented a novel image analysis technique for the detection of DR-associated lesions. In the last twenty years, there has been significant progress in image processing for ophthalmology, enabling the automated diagnosis of several diseases such as DR [66, 67], AMD [68], glaucoma [69], and cataract [70]. These diagnostic tools hold potential for use in large-scale screening programs, offering the advantage of saving resources, ensuring compliance, and eliminating observer fatigue or bias.

ML and DL methods have led to the development of various automated techniques for classifying the progression of AMD based on OCT in recent years. DL methods are considered to be a cutting-edge technology and have exhibited promising outcomes in various applications. Convolutional neural networks (CNN) are a popular method used in AMD classification, as they are capable of detecting features using convolutional kernels and classifying them with fully connected (FC) layers.

Other CAD system utilzed in AMD diagnosis is designed to distinguish between non-AMD diseases and AMD. ML, a subset of AI, is particularly motivated by human learning [71]. ML methods search for shapes in training data and use these shapes to classify subsets of data. When new, previously unseen data is presented, the algorithm can determine its category. Learning based on extracted features, either unsupervised or supervised, can be accomplished by training the algorithm using examples or by triggering it to learn from previous examples.

Various ML methods have been explored to categorize OCT discoveries associated with AMD development. For instance, in a study by Hwang *et al* [4], three CNN models (VGG16, ResNet50, and InceptionV3) were employed to distinguish between normal, dry AMD, active wet, and inactive wet cases. The accuracies achieved by the models were 91.40%, 90.73%, and 92.67%, respectively. These CNN architectures have been proven to be effective in image recognition and can be adapted for better acknowledgement accuracy by adapting parameters such as epoch, batch size, and learning rate. To evaluate the performance of AI methods in comparison to ophthalmologists' predictions based on process of validating the results of a diagnostic test, a confusion matrix was employed. Hwang *et al* [4] reported a sensitivity of 99.38%, 85.64%, 97.11%, and 88.53% for normal, dry AMD, inactive wet AMD, and active wet AMD, respectively, using InceptionV3, which was the best-performing CNN model among the three models tested. In terms of specificity, InceptionV3 achieved 99.70%, 99.57%, 91.82%, and 98.99% for normal, dry AMD, inactive wet AMD, and active wet AMD, respectively. Another method used to classify OCT findings related to AMD progression is the Bag of Words approach. In a study by Venhuizen *et al* [3], an ML approach was presented that characterized between normal retina and four subclasses of AMD, namely CNV, GA, early, and intermediate. They first created a dictionary of descriptive graphical words for AMD categorization using a training set of OCT volumes. At random, they selected patches from the detected salient regions which was used to train each OCT volume. The OCT scan data was used to categorize the patches into five sets based on the severity of AMD. By utilizing the data from the OCT, the patches were divided into five sets based on the seriousness of AMD. The k-means method was applied to partition each patch into k clusters based on the cluster centroid. With this system, a sensitivity of 98.2% and a specificity of 91.2% were reached. In additional research, An *et al* [5] proposed a three-way classification system using a pre-trained VGG16 CNN to distinguish between the healthy retina, active wet AMD, and inactive wet AMD. The authors achieved an accuracy of 93.38% using this system. Motozawa *et al* [6] developed a two-tier classification system using CNN models for OCT image classification. In the first tier, OCT images were classified as normal or AMD using a CNN model. In the second tier, AMD images were further classified into those with exudative changes (presence of fluids) and those without exudative changes (absence of fluids) using another CNN model. The authors also generated a heat map using class activation mapping to highlight the regions of interest in the images that were emphasized by the CNN models. The study compared the performance of the second-tier CNN model with transfer learning and a single CNN model and evaluated their speed and stability of learning. However, the study did not report sensitivity or specificity values. In order to preserve the image quality, the CNN models were trained using cropped images, and to determine the classification of the original image, three cropped images were combined using CNN models. The first model achieved an accuracy of 99%, a sensitivity of 100%, and a specificity of 91.8%, while the second model had an accuracy of 93.9%, a sensitivity of 100%, and a specificity of 91.8%. Treder *et al* [7] also utilized a pre-trained InceptionV3 CNN to detect and differentiate exudative AMD from healthy retinas based on OCT images. A study utilized a pre-trained Inception-v3 network to recognize images from the ImageNet dataset, followed by developing a DCNN model by training its initial layers on around 1 million similar images that were categorized into roughly 1000 categories, such as strawberries, zebras, and bananas. Subsequently, a classifier was built to identify exudative AMD. To achieve this, the last layer of the DCNN model was modified to function on OCT images and diagnose exudative AMD. The results showed an average AMD score of 99.7% and an average healthy score of 92.03%, with sensitivity and specificity values of 100% and 92%, respectively. Li *et al* [63] proposed a CNN-based classification method using a pre-trained VGG16 to differentiate between normal, drusen, CNV, and diabetic macular edema (DME) in retinal OCT images. Image acquisition preceded image grading and labeling, followed by image preprocessing techniques such as image normalization. They did not perform image denoising to prevent overfitting and improve the classifier's generalization ability. The last fully connected layer of the VGG-16 network was modified to adjust it to the four output classes using a deep transfer learning approach. The accuracy achieved by their system was 98.6%. However, it lacked the ability to differentiate between various phases of AMD with considerable features. The developed method demonstrated impressive performance, achieving 100% specificity and sensitivity in categorize normal from CNV, and 98.8% sensitivity and specificity in categorize normal from DME. While observing drusen from normal, the model achieved 98.4% sensitivity and 100% specificity. However, it should be noted that distinguishing between different stages of AMD with significant detail remains a challenge, as CNV can be either inactive or active, and different stages of AMD have dissimilar predictions. Moreover, the literature can be impaired from certain limitations. Firstly, traditional DL CNNs are used to separate between AMD classes and normal retina, without incorporating medical landmarks such as retina abnormalities. Secondly, there is currently no computer-aided diagnosis (CAD) system available that can accurately distinguish among normal retina and all five AMD categories.

13.6 Discussion and future direction

In this chapter, the focus is on the application of ML techniques, particularly DL, for diagnosing AMD in a quantitative and qualitative way. ML classification methods have demonstrated high specificity and sensitivity and hold great ability in differentiating AMD from further eye conditions. ML classifiers have provided promising results in terms of diagnostic accuracy for many other diseases [72–84]. The utilization of AI can significantly enhance teleophthalmology practices, particularly in areas where patients may not have immediate access to ophthalmologists [85]. In rural areas, ML can aid in early AMD identification without requiring confirmation from a clinician and can also decrease transportation expenses for both patients and healthcare providers. Additionally, ML provides a chance for physicians in urban areas to improve efficiency and alleviate patient load [86].

Despite AMD being one of the leading causes of vision loss in countries with a population of over a million people, no studies were found on the use of machine learning for AMD detection in Africa or the Middle East. The studies that were incorporated in this analysis were carried out across different regions, including Western Europe, the United States, and Asia. Therefore, there is a need to explore the potential use of machine learning for AMD detection in these regions to improve early diagnosis and treatment.

OCT-derived biomarkers have shown that GA is the most readily detectable form of AMD, while machine learning classifiers have struggled with detecting intraretinal/subretinal fluid and PED. OCT can provide a greater variety of biomarkers beyond drusen and GA. While the techniques for drusen and GA segmentation are almost identical, there are differences in how intra/subretinal fluid, PED, and CAD tools are detected. A potential explanation for this contrast could be the varying availability and accessibility of image processing methods used for identifying drusen and GA, as these features tend to have a consistent appearance in OCT scans. Comparing algorithms can be challenging since there is no standardization in quality assessment. However, CAD tools seem to require the most comprehensive implementation. In truth, detecting pathology alone is less likely to result in errors than detecting and quantifying it together.

A future meta-analysis could provide valuable insights into how AI procedures can be utilized to categorize all classes of AMD, involving wet AMD, with dry AMD. Moreover, further research could explore how AI algorithms can differentiate between AMD and non-AMD diseases, which can be particularly challenging as some types of AMD share similarities with non-AMD diseases in terms of their characteristics. For instance, the morphological characteristics of GA on OCT can resemble those of Stargardt disease, an inherited macular degeneration that often presents in individuals under 30 years of age and shows gradual macular atrophy, like GA. Furthermore, some individuals with Stargardt disease may develop AMD later in life. However, age is a key criterion that retina specialists use to diagnose Stargardt disease. In some cases, family history can be helpful in diagnosing Stargardt disease, but it is not always available due to the possibility of "de novo" mutations. Other ophthalmic imaging techniques such as fundus photography and OCTA can be employed in combination with OCT to assist in the diagnosis of cases with similar characteristics in both AMD and non-AMD diseases. The Macula Society has established a uniform system for classifying neovascular AMD and its related features using advanced imaging technologies, particularly OCT imaging and OCT angiography. These techniques enable the precise evaluation of each component, whether neovascular or not, by providing detailed 3-dimensional analyses of the vascular anatomical structure of neovascular AMD lesions and direct imaging of anatomic features.

The aim of this chapter is to include a complete understanding of AMD, highlighting the latest advancements in the field that can help improve its diagnosis and treatment. Esteemed experts have shared their knowledge, expertise, and insights, making this survey a valuable resource for ophthalmologists seeking guidance on managing AMD. Ultimately, this survey serves as a critical tool in advancing our understanding and improving our ability to diagnose and treat this debilitating disease.

Our findings hold significant implications for both public health and clinical practice. Employing AI systems in the workplace can facilitate early detection of AMD, allowing for prompt interventions to prevent its progression to advanced stages. This can potentially reduce the burden of AMD-related vision loss on individuals, families, and healthcare systems. In addition, AI-based screening can be especially valuable in remote areas where access to qualified ophthalmologists is limited. By enabling more efficient and accurate detection of eye disorders, AI can help ensure timely diagnosis and treatment, ultimately improving the health outcomes of individuals in these underserved regions.

It is important to acknowledge that this study has some limitations. One of the limitations is that a few of the studies included in the analysis had small sample sizes, which may have impacted the reliability of the AI performance. Furthermore, the definition of AMD varied among the studies included, however, despite this heterogeneity, subgroup analyses were conducted to explore potential differences in performance.

13.7 Conclusion

To summarize, the usage of OCT-derived biomarkers representing automated recognition of AMD shows promise, but there is significant variation in the quality and type of validation methods used. Most of the testing is limited to preselected individuals, and there is a need for standardized validation procedures. Clinicians and researchers would benefit from the development of algorithms that can simultaneously analyze multiple AMD biomarkers. Overall, further advancements in this technology will help improve the diagnosis and management of AMD, ultimately leading to better patient outcomes.

In addition to the retina [87–91], this work could also be applied to various other applications in medical imaging, such as the prostate [92–96], the kidney [97–115], the heart [116–133], the lung [134–181], the brain [182–204], the vascular system [205–215], the bladder [216–220], the liver [221, 222], the head and neck [223, 224], and injury prediction [225], as well as several non-medical applications [226–232].

References

[1] Bressler N M Age-related macular degeneration is the leading cause of blindness *JAMA* **291** 1900–1 2004

[2] Pascolini D *et al* 2004 2002 Global update of available data on visual impairment: a compilation of population-based prevalence studies, *Ophthal. Epidemiol.* **11** 67–115

[3] Venhuizen F G, van Ginneken B, van Asten F, van Grinsven M J, Fauser S, Hoyng C B, Theelen T and Sanchez C I 2017 Automated staging of age-related macular degeneration using optical coherence tomography *Invest. Ophthalmol. Vis. Sci.* **58** 2318–28

[4] Hwang D-K, Hsu C-C, Chang K-J, Chao D, Sun C-H, Jheng Y-C, Yarmishyn A A, Wu J-C, Tsai C-Y and Wang M-L *et al* 2019 Artificial intelligence-based decision-making for age-related macular degeneration *Theranostics* **9** 232

[5] An G, Yokota H, Motozawa N, Takagi S, Mandai M, Kitahata S, Hirami Y, Takahashi M, Kurimoto Y and Akiba M 2019 Deep learning classification models built with two-step transfer learning for age related macular degeneration diagnosis *2019 41st Annual Int. Conf. of the IEEE Engineering in Medicine and Biology Society (EMBC)* (Piscataway, NJ: IEEE) 2049–52

[6] Motozawa N, An G, Takagi S, Kitahata S, Mandai M, Hirami Y, Yokota H, Akiba M, Tsujikawa A and Takahashi M *et al* 2019 Optical coherence tomography-based deep-learning models for classifying normal and age-related macular degeneration and exudative and non-exudative age-related macular degeneration changes *Ophthalmol. Ther.* **8** 527–39

[7] Treder M, Lauermann J L and Eter N 2018 Automated detection of exudative age-related macular degeneration in spectral domain optical coherence tomography using deep learning *Graefe's Arch. Clin. Exp. Ophthalmol.* **256** 259–65

[8] Ambati J, Ambati B K, Yoo S H, Ianchulev S and Adamis A P 2003 Age-related macular degeneration: etiology, pathogenesis, and therapeutic strategies *Surv. Ophthalmol.* **48** 257–93

[9] Lim L S, Mitchell P, Seddon J M, Holz F G and Wong T Y 2012 Age-related macular degeneration *Lancet* **379** 1728–38

[10] Yannuzzi L A, Negrao S, Tomohiro I, Carvalho C, Rodriguez-Coleman H, Slakter J, Freund K B, Sorenson J, Orlock D and Borodoker N 2012 Retinal angiomatous proliferation in age–related macular degeneration *Retina* **32** 416–34

[11] Chakravarthy U, Wong T Y, Fletcher A, Piault E, Evans C, Zlateva G, Buggage R, Pleil A and Mitchell P 2010 Clinical risk factors for agerelated macular degeneration: a systematic review and meta-analysis *BMC Ophthalmol.* **10** 1–13

[12] Klein R, Klein B E and Linton K L 1992 Prevalence of age-related maculopathy: the beaver dam eye study *Ophthalmology* **99** 933–43

[13] Leibowitz H M, Krueger D, Maunder L R, Milton R, Kini M, Kahn H, Nickerson R, Pool J, Colton T and Ganley J *et al* 1980 The framingham eye study monograph: an ophthalmological and epidemiological study of cataract, glaucoma, diabetic retinopathy, macular degeneration, and visual acuity in a general population of 2631 adults, 1973–1975 *Surv. Ophthalmol.* **24** 335–610

[14] Chou R, Dana T, Bougatsos C, Grusing S and Blazina I 2016 Screening for impaired visual acuity in older adults: updated evidence report and systematic review for the us preventive services task force *JAMA* **315** 915–33

[15] Adams M K, Simpson J A, Aung K Z, Makeyeva G A, Giles G G, English D R, Hopper J, Guymer R H, Baird P N and Robman L D 2011 Abdominal obesity and age-related macular degeneration *Am. J. Epidemiol.* **173** 1246–55

[16] Dasari B, Prasanthi J R, Marwarha G, Singh B B and Ghribi O 2011 Cholesterol-enriched diet causes age-related macular degeneration-like pathology in rabbit retina *BMC Ophthalmol.* **11** 1–11
[17] Sandberg M A, Tolentino M J, Miller S, Berson E L and Gaudio A R 1993 Hyperopia and neovascularization in age-related macular degeneration *Ophthalmology* **100** 1009–13
[18] Khan J, Shahid H, Thurlby D, Bradley M, Clayton D, Moore A, Bird A and Yates J 2006 Age related macular degeneration and sun exposure, iris colour, and skin sensitivity to sunlight *Br. J. Ophthalmol.* **90** 29–32
[19] Feskanich D, Cho E, Schaumberg D A, Colditz G A and Hankinson S E 2008 Menopausal and reproductive factors and risk of age-related macular degeneration *Arch. Ophthalmol.* **126** 519–24
[20] Chong E W-T, Kreis A J, Wong T Y, Simpson J A and Guymer R H 2008 Alcohol consumption and the risk of age-related macular degeneration: a systematic review and meta-analysis *Am. J. Ophthalmol.* **145** 707–15
[21] Gopinath B, Flood V M, Rochtchina E, Wang J J and Mitchell P 2013 Homocysteine, folate, vitamin b-12, and 10-y incidence of age-related macular degeneration *Am. J. Clin. Nutr.* **98** 129–35
[22] Millen A E, Voland R, Sondel S A, Parekh N, Horst R L, Wallace R B, Hageman G S, Chappell R, Blodi B A and Klein M L *et al* 2011 Vitamin D status and early age-related macular degeneration in postmenopausal women *Arch. Ophthalmol.* **129** 481–9
[23] Wong W L, Su X, Li X, Cheung C M G, Klein R, Cheng C-Y and Wong T Y 2014 Global prevalence of age-related macular degeneration and disease burden projection for 2020 and 2040: a systematic review and meta-analysis *Lancet Global Health* **2** e106–16
[24] Victor A A 2019 The role of imaging in age-related macular degeneration *Visual Impairment and Blindness-What We Know and What We Have to Know* (IntechOpen)
[25] Ooto S *et al* 2011 Effects of age, sex, and axial length on the threedimensional profile of normal macular layer structures *Invest. Ophthalmol. Vis. Sci.* **52** 8769–79
[26] Keane P A, Patel P J, Liakopoulos S, Heussen F M, Sadda S R and Tufail A 2012 Evaluation of age-related macular degeneration with optical coherence tomography *Surv. Ophthalmol.* **57** 389–414
[27] Ahlers C, Gotzinger E, Pircher M, Golbaz I, Prager F, Schütze C, Baumann B, Hitzenberger C K and Schmidt-Erfurth U 2010 Imaging of the retinal pigment epithelium in age-related macular degeneration using polarization-sensitive optical coherence tomography *Invest. Ophthalmol. Vis. Sci.* **51** 2149–57
[28] Ma J, Desai R, Nesper P, Gill M, Fawzi A and Skondra D 2017 Optical coherence tomographic angiography imaging in age-related macular degeneration *Ophthalmol. Eye Dis.* **9** 1179172116686075
[29] Nagiel A, Sadda S R and Sarraf D 2015 A promising future for optical coherence tomography angiography *JAMA Ophthalmol.* **133** 629–30
[30] Stahl A 2020 The diagnosis and treatment of age-related macular degeneration *Dtsch. Arztebl. Int.* **117** 513
[31] Bird A C, Bressler N M, Bressler S B, Chisholm I H, Coscas G, Davis M D, de Jong P T, Klaver C, Klein B and Klein R *et al* 1995 An international classification and grading system for age-related maculopathy and age-related macular degeneration *Surv. Ophthalmol.* **39** 367–74
[32] Joachim N, Mitchell P, Burlutsky G, Kifley A and Wang J J 2015 The incidence and progression of age-related macular degeneration over 15 years: the blue mountains eye study *Ophthalmology* **122** 2482–9

[33] Vitale S, Agron E, Clemons T E, Keenan T D and Domalpally A 2020 Association of 2-year progression along the AREDS AMD scale and development of late age-related macular degeneration or loss of visual acuity: AREDS report 41 *JAMA Ophthalmol.* **138** 610–7

[34] Coleman H R, Chan C-C, Ferris F L and Chew E Y 2008 Age-related macular degeneration *Lancet* **372** 1835–45

[35] Group A-R E D S R *et al* 2005 The age-related eye disease study severity scale for age-related macular degeneration: AREDS report no. 17 *Arch. Ophthalmol.* **123** 1484–98

[36] Kim S G, Lee S C, Seong Y S, Kim S W and Kwon O W 2003 Optical coherence tomography *Yonsei Med. J.* **44** 821–7

[37] Podoleanu A G 2013 Optical sources for optical coherence tomography (OCT) *Lasers for Medical Applications* (Amsterdam: Elsevier) pp 253–85

[38] Fujimoto J G, Pitris C, Boppart S A and Brezinski M E 2000 Optical coherence tomography: an emerging technology for biomedical imaging and optical biopsy *Neoplasia* **2** 9–25

[39] Hee M R, Baumal C R, Puliafito C A, Duker J S, Reichel E, Wilkins J R, Coker J G, Schuman J S, Swanson E A and Fujimoto J G 1996 Optical coherence tomography of age-related macular degeneration and choroidal neovascularization *Ophthalmology* **103** 1260–70

[40] Pieroni C, Witkin A, Ko T, Fujimoto J, Chan A, Schuman J, Ishikawa H, Reichel E and Duker J 2006 Ultrahigh resolution optical coherence tomography in non-exudative age related macular degeneration *Br. J. Ophthalmol.* **90** 191–7

[41] Gorczynska I, Srinivasan V J, Vuong L N, Chen R W, Liu J J, Reichel E, Wojtkowski M, Schuman J S, Duker J S and Fujimoto J G 2009 Projection OCT fundus imaging for visualising outer retinal pathology in non-exudative age-related macular degeneration *Br. J. Ophthalmol.* **93** 603–9

[42] Sikorski B L, Bukowska D, Kaluzny J J, Szkulmowski M, Kowalczyk A and Wojtkowski M 2011 Drusen with accompanying fluid underneath the sensory retina *Ophthalmology* **118** 82–92

[43] Spaide R F and Curcio C A 2010 Drusen characterization with multimodal imaging *Retina* **30** 1441

[44] Klein R, Davis M D, Magli Y L, Segal P, Klein B E and Hubbard L 1991 The wisconsin age-related maculopathy grading system *Ophthalmology* **98** 1128–34

[45] Klein R, Meuer S M, Knudtson M D, Iyengar S K and Klein B E 2008 The epidemiology of retinal reticular drusen *Am. J. Ophthalmol.* **145** 317–26

[46] Arnold J J, Sarks S H, Killingsworth M C and Sarks J P 1995 Reticular pseudodrusen. a risk factor in age-related maculopathy *Retina* **15** 183–91

[47] Cohen S Y, Dubois L, Tadayoni R, Delahaye-Mazza C, Debibie C and Quentel G 2007 Prevalence of reticular pseudodrusen in age-related macular degeneration with newly diagnosed choroidal neovascularisation *Br. J. Ophthalmol.* **91** 354–9

[48] Finger R P, Issa P C, Kellner U, Schmitz-Valckenberg S, Fleckenstein M, Scholl H P and Holz F G 2010 Spectral domain optical coherence tomography in adult-onset vitelliform macular dystrophy with cuticular drusen *Retina* **30** 1455–64

[49] Leng T, Rosenfeld P J, Gregori G, Puliafito C A and Punjabi O S 2009 Spectral domain optical coherence tomography characteristics of cuticular drusen *Retina* **29** 988–93

[50] Schmitz-Valckenberg S, Steinberg J S, Fleckenstein M, Visvalingam S, Brinkmann C K and Holz F G 2010 Combined confocal scanning laser ophthalmoscopy and spectral-domain optical coherence tomography imaging of reticular drusen associated with age-related macular degeneration *Ophthalmology* **117** 1169–76

[51] Zweifel S A, Spaide R F, Curcio C A, Malek G and Imamura Y 2010 Reticular pseudodrusen are subretinal drusenoid deposits *Ophthalmology* **117** 303–12

[52] Freeman S R, Kozak I, Cheng L, Bartsch D-U, Mojana F, Nigam N, Brar M, Yuson R and Freeman W R 2010 Optical coherence tomographyraster scanning and manual segmentation in determining drusen volume in age-related macular degeneration *Retina* **30** 431–5

[53] Gregori G, Wang F, Rosenfeld P J, Yehoshua Z, Gregori N Z, Lujan B J, Puliafito C A and Feuer W J 2011 Spectral domain optical coherence tomography imaging of drusen in nonexudative age-related macular degeneration *Ophthalmology* **118** 1373–9

[54] Holz F G, Pauleikhoff D, Klein R and Bird A C 2004 Pathogenesis of lesions in late age-related macular disease *Am. J. Ophthalmol.* **137** 504–10

[55] Sunness J S 1999 The natural history of geographic atrophy, the advanced atrophic form of age-related macular degeneration *Mol. Vis.* **5** 25

[56] Wolf-Schnurrbusch U E, Enzmann V, Brinkmann C K and Wolf S 2008 Morphologic changes in patients with geographic atrophy assessed with a novel spectral OCT–SLO combination *Investigative Ophthalmol. Vis. Sci.* **49** 3095–9

[57] Brar M, Kozak I, Cheng L, Bartsch D-U G, Yuson R, Nigam N, Oster S F, Mojana F and Freeman W R 2009 Correlation between spectraldomain optical coherence tomography and fundus autofluorescence at the margins of geographic atrophy *Am. J. Ophthalmol.* **148** 439–44

[58] Lujan B J, Rosenfeld P J, Gregori G, Wang F, Knighton R W, Feuer W J and Puliafito C A 2009 Spectral domain optical coherence tomographic imaging of geographic atrophy *Ophthal. Surg., Lasers Imaging Retina* **40** 96–101

[59] Grossniklaus H E and Green W R 2004 Choroidal neovascularization *Am. J. Ophthalmol.* **137** 496–503

[60] Green W R 1991 Clinicopathologic studies of treated choroidal neovascular membranes. a review and report of two cases *Retina* **11** 328–56

[61] Green W R *et al* 1999 Histopathology of age-related macular degeneration *Mol. Vis* **5** 1–10

[62] Green W R and Enger C 1993 Age-related macular degeneration histopathologic studies: the 1992 Lorenz E. Zimmerman lecture *Ophthalmology* **100** 1519–35

[63] Li F, Chen H, Liu Z, Zhang X and Wu Z 2019 Fully automated detection of retinal disorders by image-based deep learning *Graefe's Arch. Clin. Exp. Ophthalmol.* **257** 495–505

[64] Matsui M, Tashiro T, Matsumoto K and Yamamoto S 1973 [A study on automatic and quantitative diagnosis of fundus photographs. I. Detection of contour line of retinal blood vessel images on color fundus photographs (author's transl)] *Nippon Ganka Gakkai Zasshi* **77** 907–18

[65] Baudoin C, Lay B and Klein J 1984 Automatic detection of microaneurysms in diabetic fluorescein angiography *Rev. Epidemiol. Sante Publique* **32** 254–61

[66] Narasimha-Iyer H, Can A, Roysam B, Stewart V, Tanenbaum H L, Majerovics A and Singh H 2006 Robust detection and classification of longitudinal changes in color retinal fundus images for monitoring diabetic retinopathy *IEEE Trans. Biomed. Eng.* **53** 1084–98

[67] Sleman A A, Soliman A, Elsharkawy M, Giridharan G, Ghazal M, Sandhu H, Schaal S, Keynton R, Elmaghraby A and El-Baz A 2021 A novel 3D segmentation approach for extracting retinal layers from optical coherence tomography images *Med. Phys.* **48** pp 1584–95

[68] Quellec G, Lee K, Dolejsi M, Garvin M K, Abramoff M D and Sonka M 2010 Three-dimensional analysis of retinal layer texture: identification of fluid-filled regions in SD-OCT of the macula *IEEE Trans. Med. Imaging* **29** 1321–30

[69] Liu J, Wong D, Lim J, Li H, Tan N, Zhang Z, Wong T and Lavanya R 2009 Argali: an automatic cup-to-disc ratio measurement system for glaucoma analysis using level-set image processing *13th Int. Conf. on Biomedical Engineering* (Berlin: Springer) pp 559–62

[70] Huang W, Chan K L, Li H, Lim J H, Liu J and Wong T Y 2010 A computer assisted method for nuclear cataract grading from slit-lamp images using ranking *IEEE Trans. Med. Imaging* **30** 94–107

[71] LeCun Y, Bengio Y and Hinton G 2015 Deep learning *Nature* **521** 436–44

[72] Sharafeldeen A, Elsharkawy M, Shaffie A, Khalifa F, Soliman A, Naglah A, Khaled R, Hussein M, Alrahmawy M and Elmougy S *et al* 2022 Thyroid cancer diagnostic system using magnetic resonance imaging *2022 26th Int. Conf. on Pattern Recognition (ICPR)* (Piscataway, NJ: IEEE) pp 4365–70

[73] Sharafeldeen A, Elsharkawy M, Khalifa F, Soliman A, Ghazal M, AlHalabi M, Yaghi M, Alrahmawy M, Elmougy S and Sandhu H *et al* 2021 Precise higher-order reflectivity and morphology models for early diagnosis of diabetic retinopathy using OCT images *Sci. Rep.* **11** 1–16

[74] Elsharkawy M, Sharafeldeen A, Soliman A, Khalifa F, Widjajahakim R, Switala A, Elnakib A, Schaal S, Sandhu H S and Seddon J M *et al* 2021 Automated diagnosis and grading of dry age-related macular degeneration using optical coherence tomography imaging *Invest. Ophthalmol. Vis. Sci.* **62** 107–7

[75] Haggag S, Elnakib A, Sharafeldeen A, Elsharkawy M, Khalifa F, Farag R K, Mohamed M A, Sandhu H S, Mansoor W and Sewelam A *et al* 2022 A computer-aided diagnostic system for diabetic retinopathy based on local and global extracted features *Appl. Sci.* **12** 8326

[76] Elsharkawy M, Elrazzaz M, Sharafeldeen A, Alhalabi M, Khalifa F, Soliman A, Elnakib A, Mahmoud A, Ghazal M and El-Daydamony E *et al* 2022 The role of different retinal imaging modalities in predicting progression of diabetic retinopathy: a survey *Sensors* **22** 3490

[77] Elsharkawy M, Sharafeldeen A, Soliman A, Khalifa F, Ghazal M, El-Daydamony E, Atwan A, Sandhu H S and El-Baz A 2022 Diabetic retinopathy diagnostic cad system using 3D-OCT higher order spatial appearance model *2022 IEEE 19th Int. Symp. on Biomedical Imaging (ISBI)* (Piscataway, NJ: IEEE) pp 1–4

[78] Elsharkawy M, Sharafeldeen A, Taher F, Shalaby A, Soliman A, Mahmoud A, Ghazal M, Khalil A, Alghamdi N S and Razek A A K A *et al* 2021 Early assessment of lung function in coronavirus patients using invariant markers from chest x-rays images *Sci. Rep.* **11** 12095

[79] Sandhu H S, Elmogy M, Sharafeldeen A T, Elsharkawy M, ElAdawy N, Eltanboly A, Shalaby A, Keynton R and El-Baz A 2020 Automated diagnosis of diabetic retinopathy using clinical biomarkers, optical coherence tomography, and optical coherence tomography angiography *Am. J. Ophthalmol.* **216** 201–6

[80] Farahat I S, Sharafeldeen A, Elsharkawy M, Soliman A, Mahmoud A, Ghazal M, Taher F, Bilal M, Abdel Razek A A K and Aladrousy W *et al* 2022 The role of 3D CT imaging in the accurate diagnosis of lung function in coronavirus patients *Diagnostics* **12** 696

[81] Elsharkawy M, Sharafeldeen A, Soliman A, Khalifa F, Ghazal M, El-Daydamony E, Atwan A, Sandhu H S and El-Baz A 2022 A novel computer-aided diagnostic system for early detection of diabetic retinopathy using 3D-OCT higher-order spatial appearance model *Diagnostics* **12** 461

[82] Sharafeldeen A, Elsharkawy M, Khaled R, Shaffie A, Khalifa F, Soliman A, Abdel Razek A A k, Hussein M M, Taman S and Naglah A *et al* 2022 Texture and shape analysis of diffusion-weighted imaging for thyroid nodules classification using machine learning *Med. Phys.* **49** 988–99

[83] Sharafeldeen A, Elsharkawy M, Alghamdi N S, Soliman A and El-Baz A 2021 Precise segmentation of covid-19 infected lung from CT images based on adaptive first-order appearance model with morphological/anatomical constraints *Sensors* **21** 5482

[84] Elsharkawy M, Elrazzaz M, Ghazal M, Alhalabi M, Soliman A, Mahmoud A, El-Daydamony E, Atwan A, Thanos A and Sandhu H S *et al* 2021 Role of optical coherence tomography imaging in predicting progression of age-related macular disease: a survey *Diagnostics* **11** 2313

[85] Li H K 1999 Telemedicine and ophthalmology *Surv. Ophthalmol.* **44** 61–72

[86] Saleem S M, Pasquale L R, Sidoti P A and Tsai J C 2020 Virtual ophthalmology: telemedicine in a Covid-19 era *Am. J. Ophthalmol.* **216** 237–42

[87] Sleman A A, Soliman A, Ghazal M, Sandhu H, Schaal S, Elmaghraby A and El-Baz A 2019 Retinal layers OCT scans 3-D segmentation *2019 IEEE Int. Conf. on Imaging Systems and Techniques (IST)* (Piscataway, NJ: IEEE) pp 1–6

[88] Eladawi N, Elmogy M, Ghazal M, Helmy O, Aboelfetouh A, Riad A, Schaal S and El-Baz A 2018 Classification of retinal diseases based on OCT images *Front Biosci (Landmark Ed)* **23** 247–64

[89] ElTanboly A, Ismail M, Shalaby A, Switala A, El-Baz A, Schaal S, Gimel'farb G and El-Azab M 2017 A computer-aided diagnostic system for detecting diabetic retinopathy in optical coherence tomography images *Med. Phys.* **44** 914–23

[90] Sandhu H S, El-Baz A and Seddon J M 2018 Progress in automated deep learning for macular degeneration *JAMA Ophthalmol.* **136** 1366–7

[91] Ghazal M, Ali S S, Mahmoud A H, Shalaby A M and El-Baz A 2020 Accurate detection of non-proliferative diabetic retinopathy in optical coherence tomography images using convolutional neural networks *IEEE Access* **8** 34 387–97

[92] Reda I, Ghazal M, Shalaby A, Elmogy M, AbouEl-Fetouh A, Ayinde B O, AbouEl-Ghar M, Elmaghraby A, Keynton R and El-Baz A 2018 A novel ADCS-based CNN classification system for precise diagnosis of prostate cancer *2018 24th Int. Conf. on Pattern Recognition (ICPR)* (Piscataway, NJ: IEEE) pp 3923–8

[93] Reda I, Khalil A, Elmogy M, Abou El-Fetouh A, Shalaby A, Abou El-Ghar M, Elmaghraby A, Ghazal M and El-Baz A 2018 Deep learning role in early diagnosis of prostate cancer *Technol. Cancer Res. Treat.* **17** 1533034618775530

[94] Reda I, Ayinde B O, Elmogy M, Shalaby A, El-Melegy M, El-Ghar M A, El-fetouh A A, Ghazal M and El-Baz A 2018 A new CNN-based system for early diagnosis of prostate cancer *2018 IEEE 15th Int. Symp. on Biomedical Imaging (ISBI 2018)* (Piscataway, NJ: IEEE) pp 207–10

[95] Ayyad S M *et al* 2022 A new framework for precise identification of prostatic adenocarcinoma *Sensors* **22** 1848

[96] Hammouda K, Khalifa F, El-Melegy M, Ghazal M, Darwish H E, El-Ghar M A and El-Baz A 2021 A deep learning pipeline for grade groups classification using digitized prostate biopsy specimens *Sensors* **21** 6708

[97] Shehata M, Shalaby A, Switala A E, El-Baz M, Ghazal M, Fraiwan L, Khalil A, El-Ghar M A, Badawy M and Bakr A M *et al* 2020 A multimodal computer-aided diagnostic system for precise identification of renal allograft rejection: preliminary results *Med. Phys.* **47** 2427–40

[98] Shehata M, Khalifa F, Soliman A, Ghazal M, Taher F, Abou El-Ghar M, Dwyer A C, Gimel'farb G, Keynton R S and El-Baz A 2018 Computer-aided diagnostic system for early detection of acute renal transplant rejection using diffusion-weighted MRI *IEEE Trans. Biomed. Eng.* **66** 539–52

[99] Hollis E, Shehata M, Abou El-Ghar M, Ghazal M, El-Diasty T, Merchant M, Switala A E and El-Baz A 2017 Statistical analysis of adcs and clinical biomarkers in detecting acute renal transplant rejection *Br. J. Radiol.* **90** 20170125

[100] Shehata M, Alksas A, Abouelkheir R T, Elmahdy A, Shaffie A, Soliman A, Ghazal M, Abu Khalifeh H, Salim R and Abdel Razek A A K *et al* 2021 A comprehensive computer-assisted diagnosis system for early assessment of renal cancer tumors *Sensors* **21** 4928

[101] Khalifa F, Beache G M, El-Ghar M A, El-Diasty T, Gimel'farb G, Kong M and El-Baz A 2013 Dynamic contrast-enhanced MRI-based early detection of acute renal transplant rejection *IEEE Trans. Med. Imaging* **32** 1910–27

[102] Khalifa F, El-Ghar M A, Abdollahi B, Frieboes H, El-Diasty T and El-Baz A 2013 A comprehensive non-invasive framework for automated evaluation of acute renal transplant rejection using DCE-MRI *NMR Biomed.* **26** 1460–70

[103] Khalifa F, Elnakib A, Beache G M, Gimel'farb G, El-Ghar M A, Sokhadze G, Manning S, McClure P and El-Baz A 2011 3D kidney segmentation from CT images using a level set approach guided by a novel stochastic speed function *Proc. of Int. Conf. Medical Image Computing and Computer-Assisted Intervention, (MICCAI'11) (Toronto)* pp 587–94

[104] Shehata M, Khalifa F, Hollis E, Soliman A, Hosseini-Asl E, El-Ghar M A, El-Baz M, Dwyer A C, El-Baz A and Keynton R 2016 A new non-invasive approach for early classification of renal rejection types using diffusion-weighted MRI *IEEE Int. Conf. on Image Processing (ICIP), 2016* (Piscataway, NJ: IEEE) pp 136–40

[105] Khalifa F, Soliman A, Takieldeen A, Shehata M, Mostapha M, Shaffie A, Ouseph R, Elmaghraby A and El-Baz A 2016 Kidney segmentation from CT images using a 3D NMF-guided active contour model *IEEE 13th Int. Symp. on Biomedical Imaging (ISBI), 2016* (Piscataway, NJ: IEEE) pp 432–5

[106] Shehata M, Khalifa F, Soliman A, Takieldeen A, El-Ghar M A, Shaffie A, Dwyer A C, Ouseph R, El-Baz A and Keynton R 2016 3D diffusion MRI-based cad system for early diagnosis of acute renal rejection *2016 IEEE 13th Int. Symp. on Biomedical Imaging (ISBI)* (Piscataway, NJ: IEEE) pp 1177–80

[107] Shehata M, Khalifa F, Soliman A, Alrefai R, El-Ghar M A, Dwyer A C, Ouseph R and El-Baz A 2015 A level set-based framework for 3D kidney segmentation from diffusion mr images *IEEE Int. Conf. on Image Processing (ICIP), 2015* (Piscataway, NJ: IEEE) pp 4441–5

[108] Shehata M, Khalifa F, Soliman A, El-Ghar M A, Dwyer A C, Gimel'farb G, Keynton R and El-Baz A 2016 A promising non-invasive cad system for kidney function assessment *Int. Conf. on Medical Image Computing and Computer-Assisted Intervention* (Berlin: Springer) pp 613–21

[109] Khalifa F, Soliman A, Elmaghraby A, Gimel'farb G and El-Baz A 2017 3d kidney segmentation from abdominal images using spatial-appearance models *Computat. Math. Methods Med.* **2017** 1–10

[110] Hollis E, Shehata M, Khalifa F, El-Ghar M A, El-Diasty T and El-Baz A 2016 Towards non-invasive diagnostic techniques for early detection of acute renal transplant rejection: a review *Egypt. J. Radiol. Nucl. Med.* **48** 257–69

[111] Shehata M, Khalifa F, Soliman A, El-Ghar M A, Dwyer A C and El-Baz A 2017 Assessment of renal transplant using image and clinical-based biomarkers *Proc. of 13th Annual Scientific Meeting of American Society for Diagnostics and Interventional Nephrology (ASDIN'17) (New Orleans, LA)*

[112] 2016 Early assessment of acute renal rejection *Proc. of 12th Annual Scientific Meeting of American Society for Diagnostics and Interventional Nephrology (ASDIN'16) (Pheonix, AZ)* February 19–21, 2016

[113] Eltanboly A, Ghazal M, Hajjdiab H, Shalaby A, Switala A, Mahmoud A, Sahoo P, El-Azab M and El-Baz A 2019 Level sets-based image segmentation approach using statistical shape priors *Appl. Math. Comput.* **340** 164–79

[114] Shehata M, Mahmoud A, Soliman A, Khalifa F, Ghazal M, El-Ghar M A, El-Melegy M and El-Baz A 2018 3d kidney segmentation from abdominal diffusion MRI using an appearance-guided deformable boundary *PLoS One* **13** e0200082

[115] Abdeltawab H, Shehata M, Shalaby A, Khalifa F, Mahmoud A, El-Ghar M A, Dwyer A C, Ghazal M, Hajjdiab H and Keynton R *et al* 2019 A novel CNN-based cad system for early assessment of transplanted kidney dysfunction *Sci. Rep.* **9** 5948

[116] Hammouda K, Khalifa F, Abdeltawab H, Elnakib A, Giridharan G, Zhu M, Ng C, Dassanayaka S, Kong M and Darwish H *et al* 2020 A new framework for performing cardiac strain analysis from cine MRI imaging in mice *Sci. Rep.* **10** 1–15

[117] Abdeltawab H, Khalifa F, Hammouda K, Miller J M, Meki M M, Ou Q, El-Baz A and Mohamed T 2021 Artificial intelligence based framework to quantify the cardiomyocyte structural integrity in heart slices *Cardiovasc. Eng. Technol.* **13** 170–80

[118] Khalifa F, Beache G M, Elnakib A, Sliman H, Gimel'farb G, Welch K C and El-Baz A 2013 A new shape-based framework for the left ventricle wall segmentation from cardiac first-pass perfusion MRI *Proc. of IEEE Int. Symp. on Biomedical Imaging: From Nano to Macro, (ISBI'13) (San Francisco, CA)* 41–4

[119] 2012 A new nonrigid registration framework for improved visualization of transmural perfusion gradients on cardiac first–pass perfusion MRI *Proc. of IEEE Int. Symp. on Biomedical Imaging: From Nano to Macro, (ISBI'12) (Barcelona)* pp 828–31

[120] Khalifa F, Beache G M, Firjani A, Welch K C, Gimel'farb G and El-Baz A 2012 A new nonrigid registration approach for motion correction of cardiac first-pass perfusion MRI *Proc. of IEEE Int. Conf. on Image Processing, (ICIP'12) (Lake Buena Vista, FL)* 1665–8

[121] Khalifa F, Beache G M, Gimel'farb G and El-Baz A 2012 A novel CAD system for analyzing cardiac first-pass MR images *Proc. of IAPR Int. Conf. on Pattern Recognition (ICPR'12) (Tsukuba Science City, Japan)* pp 77–80

[122] A novel approach for accurate estimation of left ventricle global indexes from short-axis cine MRI *Proc. of IEEE Int. Conf. on Image Processing, (ICIP'11)(September 11–14, 2011) (Brussels)* pp 2645–9

[123] Khalifa F, Beache G M, Gimel'farb G, Giridharan G A and El-Baz A 2011 A new image-based framework for analyzing cine images *Handbook of Multi Modality State-of-the-Art*

Medical Image Segmentation and Registration Methodologies ed A El-Baz, U R Acharya, M Mirmedhdi and J S Suri (New York: Springer) ch 2 pp 69–98

[124] Accurate automatic analysis of cardiac cine images *IEEE Trans. Biomed. Eng* **59** 445–55 2012

[125] Khalifa F, Beache G M, Nitzken M, Gimel'farb G, Giridharan G A and El-Baz A 2011 Automatic analysis of left ventricle wall thickness using short-axis cine CMR images *Proc. of IEEE Int. Symp. on Biomedical Imaging: From Nano to Macro, (ISBI'11) (Chicago, IL)* pp 1306–9

[126] Nitzken M, Beache G, Elnakib A, Khalifa F, Gimel'farb G and El-Baz A 2012 Accurate modeling of tagged CMR 3D image appearance characteristics to improve cardiac cycle strain estimation *Image Processing (ICIP), 2012 19th IEEE Int. Conf. on (Orlando, FL)* pp 521–4

[127] Improving full-cardiac cycle strain estimation from tagged CMR by accurate modeling of 3D image appearance characteristics *Biomedical Imaging (ISBI), 2012 9th IEEE Int. Symp. on(May 2012) (Barcelona, Spain)* (IEEE) pp 462–5 (Selected for oral presentation)

[128] Nitzken M J, El-Baz A S and Beache G M 2012 Markov-gibbs random field model for improved full-cardiac cycle strain estimation from tagged CMR *J. Cardiovasc. Magn. Resonan.* **14** 1–2

[129] Sliman H, Elnakib A, Beache G, Elmaghraby A and El-Baz A 2014 Assessment of myocardial function from cine cardiac MRI using a novel 4D tracking approach *J. Comput. Sci. Syst. Biol.* **7** 169–73

[130] Sliman H, Elnakib A, Beache G M, Soliman A, Khalifa F, Gimel'farb G, Elmaghraby A and El-Baz A 2014 A novel 4D PDE-based approach for accurate assessment of myocardium function using cine cardiac magnetic resonance images *Proc. of IEEE Int. Conf. on Image Processing (ICIP'14) (Paris)* pp 3537–41

[131] Sliman H, Khalifa F, Elnakib A, Beache G M, Elmaghraby A and El-Baz A 2013 A new segmentation-based tracking framework for extracting the left ventricle cavity from cine cardiac MRI *Proceedings of IEEE Int. Conf. on Image Processing, (ICIP'13) (Melbourne)* pp 685–9

[132] Sliman H, Khalifa F, Elnakib A, Soliman A, Beache G M, Elmaghraby A, Gimel'farb G and El-Baz A 2013 Myocardial borders segmentation from cine MR images using bi-directional coupled parametric deformable models *Med. Phys.* **40** 1–13

[133] Sliman H, Khalifa F, Elnakib A, Soliman A, Beache G M, Gimel'farb G, Emam A, Elmaghraby A and El-Baz A 2013 Accurate segmentation framework for the left ventricle wall from cardiac cine MRI *Proc. of Int. Symp. on Computational Models for Life Science, (CMLS'13) (Sydney)* 287–96

[134] Abdollahi B, Civelek A C, Li X-F, Suri J and El-Baz A 2014 PET/CT nodule segmentation and diagnosis: a survey *Multi Detector* ed C T Imaging, L Saba and J S Suri (London: Taylor and Francis) ch 30 pp 639–51

[135] Abdollahi B, El-Baz A and Amini A A 2011 A multi-scale non-linear vessel enhancement technique *Engineering in Medicine and Biology Society, EMBC, 2011 Annual Int. Conf. of the IEEE* (Piscataway, NJ: IEEE) 3925–9

[136] Abdollahi B, Soliman A, Civelek A, Li X-F, Gimel'farb G and El-Baz A 2012 A novel gaussian scale space-based joint MGRF framework for precise lung segmentation *Proceedings of IEEE Int. Conf. on Image Processing, (ICIP'12)* (Piscataway, NJ: IEEE) pp 2029–32

[137] novel 3D joint MGRF framework for precise lung segmentation *Machine Learning in Medical Imaging* (Springer) 2012 pp 86–93

[138] Ali A M, El-Baz A S and Farag A A 2007 A novel framework for accurate lung segmentation using graph cuts *Proceedings of IEEE Int. Symp. on Biomedical Imaging: From Nano to Macro, (ISBI'07)* (Piscataway, NJ: IEEE) 908–11

[139] El-Baz A, Beache G M, Gimel'farb G, Suzuki K and Okada K 2013 Lung imaging data analysis *Int. J. Biomed. Imaging* **2013** 1–2

[140] El-Baz A, Beache G M, Gimel'farb G, Suzuki K, Okada K and Elnakib A 2013 Computer-aided diagnosis systems for lung cancer: challenges and methodologies *Int. J. Biomed. Imaging* **2013** 1–46

[141] El-Baz A, Elnakib A, Abou El-Ghar M, Gimel'farb G, Falk R and Farag A 2013 Automatic detection of 2D and 3D lung nodules in chest spiral CT scans *Int. J. Biomed. Imaging* **2013** 1–11

[142] El-Baz A, Farag A A, Falk R and La Rocca R 2003 A unified approach for detection, visualization, and identification of lung abnormalities in chest spiral CT scans *Int. Congress Series* **1256** (Amsterdam: Elsevier) pp 998–1004

[143] El-Baz A *et al* 2002 Detection, visualization and identification of lung abnormalities in chest spiral CT scan: phase-I *Proceedings of Int. Conf. on Biomedical Engineering (Cairo, Egypt) vol 12*

[144] El-Baz A, Farag A, Gimel'farb G, Falk R, El-Ghar M A and Eldiasty T 2006 A framework for automatic segmentation of lung nodules from low dose chest CT scans *Proceedings of Int. Conf. on Pattern Recognition, (ICPR'06)* 3 (Piscataway, NJ: IEEE) 611–4

[145] El-Baz A, Farag A, Gimel'farb G, Falk R and El-Ghar M A 2011 A novel level set-based computer-aided detection system for automatic detection of lung nodules in low dose chest computed tomography scans *Lung Imaging and Computer Aided Diagnosis* (London: Taylor and Francis) ch 10 221–38

[146] El-Baz A, Gimel'farb G, Abou El-Ghar M and Falk R 2012 Appearance-based diagnostic system for early assessment of malignant lung nodules *Proceedings of IEEE Int. Conf. on Image Processing, (ICIP'12)* (Piscataway, NJ: IEEE) 533–6

[147] El-Baz A, Gimel'farb G and Falk R 2011 A novel 3D framework for automatic lung segmentation from low dose CT images *Lung Imaging and Computer Aided Diagnosis* ed A El-Baz and J S Suri (London: Taylor and Francis) ch 1 pp 1–16

[148] El-Baz A, Gimel'farb G, Falk R and El-Ghar M 2010 Appearance analysis for diagnosing malignant lung nodules *Proceedings of IEEE Int. Symp. on Biomedical Imaging: From Nano to Macro (ISBI'10)* (Piscataway, NJ: IEEE) 193–6

[149] El-Baz A, Gimel'farb G, Falk R and El-Ghar M A 2011 A novel level set-based CAD system for automatic detection of lung nodules in low dose chest CT scans *Lung Imaging and Computer Aided Diagnosis* ed A El-Baz and J S Suri (London: Taylor and Francis) ch 1 pp 221–38

[150] El-Baz A, Gimel'farb G, Falk R and El-Ghar M A 2008 A new approach for automatic analysis of 3D low dose CT images for accurate monitoring the detected lung nodules *Proc. of Int. Conf. on Pattern Recognition, (ICPR'08)* (Piscataway, NJ: IEEE) pp 1–4

[151] El-Baz A, Gimel'farb G, Falk R and El-Ghar M A 2007 A novel approach for automatic follow-up of detected lung nodules *Proc. of IEEE Int. Conf. on Image Processing, (ICIP'07)* vol 5 (Piscataway, NJ: IEEE) pp V–501

[152] El-Baz A, Gimel'farb G, Falk R and El-Ghar M A 2007 A new CAD system for early diagnosis of detected lung nodules *Image Processing, 2007. ICIP 2007. IEEE Int. Conf. on* vol 2 (Piscataway, NJ: IEEE) pp II–461

[153] El-Baz A, Gimel'farb G, Falk R, El-Ghar M A and Refaie H 2008 Promising results for early diagnosis of lung cancer *Proc. of IEEE Int. Symp. on Biomedical Imaging: From Nano to Macro, (ISBI'08)* (Piscataway, NJ: IEEE) pp 1151–4

[154] El-Baz A, Gimel'farb G L, Falk R, Abou El-Ghar M, Holland T and Shaffer T 2008 A new stochastic framework for accurate lung segmentation *Proc. of Medical Image Computing and Computer-Assisted Intervention, (MICCAI'08)* pp 322–30

[155] El-Baz A, Gimel'farb G L, Falk R, Heredis D and Abou El-Ghar M 2008 A novel approach for accurate estimation of the growth rate of the detected lung nodules *Proc. of Int. Workshop on Pulmonary Image Analysis* pp 33–42

[156] El-Baz A, Gimel'farb G L, Falk R, Holland T and Shaffer T 2008 A framework for unsupervised segmentation of lung tissues from low dose computed tomography images *Proc. of British Machine Vision, (BMVC'08)* pp 1–10

[157] El-Baz A, Gimel'farb G, Falk R and El-Ghar M A 2011 3D MGRF-based appearance modeling for robust segmentation of pulmonary nodules in 3D LDCT chest images *Lung Imaging and Computer Aided Diagnosis* (London: Taylor and Francis) ch 3 51–63

[158] El-Baz A, Gimel'farb G, Falk R and El-Ghar M A 2009 Automatic analysis of 3D low dose CT images for early diagnosis of lung cancer *Pattern Recogn* **42** 1041–51

[159] El-Baz A, Gimel'farb G, Falk R, El-Ghar M A, Rainey S, Heredia D and Shaffer T 2009 Toward early diagnosis of lung cancer *Proc. of Medical Image Computing and Computer-Assisted Intervention, (MICCAI'09)* (Berlin: Springer) 682–9

[160] El-Baz A, Gimel'farb G, Falk R, El-Ghar M A and Suri J 2011 Appearance analysis for the early assessment of detected lung nodules *Lung Imaging and Computer Aided Diagnosis.* (London: Taylor and Francis) ch 17 395–404

[161] El-Baz A, Khalifa F, Elnakib A, Nitkzen M, Soliman A, McClure P, Gimel'farb G and El-Ghar M A 2012 A novel approach for global lung registration using 3D Markov Gibbs appearance model *Proc. of Int. Conf. Medical Image Computing and Computer-Assisted Intervention, (MICCAI'12) (Nice)* 114–21

[162] El-Baz A, Nitzken M, Elnakib A, Khalifa F, Gimel'farb G, Falk R and El-Ghar M A 2011 3D shape analysis for early diagnosis of malignant lung nodules *Proc. of Int. Conf. Medical Image Computing and Computer-Assisted Intervention, (MICCAI'11) (Toronto)* pp 175–82

[163] El-Baz A, Nitzken M, Gimel'farb G, Van Bogaert E, Falk R, El-Ghar M A and Suri J 2011 Three-dimensional shape analysis using spherical harmonics for early assessment of detected lung nodules *Lung Imaging and Computer Aided Diagnosis* (London: Taylor and Francis) ch 19 pp 421–38

[164] El-Baz A, Nitzken M, Khalifa F, Elnakib A, Gimel'farb G, Falk R and El-Ghar M A 2011 3D shape analysis for early diagnosis of malignant lung nodules *Proc. of Int. Conf. on Information Processing in Medical Imaging, (IPMI'11) (Monastery Irsee, Germany)* pp 772–83

[165] El-Baz A, Nitzken M, Vanbogaert E, Gimel'Farb G, Falk R and Abo El-Ghar M 2011 A novel shape-based diagnostic approach for early diagnosis of lung nodules *Biomedical Imaging: From Nano to Macro, 2011 IEEE Int. Symp. on* (Piscataway, NJ: IEEE) 137–40

[166] El-Baz A, Sethu P, Gimel'farb G, Khalifa F, Elnakib A, Falk R and El-Ghar M A 2011 Elastic phantoms generated by microfluidics technology: validation of an imaged-based

approach for accurate measurement of the growth rate of lung nodules *Biotechnol. J.* **6** 195–203

[167] 2010 A new validation approach for the growth rate measurement using elastic phantoms generated by state-of-the-art microfluidics technology *Proceedings of IEEE Int. Conf. on Image Processing (ICIP'10) (Hong Kong)* pp 4381–3

[168] El-Baz A, Sethu P, Gimel'farb G, Khalifa F, Elnakib A, Falk R and Suri M A E-G J 2011 Validation of a new imaged-based approach for the accurate estimating of the growth rate of detected lung nodules using real CT images and elastic phantoms generated by state-of-the-art microfluidics technology *Handbook of Lung Imaging and Computer Aided Diagnosis* ed A El-Baz and J S Suri (New York: Taylor and Francis) ch 1 pp 405–20

[169] El-Baz A, Soliman A, McClure P, Gimel'farb G, El-Ghar M A and Falk R 2012 Early assessment of malignant lung nodules based on the spatial analysis of detected lung nodules *Proceedings of IEEE Int. Symp. on Biomedical Imaging: From Nano to Macro, (ISBI'12)* (Piscataway, NJ: IEEE) pp 1463–6

[170] El-Baz A, Yuksel S E, Elshazly S and Farag A A 2005 Non-rigid registration techniques for automatic follow-up of lung nodules *Proceedings of Computer Assisted Radiology and Surgery, (CARS'05)* 1281 (Amsterdam: Elsevier) pp 1115–20

[171] El-Baz A S and Suri J S 2011 *Lung Imaging and Computer Aided Diagnosis* (Boca Raton, FL: CRC Press)

[172] Soliman A, Khalifa F, Dunlap N, Wang B, El-Ghar M and El-Baz A 2016 An iso-surfaces based local deformation handling framework of lung tissues *Biomedical Imaging (ISBI), 2016 IEEE 13th Int. Symp. on* (Piscataway, NJ: IEEE) pp 1253–9

[173] Soliman A, Khalifa F, Shaffie A, Dunlap N, Wang B, Elmaghraby A and El-Baz A 2016 Detection of lung injury using 4D-CT chest images *Biomedical Imaging (ISBI), 2016 IEEE 13th Int. Symp. on* (Piscataway, NJ: IEEE) pp 1274–7

[174] Soliman A, Khalifa F, Shaffie A, Dunlap N, Wang B, Elmaghraby A, Gimel'farb G, Ghazal M and El-Baz A 2017 A comprehensive framework for early assessment of lung injury *Image Processing (ICIP), 2017 IEEE Int. Conf. on* (Piscataway, NJ: IEEE) pp 3275–9

[175] Shaffie A, Soliman A, Ghazal M, Taher F, Dunlap N, Wang B, Elmaghraby A, Gimel'farb G and El-Baz A 2017 A new framework for incorporating appearance and shape features of lung nodules for precise diagnosis of lung cancer *Image Processing (ICIP), 2017 IEEE Int. Conf. on* (Piscataway, NJ: IEEE) pp 1372–6

[176] Soliman A, Khalifa F, Shaffie A, Liu N, Dunlap N, Wang B, Elmaghraby A, Gimel'farb G and El-Baz A 2016 Image-based CAD system for accurate identification of lung injury *Image Processing (ICIP), 2016 IEEE Int. Conf. on* (Piscataway, NJ: IEEE) pp 121–5

[177] Soliman A, Shaffie A, Ghazal M, Gimel'farb G, Keynton R and El-Baz A 2018 A novel CNN segmentation framework based on using new shape and appearance features *2018 25th IEEE Int. Conf. on Image Processing (ICIP)* (Piscataway, NJ: IEEE) pp 3488–92

[178] Shaffie A, Soliman A, Khalifeh H A, Ghazal M, Taher F, Keynton R, Elmaghraby A and El-Baz A 2018 On the integration of CT-derived features for accurate detection of lung cancer *2018 IEEE Int. Symp. on Signal Processing and Information Technology (ISSPIT)* (Piscataway, NJ: IEEE) pp 435–40

[179] Shaffie A, Soliman A, Khalifeh H A, Ghazal M, Taher F, Elmaghraby A, Keynton R and El-Baz A 2019 Radiomic-based framework for early diagnosis of lung cancer *2019 IEEE 16th Int. Symp. on Biomedical Imaging (ISBI 2019)* (Piscataway, NJ: IEEE) pp 1293–7

[180] Shaffie A, Soliman A, Ghazal M, Taher F, Dunlap N, Wang B, Van Berkel V, Gimelfarb G, Elmaghraby A and El-Baz A 2018 A novel autoencoder-based diagnostic system for early assessment of lung cancer *2018 25th IEEE Int. Conf. on Image Processing (ICIP)* (Piscataway, NJ: IEEE) pp 1393–7

[181] Shaffie A, Soliman A, Fraiwan L, Ghazal M, Taher F, Dunlap N, Wang B, van Berkel V, Keynton R and Elmaghraby A *et al* 2018 A generalized deep learning-based diagnostic system for early diagnosis of various types of pulmonary nodules *Technol. Cancer Res. Treat.* **17** 1533033818798800

[182] Abdel Razek A A K, Alksas A, Shehata M, AbdelKhalek A, Abdel Baky K, El-Baz A and Helmy E 2021 Clinical applications of artificial intelligence and radiomics in neuro-oncology imaging *Insights Imaging* **12** 152

[183] ElNakieb Y, Ali M T, Dekhil O, Khalefa M E, Soliman A, Shalaby A, Mahmoud A, Ghazal M, Hajjdiab H and Elmaghraby A *et al* 2018 Towards accurate personalized autism diagnosis using different imaging modalities: SMRI, FMRI, and DTI *2018 IEEE Int. Symp. on Signal Processing and Information Technology (ISSPIT)* (Piscataway, NJ: IEEE) pp 447–52

[184] ElNakieb Y, Soliman A, Mahmoud A, Dekhil O, Shalaby A, Ghazal M, Khalil A, Switala A, Keynton R S and Barnes G N *et al* 2019 Autism spectrum disorder diagnosis framework using diffusion tensor imaging *2019 IEEE Int. Conf. on Imaging Systems and Techniques (IST)* (Piscataway, NJ: IEEE) pp 1–5

[185] Haweel R, Dekhil O, Shalaby A, Mahmoud A, Ghazal M, Keynton R, Barnes G and El-Baz A 2019 A machine learning approach for grading autism severity levels using task-based functional MRI *2019 IEEE Int. Conf. on Imaging Systems and Techniques (IST)* (Piscataway, NJ: IEEE) pp 1–5

[186] Dekhil O, Ali M, Haweel R, Elnakib Y, Ghazal M, Hajjdiab H, Fraiwan L, Shalaby A, Soliman A and Mahmoud A *et al* 2020 A comprehensive framework for differentiating autism spectrum disorder from neurotypicals by fusing structural MRI and resting state functional MRI *Seminars in Pediatric Neurology* (Amsterdam: Elsevier) p 100805

[187] Haweel R, Dekhil O, Shalaby A, Mahmoud A, Ghazal M, Khalil A, Keynton R, Barnes G and El-Baz A 2020 A novel framework for grading autism severity using task-based FMRI *2020 IEEE 17th Int. Symp. on Biomedical Imaging (ISBI)* (Piscataway, NJ: IEEE) pp 1404–7

[188] El-Baz A, Elnakib A, Khalifa F, El-Ghar M A and McClure P 2012 Precise segmentation of 3-D magnetic resonance angiography *IEEE Trans. Biomed. Eng.* **59** 2019–29

[189] El-Baz A, Farag A, Elnakib A, Casanova M F, Gimel'farb G, Switala A E, Jordan D and Rainey S 2011 Accurate automated detection of autism related corpus callosum abnormalities *J. Med. Syst.* **35** 929–39

[190] El-Baz A, Gimel'farb G, Falk R, El-Ghar M A, Kumar V and Heredia D 2009 A novel 3D joint Markov-Gibbs model for extracting blood vessels from PC–MRA images *Medical Image Computing and Computer-Assisted Intervention–MICCAI 2009* 5762 (Berlin: Springer) pp 943–50

[191] Elnakib A, El-Baz A, Casanova M F, Gimel'farb G and Switala A E 2010 Image-based detection of corpus callosum variability for more accurate discrimination between dyslexic and normal brains *Proc. IEEE Int. Symp. on Biomedical Imaging: From Nano to Macro (ISBI'2010)* (Piscataway, NJ: IEEE) pp 109–12

[192] Elnakib A, Casanova M F, Gimel'farb G, Switala A E and El-Baz A 2011 Autism diagnostics by centerline-based shape analysis of the corpus callosum *Proc. IEEE Int. Symp. on Biomedical Imaging: From Nano to Macro (ISBI'2011)* (Piscataway, NJ: IEEE) pp 1843–6

[193] Elnakib A, Nitzken M, Casanova M, Park H, Gimel'farb G and El-Baz A 2012 Quantification of age-related brain cortex change using 3D shape analysis *Pattern Recognition (ICPR), 2012 21st Int. Conf. on* (Piscataway, NJ: IEEE) pp 41–4

[194] Nitzken M, Casanova M, Gimel'farb G, Elnakib A, Khalifa F, Switala A and El-Baz A 2011 3D shape analysis of the brain cortex with application to dyslexia *Image Processing (ICIP), 2011 18th IEEE Int. Conf. on (Brussels)* (Piscataway, NJ: IEEE) pp 2657–60 (Selected for oral presentation. Oral acceptance rate is 10 percent and the overall acceptance rate is 35 percent)

[195] El-Gamal F E-Z A, Elmogy M M, Ghazal M, Atwan A, Barnes G N, Casanova M F, Keynton R and El-Baz A S 2017 A novel CAD system for local and global early diagnosis of Alzheimer's disease based on PIB-PET scans *2017 IEEE Int. Conf. on Image Processing (ICIP)* (Piscataway, NJ: IEEE) pp 3270–4

[196] Ismail M M, Keynton R S, Mostapha M M, ElTanboly A H, Casanova M F, Gimel'farb G L and El-Baz A 2016 Studying autism spectrum disorder with structural and diffusion magnetic resonance imaging: a survey *Front. Human Neurosci.* **10** 211

[197] Alansary A, Ismail M, Soliman A, Khalifa F, Nitzken M, Elnakib A, Mostapha M, Black A, Stinebruner K and Casanova M F *et al* 2016 Infant brain extraction in T1-weighted MR images using BET and refinement using LCDG and MGRF models *IEEE J. Biomed. Health Inform.* **20** 925–35

[198] Asl E H, Ghazal M, Mahmoud A, Aslantas A, Shalaby A, Casanova M, Barnes G, Gimel'farb G, Keynton R and El-Baz A 2018 Alzheimer's disease diagnostics by a 3d deeply supervised adaptable convolutional network *Front. Biosci. (Landmark Ed.)* **23** 584–96

[199] Dekhil O *et al* 2019 A personalized autism diagnosis cad system using a fusion of structural MRI and resting-state functional MRI data *Front. Psych.* **10** 392

[200] Dekhil O, Shalaby A, Soliman A, Mahmoud A, Kong M, Barnes G, Elmaghraby A and El-Baz A 2021 Identifying brain areas correlated with ADOS raw scores by studying altered dynamic functional connectivity patterns *Med. Image Anal.* **68** 101899

[201] Elnakieb Y A, Ali M T, Soliman A, Mahmoud A H, Shalaby A M, Alghamdi N S, Ghazal M, Khalil A, Switala A and Keynton R S *et al* 2020 Computer aided autism diagnosis using diffusion tensor imaging *IEEE Access* **8** 191 298–1308

[202] Ali M T, Elnakieb Y A, Shalaby A, Mahmoud A, Switala A, Ghazal M, Khelifi A, Fraiwan L, Barnes G and El-Baz A 2021 Autism classification using SMRI: a recursive features selection based on sampling from multi-level high dimensional spaces *2021 IEEE 18th Int. Symp. on Biomedical Imaging (ISBI)* (Piscataway, NJ: IEEE) pp 267–70

[203] Ali M T, ElNakieb Y, Elnakib A, Shalaby A, Mahmoud A, Ghazal M, Yousaf J, Abu Khalifeh H, Casanova M and Barnes G *et al* 2022 The role of structure MRI in diagnosing autism *Diagnostics* **12** 165

[204] ElNakieb Y, Ali M T, Elnakib A, Shalaby A, Soliman A, Mahmoud A, Ghazal M, Barnes G N and El-Baz A 2021 The role of diffusion tensor MR imaging (DTI) of the brain in diagnosing autism spectrum disorder: promising results *Sensors* **21** 8171

[205] Mahmoud A, El-Barkouky A, Farag H, Graham J and Farag A 2013 A non-invasive method for measuring blood flow rate in superficial veins from a single thermal image *Proc. of the IEEE Conf. on Computer Vision and Pattern Recognition Workshops* pp 354–9

[206] Elsaid N, Saied A, Kandil H, Soliman A, Taher F, Hadi M, Giridharan G, Jennings R, Casanova M and Keynton R *et al* 2021 Impact of stress and hypertension on the cerebrovasculature *Front. Biosci.-Landmark* **26** 1643

[207] Taher F, Kandil H, Gebru Y, Mahmoud A, Shalaby A, El-Mashad S and El-Baz A 2021 A novel mra-based framework for segmenting the cerebrovascular system and correlating cerebral vascular changes to mean arterial pressure *Appl. Sci.* **11** 4022

[208] Kandil H, Soliman A, Taher F, Ghazal M, Khalil A, Giridharan G, Keynton R, Jennings J R and El-Baz A 2020 A novel computer-aided diagnosis system for the early detection of hypertension based on cerebrovascular alterations *NeuroImage: Clin.* **25** 102107

[209] Kandil H, Soliman A, Ghazal M, Mahmoud A, Shalaby A, Keynton R, Elmaghraby A, Giridharan G and El-Baz A 2019 A novel framework for early detection of hypertension using magnetic resonance angiography *Sci. Rep.* **9** 1–12

[210] Gebru Y, Giridharan G, Ghazal M, Mahmoud A, Shalaby A and El-Baz A 2018 Detection of cerebrovascular changes using magnetic resonance angiography *Cardiovascular Imaging and Image Analysis* (Boca Raton, FL: CRC Press) pp 1–22

[211] Mahmoud A, Shalaby A, Taher F, El-Baz M, Suri J S and El-Baz A 2018 Vascular tree segmentation from different image modalities *Cardiovascular Imaging and Image Analysis* (Boca Raton, FL: CRC Press) pp 43–70

[212] Taher F, Mahmoud A, Shalaby A and El-Baz A 2018 A review on the cerebrovascular segmentation methods *2018 IEEE Int. Symp. on Signal Processing and Information Technology (ISSPIT)* (Piscataway, NJ: IEEE) pp 359–64

[213] Kandil H, Soliman A, Fraiwan L, Shalaby A, Mahmoud A, ElTanboly A, Elmaghraby A, Giridharan G and El-Baz A 2018 A novel mra framework based on integrated global and local analysis for accurate segmentation of the cerebral vascular system *2018 IEEE 15th Int. Symp. on Biomedical Imaging (ISBI 2018)* (Piscataway, NJ: IEEE) pp 1365–8

[214] Taher F, Soliman A, Kandil H, Mahmoud A, Shalaby A, Gimel'farb G and El-Baz A 2020 Accurate segmentation of cerebrovasculature from TOF-MRA images using appearance descriptors *IEEE Access* **8** 96139–49

[215] Precise cerebrovascular segmentation *2020 IEEE Int. Conf. on Image Processing (ICIP)* (Piscataway, NJ: IEEE) 2020 pp 394–7

[216] Hammouda K, Khalifa F, Soliman A, Ghazal M, Abou El-Ghar M, Haddad A, Elmogy M, Darwish H, Khalil A and Elmaghraby A *et al* 2019 A CNN-based framework for bladder wall segmentation using MRI *2019 Fifth Int. Conf. on Advances in Biomedical Engineering (ICABME)* (Piscataway, NJ: IEEE) pp 1–4

[217] Hammouda K, Khalifa F, Soliman A, Ghazal M, Abou El-Ghar M, Haddad A, Elmogy M, Darwish H, Keynton R and El-Baz A 2019 A deep learning-based approach for accurate segmentation of bladder wall using MR images *2019 IEEE Int. Conf. on Imaging Systems and Techniques (IST)* (Piscataway, NJ: IEEE) pp 1–6

[218] Hammouda K, Khalifa F, Soliman A, Abdeltawab H, Ghazal M, Abou El-Ghar M, Haddad A, Darwish H E, Keynton R and El-Baz A 2020 A 3D CNN with a learnable adaptive shape prior for accurate segmentation of bladder wall using MR images *2020 IEEE 17th Int. Symp. on Biomedical Imaging (ISBI)* (Piscataway, NJ: IEEE) pp 935–8

[219] Hammouda K, Khalifa F, Soliman A, Ghazal M, Abou El-Ghar M, Badawy M, Darwish H, Khelifi A and El-Baz A 2021 A multiparametric MRI-based cad system for accurate diagnosis of bladder cancer staging *Comput. Med. Imaging Graph.* **90** 101911

[220] Hammouda K, Khalifa F, Soliman A, Ghazal M, Abou El-Ghar M, Badawy M, Darwish H and El-Baz A 2021 A cad system for accurate diagnosis of bladder cancer staging using a multiparametric MRI *2021 IEEE 18th Int. Symp. on Biomedical Imaging (ISBI)* (Piscataway, NJ: IEEE) pp 1718–21

[221] Alksas A, Shehata M, Saleh G A, Shaffie A, Soliman A, Ghazal M, Khalifeh H A, Razek A A and El-Baz A 2021 A novel computer-aided diagnostic system for early assessment of hepatocellular carcinoma *2020 25th Int. Conf. on Pattern Recognition (ICPR)* (Piscataway, NJ: IEEE) pp 10 375–82

[222] Alksas A, Shehata M, Saleh G A, Shaffie A, Soliman A, Ghazal M, Khelifi A, Khalifeh H A, Razek A A and Giridharan G A *et al* 2021 A novel computer-aided diagnostic system for accurate detection and grading of liver tumors *Sci. Rep.* **11** 1–18

[223] Razek A A K A, Khaled R, Helmy E, Naglah A, AbdelKhalek A and El-Baz A 2022 Artificial intelligence and deep learning of head and neck cancer *Magn. Resonan. Imaging Clin.* **30** 81–94

[224] Naglah A, Khalifa F, Khaled R, Abdel Razek A A K, Ghazal M, Giridharan G and El-Baz A 2021 Novel MRI-based cad system for early detection of thyroid cancer using multi-input CNN *Sensors* **21** 3878

[225] Naglah A, Khalifa F, Mahmoud A, Ghazal M, Jones P, Murray T, Elmaghraby A S and El-Baz A 2018 Athlete-customized injury prediction using training load statistical records and machine learning *2018 IEEE Int. Symp. on Signal Processing and Information Technology (ISSPIT)* (Piscataway, NJ: IEEE) pp 459–64

[226] Mahmoud A H 2014 Utilizing radiation for smart robotic applications using visible, thermal, and polarization images *PhD Dissertation* University of Louisville

[227] Mahmoud A, El-Barkouky A, Graham J and Farag A 2014 Pedestrian detection using mixed partial derivative based his togram of oriented gradients *2014 IEEE Int. Conf. on Image Processing (ICIP)* (Piscataway, NJ: IEEE) pp 2334–7

[228] El-Barkouky A, Mahmoud A, Graham J and Farag A 2013 An interactive educational drawing system using a humanoid robot and light polarization *2013 IEEE Int. Conf. on Image Processing* (Piscataway, NJ: IEEE) pp 3407–11

[229] Mahmoud A H, El-Melegy M T and Farag A A 2012 Direct method for shape recovery from polarization and shading *2012 19th IEEE Int. Conf. on Image Processing* (Piscataway, NJ: IEEE) 1769–72

[230] Ghazal M A, Mahmoud A, Aslantas A, Soliman A, Shalaby A, Benediktsson J A and El-Baz A 2019 Vegetation cover estimation using convolutional neural networks *IEEE Access* **7** 132 563–176

[231] Ghazal M, Mahmoud A, Shalaby A and El-Baz A 2019 Automated framework for accurate segmentation of leaf images for plant health assessment *Environ. Monit. Assess.* **191** 491

[232] Ghazal M, Mahmoud A, Shalaby A, Shaker S, Khelifi A and El-Baz A 2020 Precise statistical approach for leaf segmentation *2020 IEEE Int. Conf. on Image Processing (ICIP)* (Piscataway, NJ: IEEE) 2985–9

IOP Publishing

Photo Acoustic and Optical Coherence Tomography Imaging, Volume 1

Diabetic retinopathy

Ayman El-Baz and Jasjit S Suri

Chapter 14

Optical coherence tomography in diabetic retinopathy: a review in application of artificial intelligence

Meysam Tavakoli

Diabetes mellitus can lead to significant microvasculature disruptions that eventually causes diabetic retinopathy (DR), or complications in the eye due to diabetes. Here, without proper screening, this disease progresses over time and eventually causes complete vision loss. One of the general ways to detect such development is using retinal optical coherence tomography (OCT) systems. In fact, OCT images provide fundamental information regarding the health of the eye. Therefore, the establishment of automatic image analysis approaches is important to provide clinicians with quantitative data that simplifies decision making. In recent years, different deep learning (DL) approaches have been presented to work on the automated image analyses, to improve performance over conventional methods, and increase repeatability. DL-based methods represent new algorithms that improve the outcomes. Results published to date show that DL networks generally achieve higher performance than conventional methods or early machine learning algorithms. Consequently, DL-based approaches for OCT image analysis have provided a fundamental advance in the area of DR detection in images from healthy eyes as well as segmentation of a variety of pathologies. This chapter provides a comprehensive review literature of current DL-based methods used in OCT images for those who are dealing with diabetes.

14.1 Introduction to optical coherence tomography

Diabetic retinopathy (DR) is a common complication of retinal vessels because of diabetes that causes progressive vision loss and blindness [1]. From a statistics

doi:10.1088/978-0-7503-2052-8ch14

viewpoint, it is estimated that the number of diabetic people will increase to 642 million by 2040 [2] and among these people the prevalence of DR ranges reportedly from 30% to 45% in which 1 in 10 suffers from vision complication because of DR [3]. Therefore, diagnosis of DR based on clinical fundus examination is a critical task. This procedure is broadly divided into two different classes: non-proliferative diabetic retinopathy (NPDR), with early signs of DR, and proliferative diabetic retinopathy (PDR), which is associated with the progress of neovascularization (NV) [3, 4].

The important challenge in diagnosing DR before complications become apparent is that diabetic people are not aware of the disease until the retinal changes have progressed to a certain level which treatment tends to be less effective. Therefore, using early DR detection systems, which could help ophthalmologists to investigate and treat the disease and make it less severe, is the more efficiently main goal of these types of studies [5, 6]. For this purpose, recently we faced a growing number of imaging modalities that can be applied in the screening, evaluating, diagnosing, and treating of DR. Here, there is an increasing request for high-resolution imaging in retinal area in the diagnosis and management of retinal diseases. In particular, the identification of disorders has been substantially improved by the introduction of optical coherence tomography (OCT). With technological advances, OCT may serve as a potential rapid, non-invasive image modality as an adjunct for evaluating microvascular changes in capillary level [7, 8]. The OCT checks living tissue by means of high-resolution tomographic cross-sections of the retina. In fact, its measurements are analogous to ultrasound mode scanning [9]. Therefore, OCT has this capability to provide important information complementary for clinical checks and certain detection of DR.

In the same direction, since accurate diagnosis is mandatory for successful treatment, the DR identification can be notably improved by establishing computer-aided diagnostic (CAD) systems based on OCT. Here, the OCT is working by capturing repeated scans at the same location to detect any variations in reflected signal from the movement of red blood cells through blood vessels in the volumetric scans [4]. In fact, without intravenous dye injection, OCT allows depth-resolved visualization of the retinal microvasculature by selecting different enface slabs from different retinal layers. Microvascular changes in the superficial, intermediate, and deep capillary plexuses can be assessed separately. The OCT works based on the analysis of the reflections of low coherence light from the retina. Its resolution with the current clinically used system is 10 μm [9]. It allows images to be captured for the retina, retinal pigment epithelial and different retinal layers (figure 14.1). It has the potential to detect most morphological changes both quantitatively and qualitatively.

14.1.1 Advantages of the optical coherence tomography

In OCT, the red blood cell mobility over time is mapped using volumetric images. The scan is repeated at each position to find the motion contrast. The only predicted motion in the retina is blood flow in arteries, therefore the degree of motion contrast

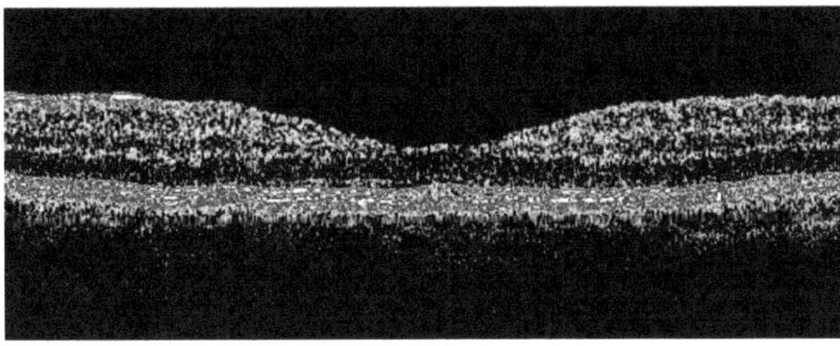

Figure 14.1. Normal OCT of a macula. Different retinal layers and structures for a normal macula.

matches flow [8]. It is non-invasive, and without the need for dye injection, OCT techniques enable the imaging and thorough assessment of alterations in the retinal microvasculature. This is crucial because OCT can be used more frequently to manage patients' eyes longitudinally than conventional retinal imaging. Furthermore, OCT scanning data is three dimensional (3D) and depth resolved. One may see and analyze individual capillary plexuses. The user can further adjust and customize the segmentation of the retinal vasculature to obtain images of other layers such as the intermediate capillary plexuses, which helps to visualize pathological features that are not available in conventional dye-based angiography. An improved software algorithm automatically generates the images of the superficial and deep capillary plexuses [4]. Moreover, better intraretinal layer segmentation with 3D OCT may be achieved by denser scanning and better resolution (2 μm in some systems). Another obvious advantage of 3D OCT images is the ability to virtually slice the data along any plane relative to the direction in which the scan is taken. In short, recent technological advances have enabled the acquisition of 3D OCT datasets. However, in spite of volumetric data improving disease diagnosis and follow-up [10], new image analysis techniques are currently needed to process dense 3D OCT datasets. Fundamental software improvements include approaches for subject segmentation of structures or volumes of interest, post-extraction of relevant data, and signal averaging to improve representation of retinal layers. Plus, an innovative way to display images can improve OCT image interpretation of pathological structures. All these methods have been developed, but most remain immature. For more details regarding the OCT advantages and advances we recommend readers to see [11].

14.1.2 Clinical signs of diabetic retinopathy in the optical coherence tomography

Before we start DR diagnosis, an important question is *what are the revealing signs of DR?* Several morphological signs of DR can be identified by OCT, including microaneurysms (MAs), different abnormal microvascular, and NV which are briefly explained below.

(1) *Microaneurysms (MAs)*: MAs are the first visible signs of DR. They appear as small and round shaped dots near tiny retinal blood vessels in retinal

images [12]. The size of MAs usually ranges from 10 to 125 μm in diameter [13]. Previous works demonstrated that increased number of MAs and their turnover are associated with a higher risk for progression of DR [14–16]. By using OCT, Thompson *et al* demonstrated that OCT could identify MAs even not shown on a dilated clinical planform [17]. Moreover, Schwartz *et al* and Ishibazawa *et al* showed that OCTA could detect MAs that are not detectable on fluorescein angiography (FA) [18, 19]. In a separate work using normal FA retinal images, Tavakoli and associates showed that the MAs were identified with more accuracy by FA images [18]. However, a number of researchers have shown that not all MAs detected by FA could be visualized with OCT, which may be influenced by blood flow turbulence within the MAs [20–24].

(2) *Macular edema*: Macular edema is a common cause of vision loss and blindness in patients with DR and can occur in any stage of the disease (see figure 14.2). Using OCT it is possible to identify, quantify and classify diabetic macular edema. It is recognized as the swelling of the macula, an area at the center of the retina. It is caused by permeability of abnormal retinal capillaries because of the fluid leakage or solutes around the macula [25, 26].

Eventually, it affects the central vision [27].

(3) *Cotton Wool (CW)*: Also called soft exudates, CW is ischemic infarctions of the nerve fiber layer. On normal retinal imaging, they appear white and located superficially. When we do OCT, they are hyperreflective, and elongated lesions, which can cause a shadow shape on the posterior layers (figure 14.3). They occur due to occlusion, or blockage of blood supply, of

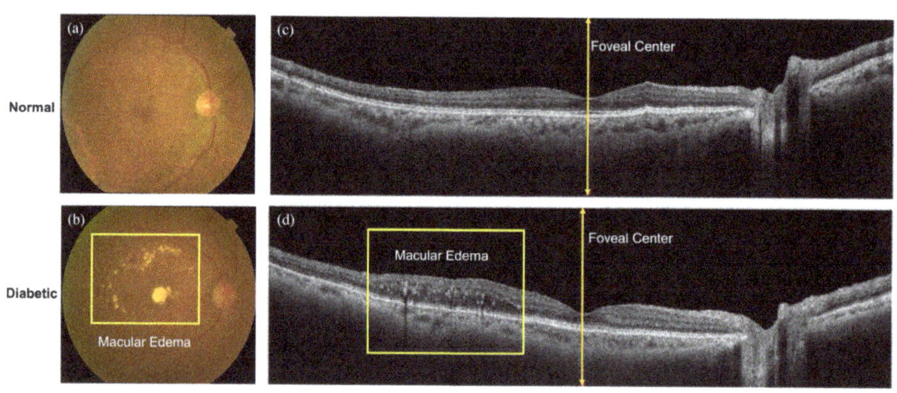

Figure 14.2. Images from a local database, MUMS-DB, without any chronic systemic diseases or eye diseases. (a) Normal color retinal image. (b) Diabetic image with different features including MAs, HEs and exudates. (c) Wide-field OCT of a horizontal scan through the center of the fovea (red line) revealing a normal retina. Images of the right eye of a 45-year-old man with signs of moderate non-proliferative diabetic retinopathy. (d) Widefield OCT of a horizontal scan through the center of the fovea reveals marked thickening of the retina at the temporal quadrant of the retina.

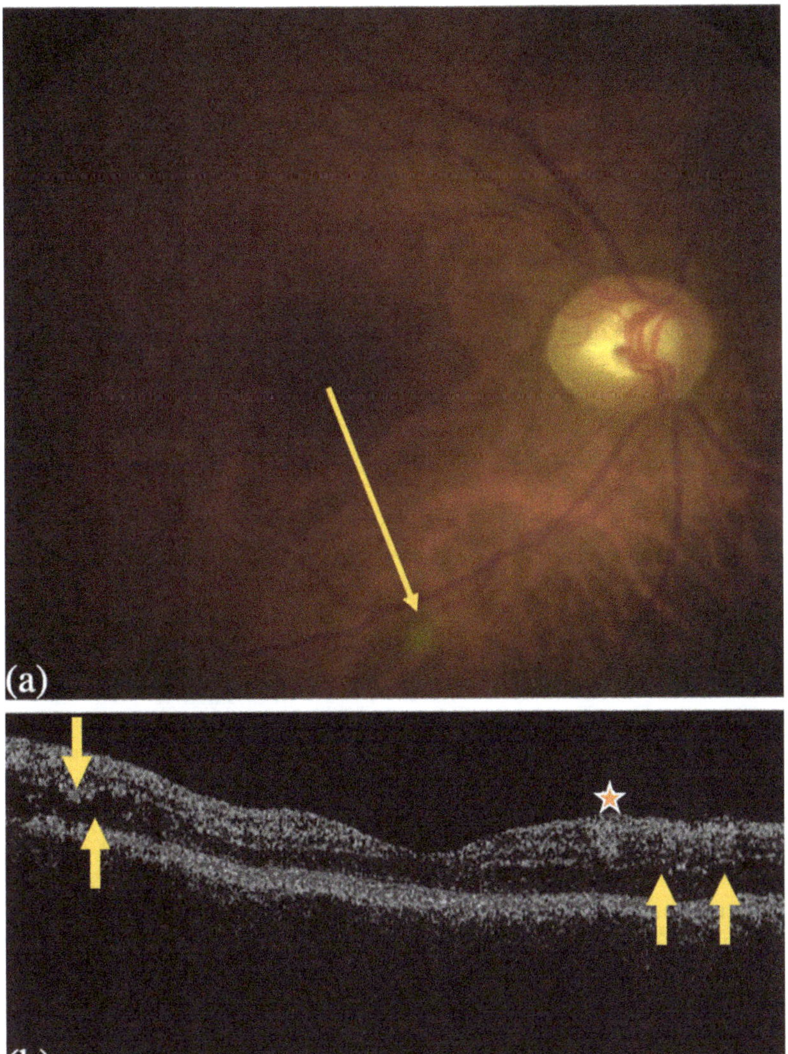

Figure 14.3. Proliferative diabetic retinopathy case with OCT structures. (a) Original image with different signs of DR such as neovascularization, hemorrhage, and clinically significant macular edema. (b) OCT of macular edema. (c) Combination of neovascularization, and hemorrhage at the superior vascular arcade shadowing the posterior layers. (d) Epipapillar neovascularization in the optic disk in the posterior layers. (e) High hemorrhage completely shadowing the posterior layers.

arterioles and cause disruption and damage to areas of tissue [28]. The reduced blood flow to the retina results in ischemia of the Retinal Nerve Fiber Layer (RNFL) which changes the axoplasmic flow and causes accumulation of axoplasmic debris in the retinal ganglion cell axons. The debris accumulation arises as fleecy white lesions in the RNFL called soft exudates or CW [28, 29].

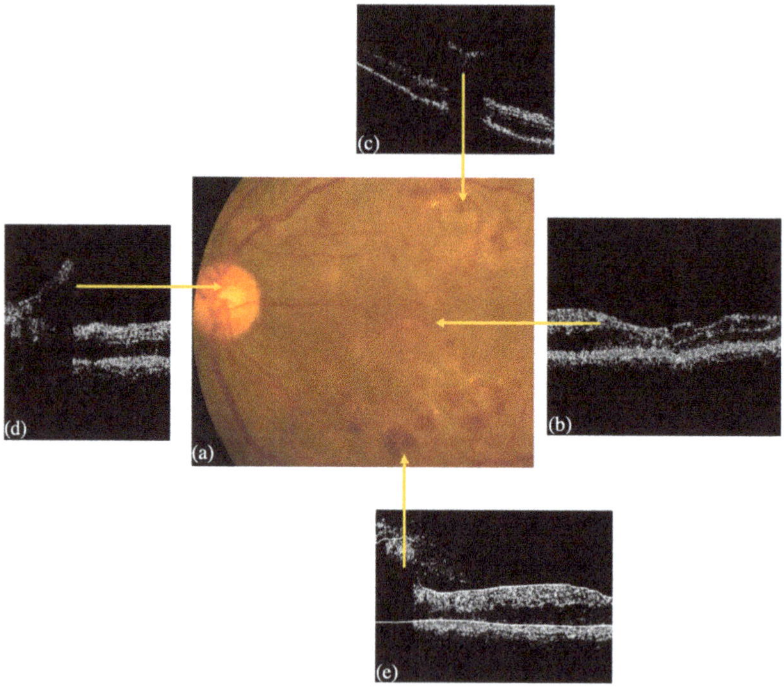

Figure 14.4. Proliferative diabetic retinopathy case with OCT structures. (a) Original image with different signs of DR such as neovascularization, hemorrhage, and clinically significant macular edema. (b) OCT of macular edema. (c) Combination of neovascularization, and hemorrhage at the superior vascular arcade shadowing the posterior layers. (d) Epipapillar neovascularization in the optic disk in the posterior layers. (e) High hemorrhage completely shadowing the posterior layers.

(4) Haemorrhages (HEs): A more severe level of DR is caused by retinal HEs or leakage of weak capillaries and ruptured MAs into the surface of the retina [30]. HEs are located pre-, intra- or subretinal in the retina. On retinal images, they have irregular shapes, such as flame shaped, when located in the nerve fiber layer, and they are rounded when placed in the deep retinal layers. They appear hyperreflective in OCT and can produce a cone-shaped shadow on the posterior layer, especially if they are located pre-retinally (figure 14.4). They are defined as a red spot with different shapes, such as 'dot', 'blot' and "flame' [30, 31] and irregular margin and uneven density [32].

(5) *Hard Exudates (EXs)*: When the lipoproteins and other proteins are leaking through abnormal retinal vessels because of the degradation or breakdown of the blood-retina barrier, hard EXs appear [33]. They become visible as small white or yellowish-white deposits with sharp borders. They are often arranged in bulks or circular rings and located in the outer layer of the retina [32].

(6) *Neovascularizations (NVs)*: The advanced stage of DR is Proliferative DR and it is characterized by the development of abnormal blood vessels—neovascularization [34]. Primarily the lack of oxygen induces the growth of

new blood vessels on the inner surface of the retina. The new blood vessels start to grow on the inner surface of the retina because of primarily the lack of oxygen. These new blood vessels push into the surrounding retinal space and are feeble and very often bleed into retinal cavity, effecting the eyesight [35, 36]. Early detection of NV is critical to facilitate early intervention. If NV results treated late it may causes substantially deteriorate patients' vision. Here, NVs are detectable using OCT [37]. Recent works showed that OCT can identify early NVs and identify the starts and morphological patterns of NVs in PDR, hence allowing classification of the lesion, offering a better understanding of the pathophysiology and helps to guide the management strategies [24, 37]. Khalid *et al* assessed the utility of wide-field OCT compared with clinical examination in diagnosing PDR [38, 39] and found that OCT has a higher detection rate of PDR than clinical tests, suggesting that this method could be applied for early detection and characterization of NV.

Figures 14.3 and 14.4 show typical retinal images of different signs and stages of DR.

14.1.3 Related conventional works

According to the presence of the aforementioned clinical signs, and the severeness/ density of these signs [40] DR is categorized into four types: mild, moderate, and severe NPDR, and PDR [32, 41, 42]. The NPDR type is introduced in terms of number of MAs and HEs as normal, mild, moderate, and sever DR. This type of DR is further divided into following levels of DR [43].

In PDR, we see NVs, which is the formation of abnormal new blood vessels. It is the severe stage of DR, in which these new abnormal vessel growths, caused by poor circulation especially from formations of MAs, protrude into the retina and even into the vitreous, the gel-like medium that gives the eye its spherical shape. Thus, both MAs and NVs are two important clinical lesions [44] and fluid in DR is classified as EXs and non-EXs [45]. MAs, in general, are a very good indication of possibly worsening conditions of DR. From a statistics point of view, DR with MAs has a 6.2% possibility to expand into PDR within a year [46]. An increased number of MAs is an important early characteristic of DR progression. Moreover, with development of ischemia, there is an increase in the possibility of PDR progress within a year. This one-year risk development increases from 11.3% to 54.8% from lower stage to advanced stage [46, 47]. In the NVs stage, patients have a 25.6% to 36.9% possibility of vision loss and blindness, if not treated properly. Furthermore, PDR eyes not treated after more than 2 years have a possibility of vision loss and blindness at 7.0% and if it is not treated for more than 4 years, the possibility of vision loss goes up to 20.9%. On the other hand, this vision loss and blindness decreases to 3.2% within 2 years of treatment and 7.4% within 4 years of treatment [47]. Diabetic people with mild DR do not need any specific treatment other than managing their DM and the related risk factors such as hypertension, anaemia, and renal malfunction. They need to be monitored closely, else it may progress to higher stages of DR [48]. In the severe and advanced stage of DR, like NV, treatment is limited [49].

For early detection of DR many of CAD systems use retinal images of the patient. Manual detection and diagnosis is an exhaustive task in both cost and lack of experts [50] when screening DR images. It would be more cost effective and helpful if the initial task of analyzing the retinal images can be automated. Automated approaches address these issues by decreasing the time, cost, and effort significantly [51]. Moreover, since image processing techniques are growing in all areas of medical science, by assisting them, especially in advanced ophthalmology, we can do automated screening [45]. For this reason, automated DR detection and classification utilizes CAD systems [12, 50]. These computer-based systems have the ability to detect any change in normal and abnormal images and systematize these changes to form a feature space. At the end, the combination of these features introduces type and stage of DR. Many CAD systems have been presented in the state of the art for early detection of DR and related lesions [48, 52–59].

Normally, each CAD system is the sequence of two algorithms, i.e., feature extraction and classification[43]. Since there are many different approaches in CAD systems, it is also necessary to evaluate their robustness and accuracy. There are many different standard methods used in assessing the effectiveness of the CAD systems [60, 61]. A common one is using Receiver Operating Characteristic (ROC) analysis and area under the curve (AUC) analysis. AUC of ROC curve is a performance measurement for an automated method, for example, a classification problem, at different threshold values. ROC is a probability curve and AUC shows degree or rate of separability. In fact, AUC tells us how much the method is able to distinguish between classes. The higher the AUC, better the method is in its prediction. By analogy, the higher the AUC, the better the method is at distinguishing between images with DR and no DR. ROC curves illustrate the tradeoff between sensitivity and specificity for a range of thresholds and enable the identification of an optimal value [62, 63]. Hence, assessment using the ROC curve is a way to evaluate the model, independent of the choice of a threshold. Here, the analytical definition when using ROC is evaluated in terms of the true positive fraction (TPF), given by sensitivity, and the false positive fraction (FPF), given by (1- specificity) [64]. Also the accuracy is determined as a measurement providing the ratio of well-classified pixels. The results for the automated method as compared to the ground truth or gold standard are calculated for each image. These metrics are defined as:

$$\begin{aligned} \text{Sensitivity(Se)} &= \frac{\text{TP}}{\text{TP} + \text{FN}} \\ \text{Specificity(Sp)} &= \frac{\text{TN}}{\text{TN} + \text{FP}} \\ \text{Accuracy(Acc)} &= \frac{\text{TP} + \text{TN}}{\text{TP} + \text{FN} + \text{TN} + \text{FP}} \end{aligned} \tag{14.1}$$

where TP is true positive, TN is true negative, FP is false positive and FN is false negative, the same as in [65, 66]. A look into the other published results has shown they used both ROC and AUC (equation (14.1)) to report the performance of CAD systems in DR screening.

There are also some authors who have reviewed multiple CAD methods and their applications [45, 46, 67–69] for DR screening. Briefly, Tavakoli *et al* [12] introduced a systematic DR screening system based on the efficacy of both manual and automated grading systems in comparison to the gold standard. Their automated approach detected DR with an accuracy of more than 90%. Abramoff *et al* [70] also presented an automated DR detection system that acquired an AUC of 0.84 of DR detection. Niemeijer *et al* [71] introduced a DR screening system using an information fusion approach with an AUC of 0.88. Quellec *et al* [53] established an automated DR screening system for MAs detection, and age-related macular degeneration and their method reached to an AUC of 0.93. Fleming *et al* [72] introduced a CAD system to detect blot HEs and provided a sensitivity of 98.60% and a specificity of 95.50%. Fleming *et al* [73] also developed another CAD system to detect MAs, which obtained a sensitivity of 85.4% and specificity of 83.1%. Perumalsamy *et al* [74] developed a CAD system and achieved an accuracy of 81.3% by comparing the performance with ophthalmologists' ideas. Patton *et al* [45] discussed image preprocessing methods, registration, landmarks and lesions segmentation, retinal topography measurements and its applications in telemedicine. Mookiah *et al* [46] reviewed the methods to locate and segment the retinal image features such as ONH, fovea, macula, vessels, hard and soft EXs, MAs, and HEs comprehensively.

From a classification viewpoint, color characteristics were utilized on a Bayesian classifier to categorize each pixel into lesion or non-lesion classes [75]. In this study, authors obtained 100% accuracy in detection of all EXs in the retinal images, and accuracy of 70% in classification of normal retinal images. By using image processing techniques and a multilayer perceptron Neural Network (NN), DR and normal retina were classified with a sensitivity of 80.2% and a specificity of 70.7% [31]. Automated detection of NPDR, based on MAs, HEs, and EXs was studied in [76]. The approach was able to correctly detect the NPDR stage with an accuracy of 81.7%. For DR screening again above lesions were used [77] and obtained the sensitivity and specificity of 74.8% and 82.7%, respectively. Different stages of DR were classified using both area and geometry of the combination of the blood vessels with a feedforward NN [78]. The average classification efficiency for this method was 84%, with sensitivity and specificity of 90% and 100%, respectively. Nayak *et al* used EXs and vessel area along with texture parameters coupled with NNs to classify NPDR, PDR, and normal [79]. They got an accuracy of 93%, sensitivity of 99%, and specificity of 100%. The feature of support vector machine (SVM) classifiers was also used by Acharya *et al* [80] to classify the retinal image into normal, mild, moderate, severe and PDR categories. They showed an average accuracy of 82%, and a sensitivity and specificity of 82% and 88%, respectively. Nicolai *et al* established an automated lesion detection system for DR screening purposes, which detected 90.1% of patients with DR and 81.3% without DR [81]. Their system demonstrated a sensitivity of 93.1% and a specificity of 71.6%.

In related OCT studies, Tan *et al* [82] automatically extracted retinal vessels to obtain the vascular network. Knowing their shapes and connections and salient points helps users to edit segmentation results. Retinal imaging is more common in current CAD systems with DR detection because it uses the same concept as traditional indirect ophthalmoscopy to form a wide field of view of the retina to adequately delineate systemic disease. However, one of the main issues is that it only captures 2D images without considering depth. Unlike this type of imaging, which is expensive because it requires highly specialized technicians and can only be assessed qualitatively, OCT is inexpensive, supports quantitative measurements, and can assess changes without human bias. The ability to detect subtle changes in retinal thickness as diabetes progresses makes OCT a powerful tool in the diagnosis of diabetic macular edema. Early detection of DR lesions can be based on OCT biomarkers such as retinal volume and total thickness and microaneurysms. Retinal thinning and decreased light reflectance are important biomarkers for detecting DR changes with OCT.

Semi-automated extraction of 9 slices from an OCT image by Yazdanpanah *et al* [83] applied energy-minimizing active contours from Chan and Vese without edge models and shape priority. However, the segmentation must be initialized manually by the user and has not been tested on normal or diseased human retinas. Kafieh [84] worked on graph-based diffusion maps to segment intraretinal slices in the OCT scans of normal controls and patients, while Ehnes [85] established graph-based segmentations of up to 11 slices, which only worked for high contrast images. Automatic choroidal segmentation by Chen *et al* takes into account the characteristics of the choroidal vessels in his OCT using enhanced depth imaging (EDI-OCT), especially the large vessels in Haller's layer adjacent to the choroidscleral junction. Satisfactory, i.e., good visibility and easy recognition of choroidal vessels, yielded results similar to manual results in normal eyes and pathologies (DR and age-related). Nam *et al* [86] used high-resolution intravascular OCT to see arterial wall microstructure, and by combining OCT intensity images with first and second order axial derivatives, luminal contours and luminal contours in fundus images.

It should be worth mentioning that most known approaches for segmentation of OCT layers are accurate enough only for images with high signal-to-noise ratio (SNR), but fail for low SNR. It is tested on normal but not disease cases and is only stratified up to eight layers. Only in high-contrast images of the normal retina can more layers be segmented. Therefore, to be robust and successful, segmentation either focuses on a specific retinal portion or requires high-resolution images [86, 87]. For age-related macular degeneration (AMD), it is better to extract the retinal pigment epithelium (RPE) and the interdigital epithelium (IC) as two separate layers for diagnosis.

14.2 Introduction to artificial intelligence

The term 'Artificial Intelligence' (AI) was first introduced in 1956 with the aid of using John McCarthy and become described because the simulation of human

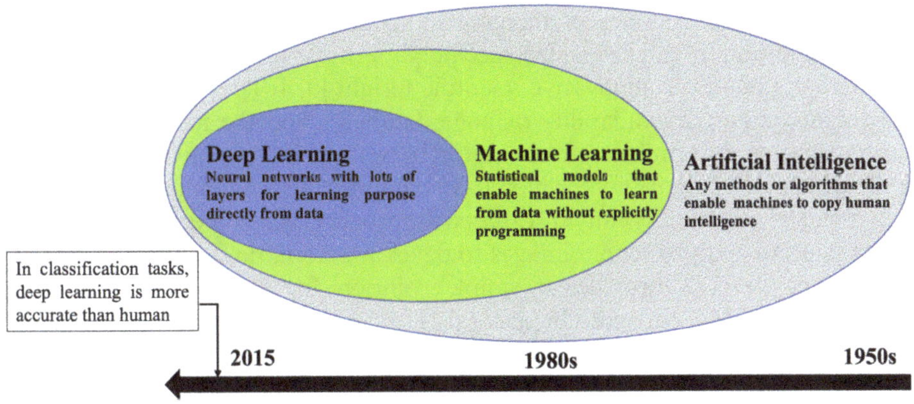

Figure 14.5. AI, ML, and DL. Relation between artificial intelligence, machine learning and deep learning.

intelligence with the aid of using machines [88]. Machine learning (ML) is a subset of AI that teaches a PC device to carry out an undertaking or expecting final results without explicitly programmed [89]. ML techniques need a set of characteristics to be measured directly from the training data (e.g., labeled lesions in retinal image). After that, based on a training database of characteristics with known labels, i.e. handcrafted features, a classifier learns to identify the correct label from the newly seen characteristics. Once a few strong classifiers have been established, the effectiveness of such ML models mostly relies on the differentiative power of the chosen characteristics which underpin the classifier performance. There are different reviews that have summarized ML approaches [46, 56, 90–94].

Deep learning (DL), a subfield of ML (figure 14.5), is on the present day area and is growing swiftly because of the advances in computation and massive records in current years. Particularly, the advent of convolutional neural networks (CNNs) delivered a sizable step forward with inside the improvement of DL for picture class and sample recognition [95, 96]. Many DL-based approaches have been established for variety of works to analyze images to establish automated CAD systems for detection of DR [97, 98]. The relation between AL, ML, and DL has been shown in the figure below [9].

In general, understanding and controlling DR or other retinal diseases has become extensively more complicated because of the large number of images and diagnoses. In fact, processing all these patient examinations seems to be a 'big data' challenge [99]. Clearly, the new era of diagnosis and clinical data mining immediately needs intelligent systems to control them sufficiently, safely and efficiently.

Recently, through the advancement of supervised learning ideas, the recent application of AI approaches for detection of DR and its related lesions have significantly increased in the research community. AI has already revealed proof-of-concept in medical science such as radiology and pathology, which like ophthalmology relies heavily on diagnostic imaging. As it's turning out, image processing is the most important application of AI in healthcare [100]. Moreover, AI is particularly suitable in assisting clinical tasks by using efficient algorithms to identify and learn features from large volumes of imaging data, and helping to reduce diagnostic errors. In addition, it can recognize specific patterns or lesions and correlate novel characteristics to obtain innovative scientific insight [9, 101].

In recent years, the availability of huge databases and the immense computing power suggested by graphics processing units (GPUs) have motivated research on deep learning (DL) algorithms, which have demonstrated magnificent performance in all medical sciences. Many DL-based approaches have been established for various tasks to analyze retinal images to develop automatic CAD systems for DR [9]. In principle, DL provides a simple solution for clinical extension of AI classification in OCT images. In general, it refers to the neural network (NN) architecture. NNs contain millions of parameters (referred to as artificial neurons) to process structures and features in an image in a feedforward procedure by extracting and processing simple structures in early layers and complex features in later ones [102]. To train the NNs for a specific classification task millions of images need to optimize the network parameters [103]. There are multiple DL-based algorithms that have been presented. Some frequently utilized deep architectures for the DR detection purpose include convolutional neural networks (CNNs), autoencoders (AEs), recurrent neural networks (RNNs) and deep belief networks (DBNs). A complete overview of these architectures is found in [104]. There are several DL-based architectures related to DR detection [105–115].

14.2.1 The advantages of deep learning compared with traditional machine learning

Prior studies on automated detection the usage of conventional ML algorithms trusted on matching of handcrafted functions designed with the aid of using highly-educated area experts. The problem with those conventional methods is that it is far more important to pick out which functions are important in every given image. As pathologies showed huge individualized versions with inside the form and size, function extraction will become a powerful task. These strategies additionally have restricted generalizability (i.e., the capacity to use ML algorithms educated on a given dataset to every other unseen dataset). On the other hand, DL has an end-to-end learning procedure with an annotated dataset. DL fashions take advantage of a couple of layers of non-linear information processing, for extraction and transformation as well as for pattern evaluation and classification. Thus, DL fashions have the ability to understand automatically the associated patterns in images compared to handcrafted approaches. Theoretically, DL may have a much better generalizability and be much less domain-specific as long as it is trained with different and diverse datasets [116–119].

In general, CNN architectures offer higher generalizability via changing the input data into different layers of abstractions and feature learning. CNNs are feedforward networks that the gaining knowledge of system takes place in, from input images to output classifications [118]. CNNs constructed with various architectures have numerous variations, however in general, the usual version of CNN consists of the enter layer, convolutional layers, pooling (or subsampling) layers, and non-linear layers. The convolutional and pooling layers are regularly grouped into modules.

Either one or fully connected layers comply with those modules. Modules are regularly stacked on top of each other to create a network. Convolution operation is a vital feature of CNNs, and it is a dot-product operation among a grid-shape set of weights and comparable grid-dependent inputs drawn from exceptional spatial localities within the input. This operation is specially beneficial for data, which has an excessive degree of spatial or different locality [120]. State-of-the-art approaches with transfer learning, 2D-CNN, 3D-CNN, multi-scale CNN, and attention frameworks have demonstrated trustable results in automated disease detection on medical images [121, 122].

14.2.2 The necessary process of developing a deep learning network

The datasets (i.e. training and tuning sets) and performance evaluation (i.e. primary and external validation sets) required to develop a DL network are shown in figure 14.6. Typically, the purpose for the training set is for the neural network to automatically learn all features, and the tuning set is a small evaluation set for monitoring real-time performance. If the network works well during training, but fits poorly to the training set, it has an overfitting problem. Changes should be made accordingly, and the learning curve should be observed to figure out the optimal stopping epoch and avoid overfitting issues [123]. A non-overlapping primary validation set (or test) is applied to test the final performance after train and tune sets have been performed. These three kinds of sets are normally separated from the

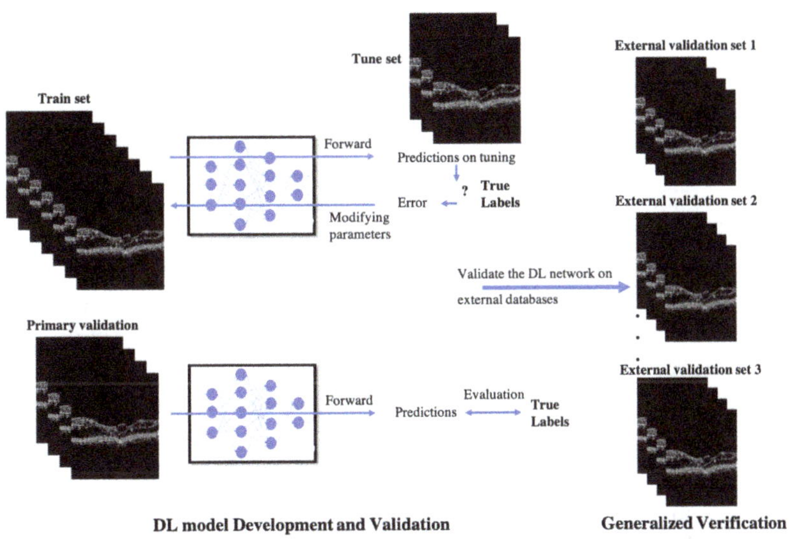

Figure 14.6. Workflow of training, validation and testing database in DL network. The train database is for learning all the features, while a tune database is a small assessment set to supervise the real-time performance. A test set or non-overlapping primary validation is applied to test the final performance after training and tuning steps. These three types of sets are usually separated from a large database based on a specific ratio. To further validate the model on external databases and prove its generalizability, other independent or unseen databases are required as external sets.

Table 14.1. Data set-up in developing deep learning model and performing assessment.

Model	Terminology	Description
	Training set	A non-overlapping set divide from the same database of tune and test sets to build different DLs.
Model		development
Tuning set	A non-	overlapping set divide from the same database of train and test sets to estimate the accuracy of various models during training, fine-tune to choose the optimum model.
Model		performance
Testing set	A non-	overlapping set divide from the same database of train and tune sets to assess the performance of the selected optimum model.
	External validation	Various databases from the database of train, tune and primary validation sets to further assess the model performance and verify its generalizability.
	k-fold	The whole database is split equally into k-folds. The (k-1)-folds are combined as train set for model development and the remaining one-fold is as the test set for final performance assessment. The process is repeated k times and the average performance will be reported.
	Leave-one-out	A special type of k-fold cross-validation when k equals the total number of data. Only one image is left out to test the model performance and the remaining images are used for training. The process is also repeated k times. Leave-one-out cross-validation is usually applied when data is very limited.

same large dataset based on some ratio. If your dataset has a limited sample size, you can develop and evaluate your DL model using k-fold cross-validation or leave-one-out cross-validation strategies. An independent or unseen dataset other than the external validation set is required to further validate the performance of the model on unseen datasets and to confirm its generalizability. A generally good performance across all validation datasets, including primary and external validation, implies a high generalizability of the DL model. For DL-based disease detection, large well-labeled datasets, good network architecture, lots of computational effort, and high generalizability are critical [96, 120]. Terms are summarized in table 14.1.

14.2.3 Lack of deep learning

In general, a major drawback to DL-based image analysis is the insufficiency of publicly available OCT databases. In better words, for the OCT, as a relatively new imaging modality, the limitation of available images presents an issue for practical implementation of DL. Here, we should have numerous input data, to train the model and reduce overfitting. Many currently existing databases are limited in the number of scans of normal cases and diseased ones (some of them are not publicly available), or contain only images obtained from just one vendor. To address this issue, The Medical Image Computing and Computer Assisted Intervention (MICCAI) Society has created benchmark studies. In 2017, the MICCAI RETOUCH Challenge organized a segmentation and detection benchmark study using a large dataset with 112 manually segmented OCT data from 3 different vendors [124]. Robust methods and algorithms were produced during the challenge, and this motivated the need for methods that generalize widely across different patient populations and different manufacturers. However, a large number of OCT databases with manual validation from multiple vendors are still required to encounter the necessities of driving DL.

Related to availability of public data, by searching in the literature a total of 11 public databases are available, among which datasets for retinal layers and anomaly segmentation purposes have been included. One of the important parts of these databases, beside the OCT scans, is that they have the ground truth class maps needed for segmentation approaches to learn and do the automatic labelling of the images. A list of these available OCT datasets that are utilized in OCT segmentation studies is shown in table 14.2. Moreover, in figure 14.7 we have shown some

Table 14.2. Publicly available OCT databases. The table includes: the dataset name, number of patients in the database, the type of acquired scans (scan type), the number of images in each dataset. List of terms: age-related macular degeneration (AMD), diabetic macular edema (DME), macular edema (ME), diabetic retinopathy (DR).

Database	Number of patients	Patients type	Type of scan	Number of images
University of Miami [125]	10	Mild NPDR	—	61/volume
Duke AMD [126]	20	AMD	0 and 90 rectangular centered at the fovea	100 B-Scans/vol
WLOA SD-OCT [127]	384	AMD	0 and 90 rectangular centered at the fovea	100 B-Scans/vol
Duke DME/AMD [128]	45	AMD/ DME	Volumetric	49 B-Scans/vol
Duke Cyst DME [129]	10	DME	—	11 B-scans/scan
Isfahan MISP [130]	13	Healthy	—	128 B-scans/vol
OCTRIMA3D [131]	10	Healthy	IR plus OCT scanning with a 30	61/vol
AFIO [132]	93	ME/ healthy	Macula centered	2497
OCTID [133]	–	AMD/ DR/ healthy	B-Scan centered at the fovea	575
AROI [134]	24	AMD	B-Scan centered at the macula	1136
OCTA-500 [135]	500	Healthy	Multi-modality	More than 350 K

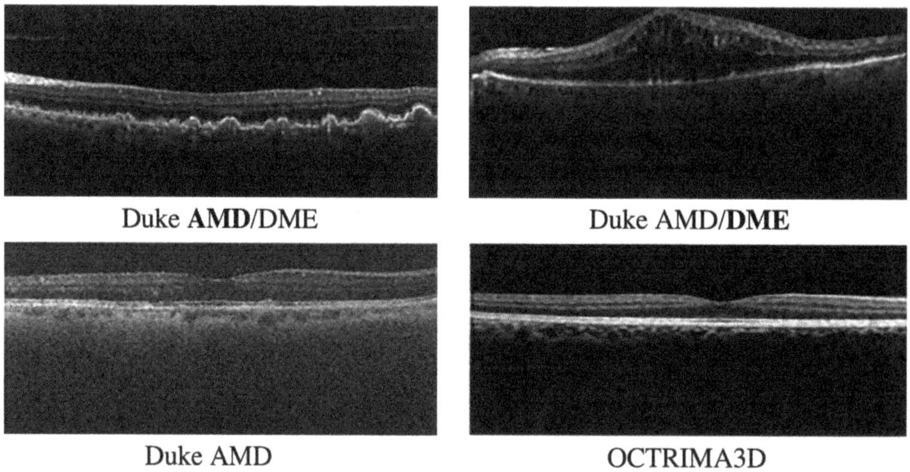

Figure 14.7. Available dataset samples. Examples of publicly available databases used in OCT image analysis studies.

examples of annotated OCT images from a publicly available dataset used to perform different segmentation tasks.

The large amount of data needed to train DL algorithms generates huge challenges. Large databases cause better training of model parameters which improves generalization. In the best-case scenario, these training databases consist of a large number validated manual annotation. However, these types of datasets are costly and therefore rare. To address this limitation, we can use small datasets to train the DL model. Here, for training the models on small data we can work on two strategies. The first approach is using B-scans with training images in the volume. The adjacent scans have similar but rather different anatomy and can be utilized as additional examples in the training step. In a similar way to computed tomography (CT), we can use unlabeled CT scan slices as additional images in addition to labeled training images, which was done in some studies like Ben-Cohen *et al* [136]. Finally, data augmentation can be used on individual scans to develop and diversify the database without obtaining new images. Here, usual transformations such as mirroring, shearing, rotation and out or inward scaling can be applied as they show the true dispersion acquired in OCT imaging. Related to this, Lee *et al* [137] applied a sliding window of size 432 × 32 for OCT images across the scans to notably increase the size of the training dataset. Morley *et al* [138] accomplished a rotation and a unique warp transformation to increase the size of the dataset to 45 times its original size. Kuwayyama *et al* [139] applied horizontal flips, rotations, and translations to increase the train data from 1,100 scans to 59 400. Similarly, Kihara *et al* [140] used data augmentation (translation, rotation, horizontal reflection) to improve the training data from 67 899 OCT-B-scans to 103 053. Gao *et al* [141] augmented the data using a mirroring operation. Devalla *et al* [142] made a comparison between the performance of DL models with and without data augmentation (rotation, horizontal flip, translation, additive white noise, multiplicative speckle noise, elastic deformation, and occlusion patches). Higher performance was presented for models trained using both real and synthetic OCT data compared to models trained using just real images. These results were ascribed to reduced overfitting and improved generalization with additional synthetic training data. Although data augmentation can produce large and different training data, it is important to bring this point to attention that larger real data with validated manual datasets have privilege to synthetic images [119, 143].

Another way to handle the above limitation is to use a transfer learning method for implementation of DL. Transfer learning is a training approach to validate some weights of a pretrained NN and properly retrain certain layers of that NN to optimize the weights for a specific task (i.e., AI classification of retinal images) [97, 144]. In fundus photography, transfer learning has been explored to conduct in retinal segmentation [145], DR and glaucoma detection [146, 147], and diabetic macular thinning evaluation [148]. Recently, transfer learning has also been used in OCT for detection of choroidal neovascularization, diabetic macular edema [144], and AMD [149]. Practically, transfer learning involves either a single layer or multiple layers, because each layer has certain weights that can be retrained. For instance, the specific number of layers needed for retraining may change, depending on the available database and specific task.

14.2.4 Deep learning related works

The new development of quantitative OCT suggests a unique opportunity to apply CAD systems and AI classification of retinal conditions. This quantitative analysis has been investigated for different assessments such as DR [150–152], AMD [153, 154] vein occlusion [155, 156] and sickle cell retinopathy [157]. DL has recently attracted enormous interest in the area of ophthalmology [144, 158, 159]. Previously, several DL systems have been established to detect DR with high sensitivity and specificity (>90%) based on both normal color retinal images [160–162] and OCT ones [144, 158, 163]. Such models could not only help in classification of different classes or stages of DR, but also make personalized therapeutic procedures for diabetic patients according to their OCT images. Most of these studies aimed to introduce: (1) developed DL models to detect the OCT patterns of DR, and (2) demonstrate the critical areas in the DL-based models in OCT images in accurate detection. Below, we briefly introduce the literature about DL-based algorithms for these tasks.

The literature review for segmentation and classification of real-world scans is mainly focused in two DL classes. The first class includes adopted well-established CNNs such as VGG [164] or AlexNet [122]. Here, for segmentation purpose, they will be converted to segmentation architecture by fine-tuning their last fully connected layers [165, 166]. In this category, some studies have introduced the use of CNN networks for classification, and dividing the OCT images into patches and giving them classes (i.e., centre or not in a boundary). Then the whole scan is classified to create a probability map on the boundary location. For this purpose, most of these studies have performed retinal layer segmentation [167–169] and fluid segmentation [170]. In the second class, we see the improvement in the field and subsequent changes in network, especially so-called encoder–decoder architecture and transfer learning model, which have been widely used for different segmentation approaches [171, 172].

Regarding the first class, Lee *et al* [137] worked on VGG network applied on a local database to classify the normal and AMD cases. Their DL model was designed with 13 convolutional layers with an increment filter and three dense layers for classification of the two categories. The dataset contained 80 839 OCT images for the training purpose and 20 163 images for the test one. The results presented a sensitivity of 84.63%, a specificity of 91.54%, and an area under the ROC curve (AUC) of 92.78% with an accuracy of 87.63%. Yu *et al* [173] proposed an automatic approach to segment hyperreflective foci using the GoogleNet [174] and ResNet [175] models for extracting feature in patches from OCT images. To evaluate the performance of their method, they used the dice coefficient, as well as precision and recall metrics. Their proposed model provided reliable segmentation results when we had small lesions in OCT images. Tan *et al* [176] used a CNN architecture to identify AMD, and achieved an accuracy of 91.17%, sensitivity of 92.66% and specificity of 88.56%. Gulshan *et al* [159] introduced a DL model to automatically detect DR in retinal images, and achieved an AUC of 0.99 with sensitivity of 90.3% and specificity of 98.1%. Although these methods could obtain promising results, they used the raw images to train the CNN from scratch requiring a large amount of training data and computation time to achieve a classification accuracy. Finally, Kermany *et al* [144] applied ImageNet, a pretrained Inception V3, to predict four different categories: normal, choroidal neovascularization (CNV), DME and Drusen. The approach was trained using a publicly available database with 108 312 images and tested in a total of 1000 images (250 per category). The highest results on the test set showed a sensitivity of 97.8%, a specificity of 97.4%, and accuracy of 96.6%. Although these works presented confident results in the classification, the output of these algorithms was highly dependent on the manual extraction of region of interests and limiting the number of images from the OCT volumes.

From the second class, encoder–decoder and transfer learning based methods, there are many studies that have utilized fully convolutional network (FCN) for segmentation tasks in OCT scans [177–180]. All of these works show that the presented FCN algorithms considerably increase the accuracy of the segmentation compared to conventional approaches. Another similar network to the FCN that follows the encoder–decoder based method is the U-Net, which was introduced by Ronneberger *et al* [181]. The U-Net structure has been broadly used for OCT image. It worked on segmenting the layers of the retina [142, 182, 183], the full retina [184], and retinal fluid [137, 183].

De Fauw *et al* [158] worked with two local databases on two-step DL-based pipelines for segmentation and classification. A deep segmentation network with a 3D U-Net architecture and a deep classification network to predict the diagnosis probability and the referral suggestions using the segmentation. The model obtained an AUC of 99.21% for the test set. For retinal layer segmentation, Guru *et al* [185] presented the use of dilated convolutions [186]. Their proposed approach was tested using the Duke Cyst DME database (table 14.2) and compared to the related literatures ReLayNet model, showing significant improvement in its accuracy. Li *et al* [187], proposed a DL architecture for early-stage DR detection that includes two parts: OrgNet, using DenseNet blocks integrated with squeeze-and-excitation blocks for feature extraction; and SegNet, containing a ReLayNet based OCT layer segmentation block, in which the dice coefficient showed competitive results. Hussain *et al* [188] used random forest method for classification of healthy and diseased retina using retinal lesions from OCT images with mean accuracy of more than 96%. Lemaitre *et al* [189], using the same types of images, presented an approach based on extracted local binary pattern features combined with dictionary learning using bag of words models for detection of DME, which obtained a sensitivity and a specificity of 81.2% and 93.7%, respectively. Alsaih *et al* [190] applied a linear SVM to classify healthy retina and DME, yielding to get 87.5% sensitivity and 87.5% specificity. Srinivasan *et al* [16] worked on two classifiers, histogram of oriented gradients descriptors and SVM, to detect DME, age-related macular degeneration (AMD), and healthy retina, which achieved 100%, 100%, and 86.67% accuracy, respectively. However, these classification methods mainly depended on features defined by doctors using their experience, causing time-consuming, poor generalization, and unfeasibility in large databases.

Lu *et al* [191] established a new system based on DL to detect and distinguish multicategorical abnormalities from OCT images automatically. They introduced a transfer learning method and used the ResNet-101 [175] model as a pretrained DL model to categorize five classes as (1) cystoid macular edema, (2) serous macular detachment, (3) epiretinal membrane, (4) macular hole, and (5) normal from OCT images. Their system's accuracy was 95.9% with mean value for AUC of 0.984. Li *et al* [163] used a transfer learning approach based on the visual geometry group 16 (VGG-16) network to categorize AMD and DME in OCT retinal images, and acquired 98.6% accuracy, 97.8% sensitivity, 99.4% specificity, and 100% AUC. Karri *et al* [192] presented another deep transfer learning approach based on an inception network to detect retinal pathologies given OCT images. According to their study, means for prediction accuracy across all validations for normal, AMD, and DME images were 99%, 89%, and 86%, respectively.

Wang *et al* [193], used multiple CNN-based DL approaches, such as ResNet, DenseNet, CliqueNet, for classification of an OCT image. In their study, two public databases were utilized for assessment, where 80% of data were employed for training. Rasti *et al* [194] introduced a multi-scale CNN ensemble method for macular OCT classification and achieved accuracy of over 98% and AUC over 0.99 in five-fold cross-validation. In general, DL-based detection of retinopathy using OCT images has already obtained acceptable performance, even comparable to ophthalmologists [144, 158]. However, as we discussed in section 2.3, the success of DL algorithms highly depends on large amount of labeled data for training step, which is costly and needs lots of human effort. To address this issue, using other image datasets and fine-tuning for pretraining is an alternative way to mitigate the need for huge data [192]. Specially, if we use a pretrained CNN, i.e., GoogLeNet, fine-tuned on OCT images for classification it would really help. A similar idea was used in [144, 163], where a pretrained VGG network was acquired and fine-tuned for detecting of retinopathy. Even with pretrained networks, several thousands of OCT images are still required to train the networks for an acceptable performance. This limits the applicability of DL-based automatic detection in our real life. To address these challenges, semi-supervised DL would be an option. It intends to utilize both labelled and unlabeled data for the training step. Ding *et al* presented a pseudo-labeling approach for semi-supervised DL, which tried to annotate unlabeled data with pseudo labels and then train the model with both the labelled and unlabeled data with the pseudo labels [195]. Yao *et al* established a temporal assembling-based semi-supervised DL approach [196] using two losses, i.e., supervised loss for labeled data and consistency loss for unlabeled data. In their study, the consistency loss computed the difference between the new prediction and the mean of old ones for unlabeled data. Mean teacher model was another popular semi-supervised DL method [197]. It worked on a teacher model based on the moving average of the weights of the original model (called student) then, a consistency loss was calculated based on the difference between the teacher and the student outputs for the unlabeled data. These were some limitations for this model: (1) the quality of pseudo labels greatly effects the model performance; (2) the consistency loss may not be suitable, especially at the early stage of training, resulting in limited performance.

In the same direction, as an extension of OCT, optical coherence tomographic angiography (OCTA) is utilized to obtain and assess the movement of blood cells in the field of vision by repeatedly acquiring images of the same retinal position to take an image of the capillary network [198]. Studies have shown that several fundus diseases, such as AMD [199], NV [200], and DR staging using different biomarkers linked to capillary changes in DR [201–203], can be detected using these images. OCTA is sensitive to the failure of vascular networks; therefore, it provides a novel way to monitor and assess the progression of DR [204]. Liu *et al* compared several ML models for DR discrimination based on OCTA scans of different segmentation layers [205]. Abdelsalam and Zahran applied an SVM to diagnose early NPDR based on multifractal geometry [206]. However, because of the special dependency of OCTA, conventional image analysis technique is not always suitable for these images. Moreover, the quality of OCTA images largely depends on some factors such as a turbid refractive medium, image noise, and artifacts of vascular projection. As a result, these studies are seriously required on feature extraction and analysis methods for OCTA images.

Compared with traditional ML algorithms, DL-based approaches show promising efficiency in analyzing medical images [118, 207, 208]. Ma *et al* [209] worked on an OCTA segmentation database and introduced split-based coarse segmentation modules for vessel segmentation. CNNs have also been used for DR classification by using OCT and OCTA [210]. Here, since the number of OCTA data with labels is much smaller than that of retinal images; better use of multilevel information and a combination of domain knowledge is the key to improve DL-based OCTA analysis methods. Related to this, Li *et al* [211] proposed a DL framework that extracted and analyzed the multilevel information in OCTA images and showed its advantage in

DR diagnosis. They presented a segmentation model based on U-Net for segmenting the boundaries of vessels.

All the above reviewed works in the state of the art have demonstrated the fact that over time, the results of automatical classification and segmentation of retinal layers, landmarks, and abnormalities with DL approaches have improved, and in this era reach a comparable efficiency to human experts. These promising and accurate approaches provide ways that simplify the task of evaluating health of the eye and monitoring its related disease. Although there have been major advances in this area, there are still opportunities for development and progress.

14.3 Limitations and future works

While above studies have shown good performance, and results, there are a number of issues that are important to mention. As we have already considered in section 2.3, DL-based approaches need a large amount of data, with enough variety in the dataset to provide generalized results to the problem. Having available public databases would simplify the development of DL algorithms. In the same way, using different datasets (cross-database) would provide a better idea about the generalization of the suggested approaches [212]. There are some studies that have worked on cross-database analysis to assess the performance of their algorithms that are different from the data used in the training process [213, 214].

The next limitation was about the architectures applied by different studies. Here, the majority of the introduced approaches used U-net network as an encoder–decoder-based algorithm, while other architectural variations have not been widely explored. We have different algorithms such as multi-scale, Bayesian, or more current architectures such as vision transformers [215–219]. The potential to use different architectures inside and beyond the encoder–decoder based method to both OCT classification and segmentation requires more attention in the future.

Another limitation is the ability for methods to consider the global context of the input image to improve the classification and segmentation prediction. In better words, some of the current approaches proposed a pixel-probability map which may not make sense from anatomical segmentation viewpoint. In this case, using regional architectures could improve segmentation of the retinal layer if we are correctly able to define the layer within OCT images when we are training the network. This architecture (called 'object hierarchy') has the potential to improve the results of anatomical outcome [172].

14.4 Conclusion

This chapter introduces a detailed survey of methods and results used for the automated detection of DR using OCT images. The DR is a complication of diabetes that damages the retina, and causes vision loss and blindness. The robust DR screening system will remarkably reduce the workload of clinicians. DL-based approaches have become the state of the art for both automatic layer and fluid segmentation of OCT images. Using DL provides better performance than previous ML and conventional approaches. The technological advances in the area of pattern recognition and computer vision, specially the improvements in DL algorithms, have had an important effect in the improvement of the automatic retinal OCT images processing [177].

By reviewing the studies in this chapter, it is obvious that valuable progress has been made in this area and the application of DL algorithms for OCT images analysis. However, the field of DL is rapidly moving forward and there are many new networks and architectures being presented. With these improvements, and with new developments in the DL area, further advances in OCT image processing tools should result, which will be beneficial for clinical practice and research.

References

[1] Flaxman S R, Bourne R R, Resnikoff S, Ackland P, Braithwaite T and Cicinelli M V *et al* 2017 Global causes of blindness and distance vision impairment 1990–2020: a systematic review and meta-analysis *Lancet Global Health* **5** e1221–34

[2] Ogurtsova K, da Rocha Fernandes J, Huang Y, Linnenkamp U, Guariguata L and Cho N H *et al* 2017 IDF diabetes atlas: global estimates for the prevalence of diabetes for 2015 and 2040 *Diabetes Res. Clin. Pract.* **128** 40–50

[3] Wong T Y, Cheung C M G, Larsen M, Sharma S and Simo´ R 2016 Erratum: Diabetic retinopathy *Nat. Rev. Dis. Primers* **2** 1

[4] Sun Z, Yang D, Tang Z, Ng D S and Cheung C Y 2021 Optical coherence tomography angiography in diabetic retinopathy: an updated review *Eye* **35** 149–61

[5] Tavakoli M, Toosi M B, Pourreza R, Banaee T and Pourreza H R 2011 Automated optic nerve head detection in fluorescein angiography fundus images *2011 IEEE Nuclear Science Symp. Conf. Record* (Piscataway, NJ: IEEE) 3057–60

[6] ElTanboly A, Ismail M, Shalaby A, Switala A, El-Baz A and Schaal S *et al* 2017 A computeraided diagnostic system for detecting diabetic retinopathy in optical coherence tomography images *Med. Phys.* **44** 914–23

[7] Spaide R F 2015 Optical coherence tomography angiography signs of vascular abnormalization with antiangiogenic therapy for choroidal neovascularization *Am. J. Ophthalmol* **160** 6–16

[8] Spaide R F, Fujimoto J G, Waheed N K, Sadda S R and Staurenghi G 2018 Optical coherence tomography angiography *Prog. Retinal Eye Res.* **64** 1–55

[9] Tavakoli M and Kelley P 2021 A comprehensive survey on computer-aided diagnostic systems in diabetic retinopathy screening *Photo Acoustic and Optical Coherence Tomography Imaging* (IOP Publishing) vol 3 pp 12-1–48

[10] Tavakoli M, Naji M, Abdollahi A and Kalantari F 2017 Attenuation correction in spect images using attenuation map estimation with its emission data *Medical Imaging 2017: Physics of Medical Imaging* vol 10132 (SPIE) pp 1279–88

[11] Gabriele M L, Wollstein G, Ishikawa H, Xu J, Kim J and Kagemann L *et al* 2010 Three dimensional optical coherence tomography imaging: advantages and advances *Prog. Retin. Eye Res.* **29** 556–79

[12] Tavakoli M, Shahri R P, Pourreza H, Mehdizadeh A, Banaee T and Toosi M H B 2013 A complementary method for automated detection of microaneurysms in fluorescein angiography fundus images to assess diabetic retinopathy *Pattern Recognit.* **46** 2740–53

[13] Walter T, Massin P, Erginay A, Ordonez R, Jeulin C and Klein J C 2007 Automatic detection of microaneurysms in color fundus images *Med. Image Anal.* **11** 555–66

[14] Tavakoli M, Mehdizadeh A, Pourreza R, Pourreza H R, Banaee T and Toosi M B 2011 Radon transform technique for linear structures detection: application to vessel detection in fluorescein angiography fundus images *2011 IEEE Nuclear Science Symp. Conf. Record* (Piscataway, NJ: IEEE) 3051–6

[15] Ribeiro M L, Nunes S G and Cunha-Vaz J G 2013 Microaneurysm turnover at the macula predicts risk of development of clinically significant macular edema in persons with mild nonproliferative diabetic retinopathy *Diabetes Care* **36** 1254–9

[16] Klein R, Meuer S M, Moss S E and Klein B E 1995 Retinal microaneurysm counts and 10-year progression of diabetic retinopathy *Arch. Ophthalmol.* **113** 1386–91

[17] Thompson I A, Durrani A K and Patel S 2019 Optical coherence tomography angiography characteristics in diabetic patients without clinical diabetic retinopathy *Eye* **33** 648–52

[18] Ishibazawa A, Nagaoka T, Takahashi A, Omae T, Tani T and Sogawa K *et al* 2015 Optical coherence tomography angiography in diabetic retinopathy: a prospective pilot study *Am. J. Ophthalmol.* **160** 35–44

[19] Schwartz D M, Fingler J, Kim D Y, Zawadzki R J, Morse L S and Park S S *et al* 2014 Phasevariance optical coherence tomography: a technique for noninvasive angiography *Ophthalmology* **121** 180–7

[20] Miwa Y, Murakami T, Suzuma K, Uji A, Yoshitake S and Fujimoto M *et al* 2016 Relationship between functional and structural changes in diabetic vessels in optical coherence tomography angiography *Sci. Rep.* **6** 1–12

[21] Yu S, Lu J, Cao D, Liu R, Liu B and Li T *et al* 2016 The role of optical coherence tomography angiography in fundus vascular abnormalities *BMC Ophthalmol.* **16** 1–7

[22] Hamada M, Ohkoshi K, Inagaki K, Ebihara N and Murakami A 2018 Visualization of microaneurysms using optical coherence tomography angiography: comparison of OCTA en face, oct B-scan, OCT en face, FA, and IA images *Japan. J. Ophthalmol.* **62** 168–75

[23] Couturier A, Mané V, Bonnin S, Erginay A, Massin P and Gaudric A *et al* 2015 Capillary plexus anomalies in diabetic retinopathy on optical coherence tomography angiography *Retina* **35** 2384–91

[24] Hwang T S, Jia Y, Gao S S, Bailey S T, Lauer A K and Flaxel C J *et al* 2015 Optical coherence tomography angiography features of diabetic retinopathy *Retina* **35** 2371

[25] Williams R, Airey M and Baxter H 2004 Epidemiology of diabetic retinopathy and macular oedema: a systematic review *Eye* **18** 963–83

[26] Giancardo L, Meriaudeau F, Karnowski T P, Li Y, Garg S and Tobin K W *et al* 2012 Exudate-based diabetic macular edema detection in fundus images using publicly available datasets *Med. Image Anal.* **16** 216–26

[27] Taylor S R, Lightman S L, Sugar E A, Jaffe G J, Freeman W R and Altaweel M M *et al* 2012 The impact of macular edema on visual function in intermediate, posterior, and panuveitis *Ocular Immunol. Inflam.* **20** 171–81

[28] McLeod D 2005 Why cotton wool spots should not be regarded as retinal nerve fibre layer infarcts *Br. J. Ophthalmol.* **89** 229–37

[29] Chui T Y, Thibos L N, Bradley A and Burns S A 2009 The mechanisms of vision loss associated with a cotton wool spot *Vis. Res.* **49** 2826–34

[30] Seoud L, Hurtut T, Chelbi J, Cheriet F and Langlois J P 2015 Red lesion detection using dynamic shape features for diabetic retinopathy screening *IEEE Trans. Med. Imaging* **35** 1116–26

[31] Sinthanayothin C, Boyce J F, Williamson T H, Cook H L, Mensah E and Lal S *et al* 2002 Automated detection of diabetic retinopathy on digital fundus images *Diabetic Med.* **19** 105–12
[32] Group ETDRSR *et al* 1991 Grading diabetic retinopathy from stereoscopic color fundus photographs—an extension of the modified airlie house classification: ETDRS report number 10 *Ophthalmology* **98** 786–806
[33] Cusick M, Chew E Y, Chan C C, Kruth H S, Murphy R P and Ferris F L 2003 Histopathology and regression of retinal hard exudates in diabetic retinopathy after reduction of elevated serum lipid levels *Ophthalmology* **110** 2126–33
[34] de Carlo T E, Bonini Filho M A, Baumal C R, Reichel E, Rogers A and Witkin A J *et al* 2016 Evaluation of preretinal neovascularization in proliferative diabetic retinopathy using optical coherence tomography angiography *Ophthal. Surg., Lasers Imaging Retina* **47** 115–9
[35] Vallabha D, Dorairaj R, Namuduri K and Thompson H 2004 Automated detection and classification of vascular abnormalities in diabetic retinopathy *Conf. Record of the Thirty-Eighth Asilomar Conf. on Signals, Systems and Computers, 2004* vol 2 (Piscataway, NJ: IEEE) 1625–9
[36] Patz A 1980 Studies on retinal neovascularization. Friedenwald lecture *Invest. Ophthalmol. Vis. Sci.* **19** 1133–8
[37] Pan J, Chen D, Yang X, Zou R, Zhao K and Cheng D *et al* 2018 Characteristics of neovascularization in early stages of proliferative diabetic retinopathy by optical coherence tomography angiography *Am. J. Ophthalmol.* **192** 146–56
[38] Khalid H, Schwartz R, Nicholson L, Huemer J, El-Bradey M H and Sim D A *et al* 2021 Widefield optical coherence tomography angiography for early detection and objective evaluation of proliferative diabetic retinopathy *Br. J. Ophthalmol.* **105** 118–23
[39] Tavakoli M, Taylor J N, Li C B, Komatsuzaki T and Pressé S 2017 Single molecule data analysis: an introduction *Adv. Chem. Phys.* **162** 205–305
[40] Wilkinson C, Ferris F L, Klein R E, Lee P P, Agardh C D and Davis M *et al* 2003 Proposed international clinical diabetic retinopathy and diabetic macular edema disease severity scales *Ophthalmology* **110** 1677–82
[41] Group ETDRSR *et al* 1991 Classification of diabetic retinopathy from fluorescein angiograms: ETDRS report number 11 *Ophthalmology* **98** 807–22
[42] Philip S, Fleming A D, Goatman K A, Fonseca S, Mcnamee P and Scotland G S *et al* 2007 The efficacy of automated 'disease/no disease' grading for diabetic retinopathy in a systematic screening programme *Br. J. Ophthalmol.* **91** 1512
[43] Salamat N, Missen M M S and Rashid A 2019 Diabetic retinopathy techniques in retinal images: a review *Artif. Intell. Med.* **97** 168–88
[44] Venkatesan R, Chandakkar P, Li B and Li H K 2012 Classification of diabetic retinopathy images using multi-class multiple-instance learning based on color correlogram features *2012 Annual Int. Conf. of the IEEE Engineering in Medicine and Biology Society* (Piscataway, NJ: IEEE) 1462–5
[45] Patton N, Aslam T M, MacGillivray T, Deary I J, Dhillon B and Eikelboom R H *et al* 2006 Retinal image analysis: concepts, applications and potential *Prog. Retin. Eye Res.* **25** 99–127
[46] Mookiah M R K, Acharya U R, Chua C K, Lim C M, Ng E and Laude A 2013 Computer-aided diagnosis of diabetic retinopathy: a review *Comput. Biol. Med.* **43** 2136–55

[47] Scanlon P H 2017 The english national screening programme for diabetic retinopathy 2003–2016 *Acta Diabetol.* **54** 515–25
[48] Pourreza H R, Bahreyni Toossi M H, Mehdizadeh A, Pourreza R and Tavakoli M 2009 Automatic detection of microaneurysms in color fundus images using a local radon transform method *Iran. J. Med. Phys.* **6** 13–20
[49] Ciulla T A, Amador A G and Zinman B 2003 Diabetic retinopathy and diabetic macular edema: pathophysiology, screening, and novel therapies *Diabetes Care* **26** 2653–64
[50] Tavakoli M, Nazar M and Mehdizadeh A 2020 The efficacy of microaneurysms detection with and without vessel segmentation in color retinal images *Medical Imaging 2020: Comuter-Aided Diagnosis* (International Society for Optics and Photonics) vol 11314 113143Y
[51] Hsu W, Pallawala P, Lee M L and Eong K G A 2001 The role of domain knowledge in the detection of retinal hard exudates *Proc. of the 2001 IEEE Computer Society Conf. on Computer Vision and Pattern Recognition. CVPR 2001* vol 2 (Piscataway, NJ: IEEE)
[52] Tavakoli M, Mehdizadeh A, Pourreza R, Banaee T, Bahreyni Toossi M H and Pourreza H R 2010 Early detection of diabetic retinopathy in fluorescent angiography retinal images using image processing methods *Iran. J. Med. Phys.* **7** 7–14
[53] Quellec G, Russell S R and Abra`moff M D 2010 Optimal filter framework for automated, instantaneous detection of lesions in retinal images *IEEE Trans. Med. Imaging* **30** 523–33
[54] Amel F, Mohammed M and Abdelhafid B 2012 Improvement of the hard exudates detection method used for computer-aided diagnosis of diabetic retinopathy *Int. J. Image, Graph. Signal Process.* **4** 28–34
[55] Sánchez C I, Niemeijer M, Dumitrescu A V, Suttorp-Schulten M S, Abramoff M D and van Ginneken B 2011 Evaluation of a computer-aided diagnosis system for diabetic retinopathy screening on public data *Invest. Ophthalmol. Vis. Sci.* **52** 4866–71
[56] Mansour R F 2017 Evolutionary computing enriched computer-aided diagnosis system for diabetic retinopathy: a survey *IEEE Rev. Biomed. Eng.* **10** 334–49
[57] Kumar D, Taylor G W and Wong A 2019 Discovery radiomics with clear-dr: interpretable computer aided diagnosis of diabetic retinopathy *IEEE Access* **7** 25891–6
[58] Ganesan K, Martis R J, Acharya U R, Chua C K, Min L C and Ng E *et al* 2014 Computeraided diabetic retinopathy detection using trace transforms on digital fundus images *Med. Biol. Eng. Comput.* **52** 663–72
[59] Sim D A, Keane P A, Tufail A, Egan C A, Aiello L P and Silva P S 2015 Automated retinal image analysis for diabetic retinopathy in telemedicine *Curr. Diabetes Rep.* **15** 14
[60] Tavakoli M, Kelley P, Nazar M and Kalantari F 2017 Automated fovea detection based on unsupervised retinal vessel segmentation method *2017 IEEE Nuclear Science Symp. and Medical Imaging Conf. (NSS/MIC)* (Piscataway, NJ: IEEE) 1–7
[61] Tavakoli M, Mehdizadeh A, Pourreza Shahri R and Dehmeshki J 2021 Unsupervised automated retinal vessel segmentation based on radon line detector and morphological reconstruction *IET Image Proc.* **15** 1484–98
[62] Mandrekar J N 2010 Receiver operating characteristic curve in diagnostic test assessment *J. Thorac. Oncol.* **5** 1315–6
[63] Tavakoli M, Kalantari F and Golestaneh A 2017 Comparing different preprocessing methods in automated segmentation of retinal vasculature *2017 IEEE Nuclear Science Symp. and Medical Imaging Conf. (NSS/MIC)* (Piscataway, NJ: IEEE) 1–8

[64] Tavakoli M, Nazar M, Golestaneh A and Kalantari F 2017 Automated optic nerve head detection based on different retinal vasculature segmentation methods and mathematical morphology *2017 IEEE Nuclear Science Symp. and Medical Imaging Conf. (NSS/MIC)* (Piscataway, NJ: IEEE) 1–7

[65] Marín D, Aquino A, Gegúndez-Arias M E and Bravo J M 2010 A new supervised method for blood vessel segmentation in retinal images by using gray-level and moment invariants based features *IEEE Trans. Med. Imaging* **30** 146–58

[66] Yan Z, Yang X and Cheng K T 2017 A skeletal similarity metric for quality evaluation of retinal vessel segmentation *IEEE Trans. Med. Imaging* **37** 1045–57

[67] Winder R J, Morrow P J, McRitchie I N, Bailie J and Hart P M 2009 Algorithms for digital image processing in diabetic retinopathy *Comput. Med. Imaging Graph.* **33** 608–22

[68] Teng T, Lefley M and Claremont D 2002 Progress towards automated diabetic ocular screening: a review of image analysis and intelligent systems for diabetic retinopathy *Med. Biol. Eng. Comput.* **40** 2–13

[69] Abr`amoff M D, Garvin M K and Sonka M 2010 Retinal imaging and image analysis *IEEE Rev. Biomed. Eng.* **3** 169–208

[70] Abràmoff M D, Niemeijer M, Suttorp-Schulten M S, Viergever M A, Russell S R and Van Ginneken B 2008 Evaluation of a system for automatic detection of diabetic retinopathy from color fundus photographs in a large population of patients with diabetes *Diabetes Care* **31** 193–8

[71] Niemeijer M, Abramoff M D and Van Ginneken B 2009 Information fusion for diabetic retinopathy cad in digital color fundus photographs *IEEE Trans. Med. Imaging* **28** 775–85

[72] Fleming A, Goatman K, Williams G, Philip S, Sharp P and Olson J 2008 Automated detection of blot haemorrhages as a sign of referable diabetic retinopathy *Proc. Medical Image Understanding and Analysis*

[73] Fleming A D, Philip S, Goatman K A, Olson J A and Sharp P F 2006 Automated assessment of diabetic retinal image quality based on clarity and field definition *Invest. Ophthalmol. Vis. Sci.* **47** 1120–5

[74] Perumalsamy N, Prasad N M, Sathya S and Ramasamy K 2007 Software for reading and grading diabetic retinopathy: Aravind diabetic retinopathy screening 3.0 *Diabetes Care* **30** 2302–6

[75] Wang H, Hsu W, Goh K G and Lee M L 2000 An effective approach to detect lesions in color retinal images *Proc. IEEE Conf. on Computer Vision and Pattern Recognition. CVPR 2000 (Cat. No. PR00662)* vol 2 (Piscataway, NJ: IEEE) 181–6

[76] Mookiah M R K, Acharya U R, Martis R J, Chua C K, Lim C M and Ng E *et al* 2013 Evolutionary algorithm based classifier parameter tuning for automatic diabetic retinopathy grading: a hybrid feature extraction approach *Knowl.-Based Syst.* **39** 9–22

[77] Gang L, Chutatape O and Krishnan S M 2002 Detection and measurement of retinal vessels in fundus images using amplitude modified second-order Gaussian filter *IEEE Trans. Biomed. Eng.* **49** 168–72

[78] Yun W L, Acharya U R, Venkatesh Y V, Chee C, Min L C and Ng E Y K 2008 Identification of different stages of diabetic retinopathy using retinal optical images *Inf. Sci.* **178** 106–21

[79] Nayak J, Bhat P S, Acharya R, Lim C M and Kagathi M 2008 Automated identification of diabetic retinopathy stages using digital fundus images *J. Med. Syst.* **32** 107–15

[80] Acharya U R, Lim C M, Ng E Y K, Chee C and Tamura T 2009 Computer-based detection of diabetes retinopathy stages using digital fundus images *Proc. Inst. Mech. Eng., Part H: J. Eng. Med.* **223** 545–53
[81] Larsen N, Godt J, Grunkin M, Lund-Andersen H and Larsen M 2003 Automated detection of diabetic retinopathy in a fundus photographic screening population *Invest. Ophthalmol. Vis. Sci.* **44** 767–71
[82] Tan J H, Acharya U R, Chua K C, Cheng C and Laude A 2016 Automated extraction of retinal vasculature *Med. Phys.* **43** 2311–22
[83] Yazdanpanah A, Hamarneh G, Smith B R and Sarunic M V 2010 Segmentation of intra-retinal layers from optical coherence tomography images using an active contour approach *IEEE Trans. Med. Imaging* **30** 484–96
[84] Kafieh R, Rabbani H, Abramoff M D and Sonka M 2013 Intra-retinal layer segmentation of 3d optical coherence tomography using coarse grained diffusion map *Med. Image Anal.* **17** 907–28
[85] Ehnes A, Wenner Y, Friedburg C, Preising M N, Bowl W and Sekundo W *et al* 2014 Optical coherence tomography (oct) device independent intraretinal layer segmentation *Transl. Vis. Sci. Technol.* **3** 1
[86] Nam H S, Kim C S, Lee J J, Song J W, Kim J W and Yoo H 2016 Automated detection of vessel lumen and stent struts in intravascular optical coherence tomography to evaluate stent apposition and neointimal coverage *Med. Phys.* **43** 1662–75
[87] Chen Q, Niu S, Yuan S, Fan W and Liu Q 2016 Choroidal vasculature characteristics based choroid segmentation for enhanced depth imaging optical coherence tomography images *Med. Phys.* **43** 1649–61
[88] Andresen S L 2002 John mccarthy: father of ai *IEEE Intell. Syst.* **17** 84–5
[89] Simon A, Singh Deo M, Venkatesan S and Ramesh Babu D R 2015 An overview of machine learning and its applications *Int. J. Electr. Sci. Eng. (IJESE)* **1** 22–4
[90] Faust O, Acharya R, Ng E Y K, Ng K H and Suri J S 2012 Algorithms for the automated detection of diabetic retinopathy using digital fundus images: a review *J. Med. Syst.* **36** 145–57
[91] Joshi S and Karule P 2018 A review on exudates detection methods for diabetic retinopathy *Biomed. Pharmacother.* **97** 1454–60
[92] Almotiri J, Elleithy K and Elleithy A 2018 Retinal vessels segmentation techniques and algorithms: a survey *Appl. Sci.* **8** 155
[93] Almazroa A, Burman R, Raahemifar K and Lakshminarayanan V 2015 Optic disc and optic cup segmentation methodologies for glaucoma image detection: a survey *J. Ophthalmol.* **2015** 180972
[94] Thakur N and Juneja M 2018 Survey on segmentation and classification approaches of optic cup and optic disc for diagnosis of glaucoma *Biomed. Signal Process. Control* **42** 162–89
[95] Abbasi S, Tavakoli M, Boveiri H R, Shirazi M A M, Khayami R and Khorasani H *et al* 2022 Medical image registration using unsupervised deep neural network: a scoping literature review *Biomed. Signal Process. Control* **73** 103444
[96] Ran A R, Tham C C, Chan P P, Cheng C Y, Tham Y C and Rim T H *et al* 2021 Deep learning in glaucoma with optical coherence tomography: a review *Eye* **35** 188–201
[97] Tavakoli M, Mehdizadeh A, Aghayan A, Shahri R P, Ellis T and Dehmeshki J 2021 Automated microaneurysms detection in retinal images using radon transform and

supervised learning: application to mass screening of diabetic retinopathy *IEEE Access* **9** 67302–14

[98] Tavakoli M and Nazar M 2020 Comparison different vessel segmentation methods in automated microaneurysms detection in retinal images using convolutional neural networks *arXiv preprint* arXiv:2005.09097

[99] Obermeyer Z and Lee T H 2017 Lost in thought: the limits of the human mind and the future of medicine *New Engl. J. Med.* **377** 1209

[100] Jiang F, Jiang Y, Zhi H, Dong Y, Li H and Ma S *et al* 2017 Artificial intelligence in healthcare: past, present and future *Stroke Vascul. Neurol.* **2** 230–43

[101] Schmidt-Erfurth U, Sadeghipour A, Gerendas B S, Waldstein S M and Bogunović H 2018 Artificial intelligence in retina *Prog. Retin. Eye Res.* **67** 1–29

[102] Kheradpisheh S R, Ghodrati M, Ganjtabesh M and Masquelier T 2016 Deep networks can resemble human feed-forward vision in invariant object recognition *Sci. Rep.* **6** 1–24

[103] Cox D D and Dean T 2014 Neural networks and neuroscience-inspired computer vision *Curr. Biol.* **24** R921–9

[104] Litjens G, Kooi T, Bejnordi B E, Setio A A A, Ciompi F and Ghafoorian M *et al* 2017 A survey on deep learning in medical image analysis *Med. Image Anal.* **42** 60–88

[105] Khojasteh P, Aliahmad B and Kumar D K 2018 Fundus images analysis using deep features for detection of exudates, hemorrhages and microaneurysms *BMC Ophthalmol.* **18** 1–13

[106] Parmar R, Lakshmanan R, Purushotham S and Soundrapandiyan R 2019 Detecting diabetic retinopathy from retinal images using cuda deep neural network *Intelligent Pervasive Computing Systems for Smarter Healthcare* (Wiley) pp 379–96

[107] Quellec G, Charri`ere K, Boudi Y, Cochener B and Lamard M 2017 Deep image mining for diabetic retinopathy screening *Med. Image Anal.* **39** 178–93

[108] Zhou L, Zhao Y, Yang J, Yu Q and Xu X 2017 Deep multiple instance learning for automatic detection of diabetic retinopathy in retinal images *IET Image Proc.* **12** 563–71

[109] Gargeya R and Leng T 2017 Automated identification of diabetic retinopathy using deep learning *Ophthalmology* **124** 962–9

[110] Son J, Shin J Y, Kim H D, Jung K H, Park K H and Park S J 2020 Development and validation of deep learning models for screening multiple abnormal findings in retinal fundus images *Ophthalmology* **127** 85–94

[111] Shankar K, Sait A R W, Gupta D, Lakshmanaprabu S, Khanna A and Pandey H M 2020 Automated detection and classification of fundus diabetic retinopathy images using synergic deep learning model *Pattern Recogn. Lett.* **133** 210–6

[112] Abràmoff M D, Lou Y, Erginay A, Clarida W, Amelon R and Folk J C *et al* 2016 Improved automated detection of diabetic retinopathy on a publicly available dataset through integration of deep learning *Invest. Ophthalmol. Vis. Sci.* **57** 5200–6

[113] Wang Z, Yin Y, Shi J, Fang W, Li H and Wang X 2017 Zoom-in-net: deep mining lesions for diabetic retinopathy detection *Int. Conf. on Medical Image Computing and Computer-Assisted Intervention* (Berlin: Springer) 267–75

[114] Lam C, Yu C, Huang L and Rubin D 2018 Retinal lesion detection with deep learning using image patches *Invest. Ophthalmol. Vis. Sci.* **59** 590–6

[115] Sayres R, Taly A, Rahimy E, Blumer K, Coz D and Hammel N *et al* 2019 Using a deep learning algorithm and integrated gradients explanation to assist grading for diabetic retinopathy *Ophthalmology* **126** 552–64

[116] O'Mahony N, Campbell S, Carvalho A, Harapanahalli S, Hernandez G V and Krpalkova L *et al* 2019 Deep learning vs. traditional computer vision *Science and information conference* (Berlin: Springer) 128–44
[117] Wang J, Ma Y, Zhang L, Gao R X and Wu D 2018 Deep learning for smart manufacturing: methods and applications *J. Manuf. Syst.* **48** 144–56
[118] Rawat W and Wang Z 2017 Deep convolutional neural networks for image classification: a comprehensive review *Neural Comput.* **29** 2352–449
[119] Tavakoli M, Tsekouras K, Day R, Dunn K W and Presse S 2019 Quantitative kinetic models from intravital microscopy: a case study using hepatic transport *J. Phys. Chem.* B **123** 7302–12
[120] Aggarwal C C *et al* 2018 *Neural Networks and Deep Learning* (Berlin: Springer) vol 10 978–3
[121] LeCun Y, Bengio Y and Hinton G 2015 Deep learning *Nature* **521** 436–44
[122] Krizhevsky A, Sutskever I and Hinton G E 2017 Imagenet classification with deep convolutional neural networks *Commun. ACM* **60** 84–90
[123] Ying X 2019 An overview of overfitting and its solutions *J. Phys.: Conf. Ser.* **1168** 022022
[124] Bogunović H, Venhuizen F, Klimscha S, Apostolopoulos S, Bab-Hadiashar A and Bagci U *et al* 2019 Retouch: the retinal oct fluid detection and segmentation benchmark and challenge *IEEE Trans. Med. Imaging* **38** 1858–74
[125] Tian J, Varga B, Tatrai E, Fanni P, Somfai G M and Smiddy W E *et al* 2016 Performance evaluation of automated segmentation software on optical coherence tomography volume data *J. Biophotonics* **9** 478–89
[126] Chiu S J, Izatt J A, O'Connell R V, Winter K P, Toth C A and Farsiu S 2012 Validated automatic segmentation of amd pathology including drusen and geographic atrophy in SD-OCT images *Invest. Ophthalmol. Vis. Sci.* **53** 53–61
[127] Farsiu S, Chiu S J, O'Connell R V, Folgar F A, Yuan E and Izatt J A *et al* 2014 Quantitative classification of eyes with and without intermediate age-related macular degeneration using optical coherence tomography *Ophthalmology* **121** 162–72
[128] Srinivasan P P, Kim L A, Mettu P S, Cousins S W, Comer G M and Izatt J A *et al* 2014 Fully automated detection of diabetic macular edema and dry age-related macular degeneration from optical coherence tomography images *Biomed. Opt. Express* **5** 3568–77
[129] Chiu S J, Allingham M J, Mettu P S, Cousins S W, Izatt J A and Farsiu S 2015 Kernel regression based segmentation of optical coherence tomography images with diabetic macular edema *Biomed. Opt. Express* **6** 1172–94
[130] Kashefpur M, Kafieh R, Jorjandi S, Golmohammadi H, Khodabande Z and Abbasi M *et al* 2017 Isfahan misp dataset *J. Med. Signals Sens.* **7** 43
[131] Tian J, Varga B, Somfai G M, Lee W H, Smiddy W E and Cabrera DeBuc D 2015 Real-time automatic segmentation of optical coherence tomography volume data of the macular region *PLoS One* **10** e0133908
[132] Hassan T, Akram M U, Masood M F and Yasin U 2018 Biomisa retinal image database for macular and ocular syndromes *Int. Conf. Image Analysis and Recognition* (Berlin: Springer) 695–705
[133] Gholami P, Roy P, Parthasarathy M K and Lakshminarayanan V 2020 Octid: optical coherence tomography image database *Comput. Electr. Eng.* **81** 106532
[134] Melinščak M, Radmilović M, Vatavuk Z and Lončarić S 2021 Annotated retinal optical coherence tomography images (aroi) database for joint retinal layer and fluid segmentation *Automatika* **62** 375–85

[135] Li M, Zhang Y, Ji Z, Xie K, Yuan S and Liu Q *et al* 2020 IPN-V2 and OCTA-500: methodology and dataset for retinal image segmentation *arXiv preprint* arXiv:2012.07261
[136] Ben-Cohen A, Klang E, Amitai M M, Goldberger J and Greenspan H 2018 Anatomical data augmentation for CNN based pixel-wise classification *2018 IEEE 15th Int. Symp. on Biomedical Imaging (ISBI 2018)* (Piscataway, NJ: IEEE) 1096–9
[137] Lee C S, Tyring A J, Deruyter N P, Wu Y, Rokem A and Lee A Y 2017 Deep-learning based, automated segmentation of macular edema in optical coherence tomography *Biomed. Opt. Express* **8** 3440–8
[138] Morley D, Foroosh H, Shaikh S and Bagci U 2017 Simultaneous detection and quantification of retinal fluid with deep learning *arXiv preprint* arXiv:1708.05464
[139] Kuwayama S, Ayatsuka Y, Yanagisono D, Uta T, Usui H and Kato A *et al* 2019 Automated detection of macular diseases by optical coherence tomography and artificial intelligence machine learning of optical coherence tomography images *J. Ophthalmol.* **2019** 6319581
[140] Kihara Y, Heeren T F, Lee C S, Wu Y, Xiao S and Tzaridis S *et al* 2019 Estimating retinal sensitivity using optical coherence tomography with deep-learning algorithms in macular telangiectasia type 2 *JAMA Network Open* **2** e188029
[141] Gao K, Niu S, Ji Z, Wu M, Chen Q and Xu R *et al* 2019 Double-branched and area-constraint fully convolutional networks for automated serous retinal detachment segmentation in sd-oct images *Comput. Methods Prog. Biomed.* **176** 69–80
[142] Devalla S K, Renukanand P K, Sreedhar B K, Subramanian G, Zhang L and Perera S *et al* 2018 Drunet: a dilated-residual u-net deep learning network to segment optic nerve head tissues in optical coherence tomography images *Biomed. Opt. Express* **9** 3244–65
[143] Wong S C, Gatt A, Stamatescu V and McDonnell M D 2016 Understanding data augmentation for classification: when to warp? *2016 Int. Conf. on Digital Image Computing: Techniques and Applications (DICTA)* (Piscataway, NJ: IEEE) 1–6
[144] Kermany D S, Goldbaum M, Cai W, Valentim C C, Liang H and Baxter S L *et al* 2018 Identifying medical diagnoses and treatable diseases by image-based deep learning *Cell* **172** 1122–31
[145] Hemelings R, Elen B, Stalmans I, Van Keer K, De Boever P and Blaschko M B 2019 Artery– vein segmentation in fundus images using a fully convolutional network *Comput. Med. Imaging Graph.* **76** 101636
[146] Gómez-Valverde J J, Antón A, Fatti G, Liefers B, Herranz A and Santos A *et al* 2019 Automatic glaucoma classification using color fundus images based on convolutional neural networks and transfer learning *Biomed. Opt. Express* **10** 892–913
[147] Ting D S W, Pasquale L R, Peng L, Campbell J P, Lee A Y and Raman R *et al* 2019 Artificial intelligence and deep learning in ophthalmology *Br. J. Ophthalmol.* **103** 167–75
[148] Arcadu F, Benmansour F, Maunz A, Michon J, Haskova Z and McClintock D *et al* 2019 Deep learning predicts oct measures of diabetic macular thickening from color fundus photographs *Invest. Ophthalmol. Vis. Sci.* **60** 852–7
[149] Lee C S, Tyring A J, Wu Y, Xiao S, Rokem A S and DeRuyter N P *et al* 2019 Generating retinal flow maps from structural optical coherence tomography with artificial intelligence *Sci. Rep.* **9** 1–11
[150] Alam M, Zhang Y, Lim J I, Chan R V, Yang M and Yao X 2020 Quantitative optical coherence tomography angiography features for objective classification and staging of diabetic retinopathy *Retina* **40** 322–32

[151] Hsieh Y T, Alam M N, Le D, Hsiao C C, Yang C H and Chao D L *et al* 2019 Oct angiography biomarkers for predicting visual outcomes after ranibizumab treatment for diabetic macular edema *Ophthalmol. Retina* **3** 826–34

[152] Le D, Alam M, Miao B A, Lim J I and Yao X 2019 Fully automated geometric feature analysis in optical coherence tomography angiography for objective classification of diabetic retinopathy *Biomed. Opt. Express* **10** 2493–503

[153] Moult E, Choi W, Waheed N K, Adhi M, Lee B and Lu C D *et al* 2014 Ultrahigh-speed sweptsource oct angiography in exudative amd *Ophthal. Surg., Lasers Imaging Retina* **45** 496–505

[154] Zheng F, Zhang Q, Motulsky E H, de Oliveira Dias J R, Chen C L and Chu Z *et al* 2017 Comparison of neovascular lesion area measurements from different swept-source oct angiographic scan patterns in age-related macular degeneration *Invest. Ophthalmol. Vis. Sci.* **58** 5098–104

[155] Cabral D, Coscas F, Glacet-Bernard A, Pereira T, Geraldes C and Cachado F *et al* 2019 Biomarkers of peripheral nonperfusion in retinal venous occlusions using optical coherence tomography angiography *Transl. Vis. Sci. Technol.* **8** 7

[156] Samara W A, Shahlaee A, Sridhar J, Khan M A, Ho A C and Hsu J 2016 Quantitative optical coherence tomography angiography features and visual function in eyes with branch retinal vein occlusion *Am. J. Ophthalmol.* **166** 76–83

[157] Alam M, Thapa D, Lim J I, Cao D and Yao X 2017 Quantitative characteristics of sickle cell retinopathy in optical coherence tomography angiography *Biomed. Opt. Express* **8** 1741–53

[158] De Fauw J, Ledsam J R, Romera-Paredes B, Nikolov S, Tomasev N and Blackwell S *et al* 2018 Clinically applicable deep learning for diagnosis and referral in retinal disease *Nat. Med.* **24** 1342–50

[159] Gulshan V, Peng L, Coram M, Stumpe M C, Wu D and Narayanaswamy A *et al* 2016 Development and validation of a deep learning algorithm for detection of diabetic retinopathy in retinal fundus photographs *JAMA* **316** 2402–10

[160] Tavakoli M 2022 Automated optic disk detection in fundus images using a combination of deep learning and local histogram matching *Proc. SPIE* **2036** 120360I–1

[161] Tavakoli M, Jazani S and Nazar M 2020 Automated detection of microaneurysms in color fundus images using deep learning with different preprocessing approaches *Medical Imaging 2020: Imaging Informatics for Healthcare, Research, and Applications* vol 11318 (International Society for Optics and Photonics) p 113180E

[162] Tavakoli M and Nazar M 2020 Comparison different vessel segmentation methods in automated microaneurysms detection in retinal images using convolutional neural networks *SPIE* **11317** 113171P

[163] Li F, Chen H, Liu Z, Zhang X and Wu Z 2019 Fully automated detection of retinal disorders by image-based deep learning *Graefe's Arch. Clin. Exp. Ophthalmol.* **257** 495–505

[164] Simonyan K and Zisserman A 2014 Very deep convolutional networks for large-scale image recognition *arXiv preprint* arXiv:1409.1556

[165] Ganin, Y and Lempitsky V 2014 Fields: neural network nearest neighbor fields for image transforms *Asian Conf. on Computer Viision* (Cham: Springer) pp 536–51

[166] Ciresan D, Giusti A, Gambardella L and Schmidhuber J 2012 Deep neural networks segment neuronal membranes in electron microscopy images *NIPS'12: Proc. of the 25th Int. Conf. on Neural Information Processing Systems* vol 2 2843–51

[167] Fang L, Cunefare D, Wang C, Guymer R H, Li S and Farsiu S 2017 Automatic segmentation of nine retinal layer boundaries in oct images of non-exudative amd patients using deep learning and graph search *Biomed. Opt. Express* **8** 2732–44

[168] Shah A, Zhou L, Abràmoff M D and Wu X 2018 Multiple surface segmentation using convolution neural nets: application to retinal layer segmentation in oct images *Biomed. Opt. Express* **9** 4509–26

[169] Hamwood J, Alonso-Caneiro D, Read S A, Vincent S J and Collins M J 2018 Effect of patch size and network architecture on a convolutional neural network approach for automatic segmentation of oct retinal layers *Biomed. Opt. Express* **9** 3049–66

[170] Rashno A, Koozekanani D D and Parhi K K 2018 Oct fluid segmentation using graph shortest path and convolutional neural network *2018 40th Annual Int. Conf. of the IEEE Engineering in Medicine and Biology Society (EMBC)* (Piscataway, NJ: IEEE) 3426–9

[171] Lateef F and Ruichek Y 2019 Survey on semantic segmentation using deep learning techniques *Neurocomputing* **338** 321–48

[172] Viedma I A, Alonso-Caneiro D, Read S A and Collins M J 2022 Deep learning in retinal optical coherence tomography (OCT): A comprehensive survey *Neurocomputing* **507** 247–64

[173] Yu C, Xie S, Niu S, Ji Z, Fan W and Yuan S *et al* 2019 Hyper-reflective foci segmentation in sdoct retinal images with diabetic retinopathy using deep convolutional neural networks *Med. Phys.* **46** 4502–19

[174] Szegedy C, Liu W, Jia Y, Sermanet P, Reed S and Anguelov D *et al* 2015 Going deeper with convolutions *Proc. of the IEEE Conf. on Computer Vision and Pattern Recognition* pp 1–9

[175] He K, Zhang X, Ren S and Sun J 2016 Deep residual learning for image recognition *Proc. of the IEEE Conf. on Computer Vision and Pattern Recognition* pp 770–8

[176] Tan J H, Bhandary S V, Sivaprasad S, Hagiwara Y, Bagchi A and Raghavendra U *et al* 2018 Age-related macular degeneration detection using deep convolutional neural network *Future Gener. Comput. Syst.* **87** 127–35

[177] Xu Y, Yan K, Kim J, Wang X, Li C and Su L *et al* 2017 Dual-stage deep learning framework for pigment epithelium detachment segmentation in polypoidal choroidal vasculopathy *Biomed. Opt. Express* **8** 4061–76

[178] Kiaee F, Fahimi H and Rabbani H 2018 Intra-retinal layer segmentation of optical coherence tomography using 3d fully convolutional networks *2018 25th IEEE Int. Conf. on Image Processing (ICIP)* (Piscataway, NJ: IEEE) 2795–9

[179] Pekala M, Joshi N, Liu T A, Bressler N M, DeBuc D C and Burlina P 2019 Deep learning based retinal oct segmentation *Comput. Biol. Med.* **114** 103445

[180] Venhuizen F G, van Ginneken B, Liefers B, van Asten F, Schreur V and Fauser S *et al* 2018 Deep learning approach for the detection and quantification of intraretinal cystoid fluid in multivendor optical coherence tomography *Biomed. Opt. Express* **9** 1545–69

[181] Ronneberger O, Fischer P and Brox T 2015 U-net: convolutional networks for biomedical image segmentation *Int. Conf. on Medical image computing and computer-assisted intervention* (Berlin: Springer) 234–41

[182] Apostolopoulos S, Zanet S D, Ciller C, Wolf S and Sznitman R 2017 Pathological OCT retinal layer segmentation using branch residual u-shape networks *Int. Conf. on Medical Image Computing and Computer-Assisted Intervention* (Berlin: Springer) 294–301

[183] Roy A G, Conjeti S, Karri S P K, Sheet D, Katouzian A and Wachinger C *et al* 2017 Relaynet: retinal layer and fluid segmentation of macular optical coherence tomography using fully convolutional networks *Biomed. Opt. Express* **8** 3627–42
[184] Venhuizen F G, van Ginneken B, Liefers B, van Grinsven M J, Fauser S and Hoyng C *et al* 2017 Robust total retina thickness segmentation in optical coherence tomography images using convolutional neural networks *Biomed. Opt. Express* **8** 3292–316
[185] Guru Pradeep Reddy T, Ashritha K S, Prajwala T, Girish G, Kothari A R and Koolagudi S G *et al* 2020 Retinal-layer segmentation using dilated convolutions *Proc. of 3rd Int. Conf. on Computer Vision and Image Processing* (Berlin: Springer) 279–92
[186] Yu F and Koltun V 2015 Multi-scale context aggregation by dilated convolutions *arXiv preprint* arXiv:1511.07122
[187] Li F, Chen H, Liu Z, Zhang X, Jiang M and Wu Z *et al* 2019 Deep learning-based automated detection of retinal diseases using optical coherence tomography images *Biomed. Opt. Express* **10** 6204–26
[188] Hussain M A, Bhuiyan A D, Luu C and Theodore Smith R H *et al* 2018 Classification of healthy and diseased retina using sd-oct imaging and random forest algorithm *PLoS One* **13** e0198281
[189] Lemaître G, Rastgoo M, Massich J, Cheung C Y, Wong T Y and Lamoureux E *et al* 2016 Classification of SD-OCT volumes using local binary patterns: experimental validation for dme detection *J. Ophthalmol.* **2016** 3298606
[190] Alsaih K, Lemaître G, Rastgoo M, Massich J, Sidibé D and Meriaudeau F 2017 Machine learning techniques for diabetic macular edema (dme) classification on sd-oct images *Biomed. Eng. Online* **16** 1–12
[191] Lu W, Tong Y, Yu Y, Xing Y, Chen C and Shen Y 2018 Deep learning-based automated classification of multi-categorical abnormalities from optical coherence tomography images *Transl. Vis. Sci. Technol.* **7** 41
[192] Karri S P K, Chakraborty D and Chatterjee J 2017 Transfer learning based classification of optical coherence tomography images with diabetic macular edema and dry age-related macular degeneration *Biomed. Opt. Express* **8** 579–92
[193] Wang D and Wang L 2019 On oct image classification via deep learning *IEEE Photon. J.* **11** 1–14
[194] Rasti R, Rabbani H, Mehridehnavi A and Hajizadeh F 2017 Macular oct classification using a multi-scale convolutional neural network ensemble *IEEE Trans. Med. Imaging* **37** 1024–34
[195] Ding G, Zhang S, Khan S, Tang Z, Zhang J and Porikli F 2019 Feature affinity-based pseudo labeling for semi-supervised person re-identification *IEEE Trans. Multimedia* **21** 2891–902
[196] Yao Y, Deng J, Chen X, Gong C, Wu J and Yang J 2020 Deep discriminative CNN with temporal ensembling for ambiguously-labeled image classification *Proc. of the AAAI Conf. on Artificial Intelligence* vol 34 pp 12669–76
[197] Tarvainen A and Valpola H 2017 Mean teachers are better role models: weight-averaged consistency targets improve semi-supervised deep learning results ArXiv:1703.01780
[198] Sambhav K, Grover S and Chalam K V 2017 The application of optical coherence tomography angiography in retinal diseases *Surv. Ophthalmol.* **62** 838–66

[199] Phasukkijwatana N, Tan A C, Chen X, Freund K B and Sarraf D 2017 Optical coherence tomography angiography of type 3 neovascularisation in age-related macular degeneration after antiangiogenic therapy *Br. J. Ophthalmol.* **101** 597–602

[200] Jia Y, Bailey S T, Wilson D J, Tan O, Klein M L and Flaxel C J *et al* 2014 Quantitative optical coherence tomography angiography of choroidal neovascularization in age-related macular degeneration *Ophthalmology* **121** 1435–44

[201] Zang P, Hormel T T, Wang X, Tsuboi K, Huang D and Hwang T S *et al* 2022 A diabetic retinopathy classification framework based on deep-learning analysis of oct angiography *Transl. Vision Sci. Technol.* **11** 10

[202] Makita S, Hong Y, Yamanari M, Yatagai T and Yasuno Y 2006 Optical coherence angiography *Opt. Express* **14** 7821–40

[203] Jia Y, Bailey S T, Hwang T S, McClintic S M, Gao S S and Pennesi M E *et al* 2015 Quantitative optical coherence tomography angiography of vascular abnormalities in the living human eye *Proc. Natl. Acad. Sci.* **112** E2395–402

[204] Liu G, Xu D and Wang F 2018 New insights into diabetic retinopathy by oct angiography *Diabetes Res. Clin. Pract.* **142** 243–53

[205] Liu Z, Wang C, Cai X, Jiang H and Wang J 2021 Discrimination of diabetic retinopathy from optical coherence tomography angiography images using machine learning methods *IEEE Access* **9** 51689–94

[206] Abdelsalam M M and Zahran M 2021 A novel approach of diabetic retinopathy early detection based on multifractal geometry analysis for octa macular images using support vector machine *IEEE Access* **9** 22844–58

[207] Ker J, Wang L, Rao J and Lim T 2017 Deep learning applications in medical image analysis *IEEE Access* **6** 9375–89

[208] Ma N, Zhang X, Zheng H T and Sun J 2018 Shufflenet v2: practical guidelines for efficient CNN architecture design *Proc. of the European Conf. on Computer Vision (ECCV)* pp 116–31

[209] Ma Y, Hao H, Xie J, Fu H, Zhang J and Yang J *et al* 2020 Rose: a retinal oct-angiography vessel segmentation dataset and new model *IEEE Trans. Med. Imaging* **40** 928–39

[210] Zang P, Gao L, Hormel T T, Wang J, You Q and Hwang T S *et al* 2020 Dcardnet: diabetic retinopathy classification at multiple levels based on structural and angiographic optical coherence tomography *IEEE Trans. Biomed. Eng.* **68** 1859–70

[211] Li Q, Zhu X, Sun G, Zhang L, Zhu M and Tian T *et al* 2022 Diagnosing diabetic retinopathy in octa images based on multilevel information fusion using a deep learning framework *Comput. Math. Methods Med.* **2022** 4316507

[212] Tommasi T, Patricia N, Caputo B and Tuytelaars T 2017 A deeper look at dataset bias *Domain Adaptation in Computer Vision Applications* (Berlin: Springer) pp 37–55

[213] Li Q, Li S, He Z, Guan H, Chen R and Xu Y *et al* 2020 Deep retina: layer segmentation of retina in OCT images using deep learning *Transl. Vis. Sci. Technol.* **9** 61

[214] Cao J, Liu X, Zhang Y and Wang M 2020 A multi-task framework for topology-guaranteed retinal layer segmentation in OCT images *2020 IEEE Int. Conf. on Systems, Man, and Cybernetics (SMC)* (Piscataway, NJ: IEEE) 3142–7

[215] Wang M, Zhu W, Shi F, Su J, Chen H and Yu K *et al* 2021 Mstganet: automatic drusen segmentation from retinal oct images *IEEE Trans. Med. Imaging* **41** 394–406

[216] Tavakoli M, Jazani S, Sgouralis I, Heo W, Ishii K and Tahara T *et al* 2020 Direct photon-byphoton analysis of time-resolved pulsed excitation data using Bayesian nonparametrics *Cell Rep. Phys. Sci.* **1** 100234

[217] Tavakoli M, Jazani S, Sgouralis I, Shafraz O M, Sivasankar S and Donaphon B *et al* 2020 Pitching single-focus confocal data analysis one photon at a time with Bayesian nonparametrics *Phys. Rev.* **10** 011021

[218] Ghaempanah H, Tavakoli M, Deevband M R, Alvar A A, Najafi M and Kelley P 2022 Electronic portal image enhancement based on nonuniformity correction in wavelet domain *Med. Phys.* **49** 4599–612

[219] Tavakoli M, Jazani S, Sgouralis I and Presse S 2019 Bayesian nonparametrics for fluorescence methods *Biophys. J.* **116** 39a

www.ingramcontent.com/pod-product-compliance
Lightning Source LLC
LaVergne TN
LVHW062221200926
848484LV00006B/1445